Beneficial Plant-Bacterial Interactions

Bernard R. Glick

Beneficial Plant-Bacterial Interactions

Second Edition

 Springer

Bernard R. Glick
Department of Biology
University of Waterloo
Waterloo, ON, Canada

ISBN 978-3-030-44370-2   ISBN 978-3-030-44368-9   (eBook)
https://doi.org/10.1007/978-3-030-44368-9

This Springer imprint is published by the registered company Springer Nature Switzerland AG.
The registered company address is: Gewerbestrasse 11, 6330 Cham, Switzerland

To Rav Avram Hersh Glick,
Born circa 1850, died 1914 (Belz, Austria-
Hungary)
And, to his great-great-great-grandchildren
Benjamin, Aviva and Gabrielle Glick
Born 2009, 2011 and 2014, respectively
(Ottawa, Canada)

Preface

In the past 15–20 years, there has been increasing interest throughout the world in studies involving plant growth-promoting bacteria (PGPB) and their positive interactions with plants. This interest is a consequence of the fact that every year more and more of these bacteria are being used as a component of sustainable agricultural practice. The interest in using PGPB is based on a number of different factors. For example, the increasing world population necessitates that agricultural output be significantly increased within a relatively short period of time without depleting our limited natural resources including prime agricultural land. Thus, the possibility exists that we can partially or even (eventually) completely replace the excessive use of chemicals (fertilizers, herbicides and pesticides) that are currently employed in agriculture.

Scientists worldwide have come to a juncture where there is an ever-increasing understanding of many of the fundamental biochemical and genetic mechanisms that are operative in plant–bacterial interactions. Since practical advances in the employment of this technology are dependent upon the development of a solid understanding of the fundamental processes involving the interactions between PGPB and plants, it is clear that we are on the verge of a paradigm shift in agriculture where we will soon utilize this technology globally on a practical and large scale. However, for PGPB technology to develop to realize its full potential, it is imperative that the fundamental research in this field continues unabated.

In the past number of years, there have been a relatively large number of books dealing with various aspects of the interactions between plants and microorganisms (both bacteria and fungi). Nearly all of these books are compilations of articles written by various specialists in this field and are almost exclusively intended for burgeoning practitioners. On the other hand, this book was written to serve as a textbook (guidebook) for a one-semester undergraduate or graduate course in plant–bacterial interactions. It is based on a course that has been offered at the University of Waterloo for the past number of years and attempts to present a broad, although admittedly biased, perspective of this area of research. The book assumes some basic knowledge of microbiology, plant biology, molecular biology and biochemistry. This notwithstanding, not everyone who uses this book will have had an extensive background in all of these disciplines. Therefore, this book is intentionally written in a simple, coherent, easy to comprehend jargon-free style in which as much of the background material as is necessary to understand the major concepts is provided. The book should therefore be useful to anyone, including the non-specialist, who is interested in developing a relatively broad fundamental perspective on plant–bacterial interactions.

Waterloo, ON, Canada   Bernard R. Glick

Acknowledgements

Individual chapters of this book were reviewed by the following scientists (in alphabetical order): Drs. Olubakola Babalola (South Africa), Zhenyu Cheng (Canada), Cinzia Forni (Italy), Eloise Foo (Australia), Elisa Gamalero (Italy), Jennifer Mesa (Spain), Francisco Nascimento (Portugal) and Gustavo Santoyo (Mexico). Their comments and criticisms were extremely useful in making the manuscript as error free as possible. While their input is highly appreciated, any remaining errors or misunderstandings are entirely my fault and responsibility.

Contents

Introduction to Plant Growth-Promoting Bacteria

1

Abstract

One way to deal with the expected increase in world population and the concomitant need to globally feed a large number of additional people is to increase agricultural productivity through the directed use of plant growth-promoting bacteria. These naturally occurring bacteria, commonly found in the soil associated with the roots of plants, positively affect the growth of plants in a number of different ways. This includes increases in plant yield, nutritional content, tolerance to various abiotic and biotic stresses, and the production of useful secondary metabolites. A better fundamental understanding of the functioning of these bacteria is essential to the eventual large-scale commercialization of plant growth-promoting bacteria.

1.1 The Problem

The Reverend Thomas Robert Malthus (1766–1834) was a British cleric and scholar who became widely known for his theories about changes in the world population. He promulgated the idea that "The power of population is indefinitely greater than the power in the earth to produce subsistence for man." That is, Malthus understood that sooner or later, the earth, which is finite, would be unable to produce enough food to feed all of the people who live here. Malthus' thinking was in direct opposition to the view that was popular in the eighteenth century Europe that society would continue to improve and was in principle "perfectible." Of course, neither Malthus nor any of his critics could have possibly predicted the enormous technological changes, including changes to agricultural and food storage technologies, that have taken place over the past 150–200 years. These changes have enabled the world's population to expand dramatically in a relatively short period of time. However, these technological changes may have lulled us into a sense of false security whereby many people in society, especially in more developed countries,

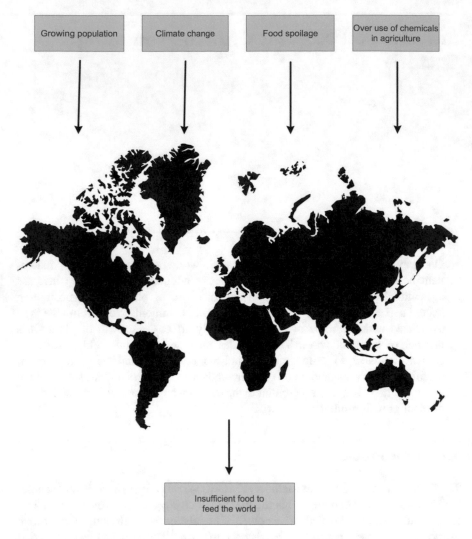

Fig. 1.1 Schematic representation of some of the limitations on the ability of the world being able to feed all of the people

believe that we have never had it so good, and as long as our policies continue to support innovation and business expansion, the good life will continue on well into the future. Unfortunately, at this juncture, the threat of insufficient food to feed the all of the world's people is once again in the headlines. This is a result of a continuously growing world population, worldwide climate change, food spoilage and wastage on an enormous scale, the unequal distribution of food resources around the world, and the overuse of energy-intensive agricultural chemicals (Fig. 1.1).

According to the United Nations World Food Program, more than 870 million of the world's people are malnourished. In addition, in the developing world,

Fig. 1.2 Estimated world population from 1950 to 2050

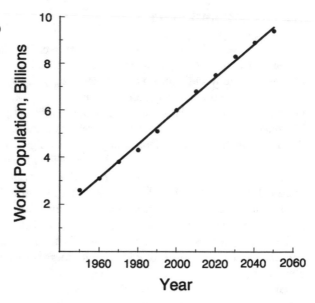

malnutrition contributes to the death of ~2.6 million children each year. At the same time, arable land is limited, and salinization and climate change limit the suitability of much the land that is not already used for agriculture. At the present time, there are around 7 billion people in the world, up from approximately 2.5 billion in 1950 and around 1 billion in the early 1800s, and it is predicted to rise to somewhere around 9.5 billion by 2050 (Fig. 1.2). That is, the increase in the world population in the next 30 or so years will be approximately equal to the number of people in the entire world in 1950. Put another way, within a period of just 100 years (1950–2050) the world population will increase nearly fourfold. Moreover, the rate of global industrialization continues to increase as does our use of and need for nonrenewable resources. Thus, we are increasingly polluting the air, water, and soil all over the planet so that some 40% of deaths worldwide have been estimated to be a consequence of that pollution. This is because, as a result of both increasing population and industrialization, the earth's atmospheric, terrestrial, and aquatic systems are no longer able to break down and tolerate the increasing amount of waste that human society produces. In fact, according to the United Nations Intergovernmental Panel on Climate Change Report released in early 2014, the overall effect of various aspects of man-made global warming is that worldwide food production will eventually be significantly reduced compared to a world without global warming.

Ironically, between 1960 and 2010, the fraction of the human population with insufficient food declined from about 60% to around 15% indicating that up until now food production has more than kept pace with population growth. However, put simply, the problem is whether we will be able to feed another ~2 billion people by 2050, approximately 30 years from now.

1.2   Possible Solutions

Given the issues that have been outlined above, it is imperative that worldwide action be taken on a variety of fronts. In the first instance, it is necessary to limit increases to the world's population, an extremely controversial subject at best. It is also essential to dramatically increase agricultural productivity within the next few decades. Next, it is necessary to be able to more efficiently deliver food from where it is produced to where it is consumed thereby decreasing much of the food spoilage that is common at present. Decreasing food wastage also entails addressing issues such as portion size in restaurants in more affluent countries. Then, efficient and productive agriculture needs to be practiced on what is now considered to be marginal land. To boost crop yields, in the short term, it will be tempting to use even more agricultural chemicals than at present. However, such an approach will certainly be counter-productive in the longer term. Thus, instead of the increased use of chemicals, it is essential that the use of transgenic plants and the widespread application of plant growth-promoting microorganisms, both bacteria and fungi, be embraced and practiced on a large scale.

Given the enormous amount of experience that scientists have had with transgenic plants over the past 35 or so years, it is time for informed scientists to take the lead and educate both the public and the politicians regarding the many potential benefits of this technology. And, where there are perceived problems with the development or use of transgenic plants, it is absolutely essential that sufficient resources of both time and money be directed to solving those problems for everyone's benefit. Transgenic plants have been produced in the field for human use since 1996, with ~2 million farmers in 30 countries planting over 190 million hectares of transgenic crops in 2017 (Table 1.1). In addition, the amount of land devoted to the growth of transgenic plants continues to grow (doubling every 5–7 years). Thus, the area devoted to transgenic crops has increased ~120-fold since 1996. Currently, the major transgenic crops worldwide include (in order of hectares planted) soybean, maize (corn), cotton, canola, tomato, squash, potato, and papaya. However, in the not too distant future, many more transgenic plants are likely to become commercially available as scientists have already engineered, and studied in controlled environments, more than 150 different plant species. It has been suggested that within the next 10–15 years, the majority of agricultural crops worldwide will be transgenic.

Despite the reluctance of most European countries to accept the use of transgenic crops, as of 2017, more land is devoted to the growing of transgenic crops in the developing than in the developed world. According to Dr. Florence Wambugu (Nairobi, Kenya), "The African continent urgently needs agricultural biotechnology, including transgenic crops, in order to improve food production. Famine provides critics with an opportunity to promote an anti-biotech message that only results in millions of people, who urgently need food, starving to death." In 2002, the Food and Agriculture Organization of the United Nations endorsed the development and use of transgenic crops. In July of 2016, a group of 110 Nobel Prize winners in science published a letter to the opponents of transgenic plants including Greenpeace

Table 1.1 Major countries planting genetically modified crops in 2017 including the area planted in millions of hectares (Mha)

Country	Area planted (Mha)
USA	75
Brazil	50
Argentina	24
Canada	13
India	11
Paraguay	3
Pakistan	3
China	2.8
South Africa	2.7
Bolivia	1.3
Uruguay	1.1
Australia	0.9
Philippines	0.6
Myanmar	0.3
Burkina Faso	0.3
Sudan	0.2
Spain	0.1
Mexico	0.1
Columbia	0.1

(a nongovernmental agency) saying that these plants were both safe and needed. They went on to say that continuing to prevent the dissemination of Golden Rice (and other transgenic plants) could be considered to be a "crime against humanity."

Finally, transgenic plants are useful for a very wide range of applications besides increasing crop yields. These include developing plants with abiotic stress resistance, insect resistance, virus resistance, herbicide resistance, pathogen resistance, better controlled fruit ripening, increased nutritional content, altered flower color, altered oil composition, increased flower lifetime, as well as using plants to produce edible vaccines and a host of potentially less expensive human therapeutic agents. Notwithstanding the enormous benefit that is realized from transgenic plants, the increased development and use of these plants will not by itself solve world hunger. Therefore, for the remainder of this book, the biochemical and genetic mechanisms that plant growth-promoting bacteria (PGPB) utilize are discussed in some detail with the idea that these naturally occurring bacteria have facilitated plant growth in nature for many millions of years.

1.3   Plant Growth-Promoting Bacteria

The interaction between bacteria and plants may be beneficial, harmful, or neutral for the plant, and sometimes the effect of a particular bacterium may vary as the soil conditions change (Fig. 1.3). Plant beneficial bacteria are often termed plant growth-promoting bacteria. Bacteria that interact with plants but have no discernable effect

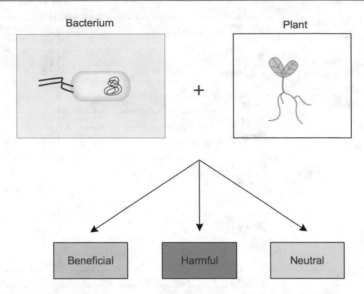

Fig. 1.3 Schematic representation of the possible effects of bacteria on plants

on the plant (although there is often a benefit to the bacteria) are termed commensal bacteria. Bacteria that are harmful to plants are generally considered to be plant pathogens (phytopathogens). PGPB are typically soil bacteria that facilitate plant growth and are often found in association with plant roots, and are sometimes found on plant leaves or flowers, or within plant tissues. It is believed that some PGPB began to develop a complex and mutually beneficial relationship with flowering plants around 80–100 million years ago. From a modern perspective, the rationale behind deliberately adding PGPB to plants (primarily in agriculture) to facilitate plant growth is that this approach copies and builds upon what has been a highly successful relationship in nature.

A large number of PGPB are commonly found in the plant rhizoplane, i.e., attached to the root surface, and the plant rhizosphere, i.e., the area immediately around the roots, generally up to 1 mm from the root surface (Fig. 1.4). This typical bacterial localization reflects the fact that most plant roots commonly exude a significant fraction of the carbon that is fixed by the plant through the process of photosynthesis and provide this fixed carbon to soil microbes that use it as a food source. Alternatively, some PGPB are endophytic (endo = inside; phyte = plant) and as such are able to enter into a plant and colonize a portion of the plant's interior. These bacteria do this without harming the plant and in fact use more or less the same mechanisms as rhizospheric bacteria to promote plant growth and development. Endophytic PGPB may be divided into two general types. First, those bacteria that bind relatively nonspecifically to plant root surfaces and then enter the plant through different points including tissue wounds, stomata (pores found in the epidermis of leaves), lenticels (raised pores in the stems of woody plants), root cracks, and germinating radicles (the part of the embryo that develops into the root system).

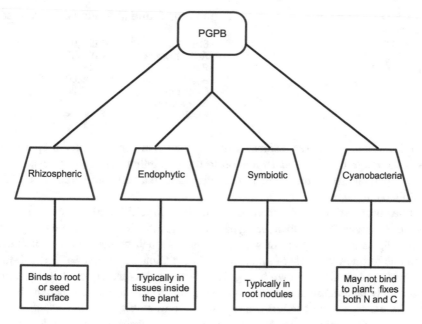

Fig. 1.4 Overview of different types of plant growth-promoting bacteria (PGPB)

Second, those bacteria that bind and infect only a very limited number of specific plants; these were formerly referred to as symbiotic bacteria (e.g., rhizobia that interact with specific legumes). Following infection, symbiotic bacteria typically form nodules on the plant root where this PGPB resides. Finally, although they are not always considered to be PGPB, many cyanobacteria are able to facilitate the growth of plants, usually aquatic plants. These cyanobacteria often fix both carbon and nitrogen; they sometimes but not always bind to plant surfaces.

The scientific literature often refers to PGPR or plant growth-promoting rhizobacteria. This older term is often still used to describe any bacterium that facilitates plant growth. However, strictly speaking, PGPR do not include endophytes, symbiotic bacteria, or cyanobacteria. Therefore, the more inclusive term PGPB is preferred, especially since, for the most part, all of these bacteria use similar mechanisms to promote plant growth and development.

1.3.1   Organisms in Soil

The natural world contains a very large number of species of different organisms, most of which are unknown and uncharacterized. For example, it has been estimated that there are ~300,000 species of plants, 30,000–40,000 species of protozoa, and 25–30 species of chromista (e.g., brown algae, diatoms, etc.). Being much smaller in size, the number of different microorganisms is very difficult to estimate so that recent estimates of the number of fungal species range from a mere 600,000 to more

Fig. 1.5 Overview of the living organisms found in a typical gram of soil

Organisms found in a typical gram of soil

Bacteria	90,000,000
Actinomycetes	4,000,000
Fungi	200,000
Algae	30,000
Protozoa	5,000
Nematodes	30
Earthworms	<1

than 5 million. Moreover, estimates of the total number of bacterial species in existence range from around 10 million to up to 1 billion. Given this perspective, it is not particularly surprising that the soil is home to a very large proportion of the world's biodiversity.

It has been estimated that there are more than 94 million organisms in a single gram of soil, with most of them being bacteria (Fig. 1.5). Of course, there really is no such thing as a typical gram of soil, and soils and their contents vary enormously from one location to another (with some soils containing between 10- and 100-fold the levels of organisms shown in Fig. 1.5). Nevertheless, nearly all soils contain a surprisingly large number of microorganisms, the vast majority of which are bacteria. The main groups of PGPB can be found along with the phyla Cyanobacteria, Actinobacteria, Bacteroidetes, Firmicutes, and Proteobacteria. In addition to the very large number of bacteria that are typically found in soil, it has been estimated that many soils may contain an equal or even greater number of number of viruses, largely bacterial and plant viruses. Unfortunately, it is technically extremely difficult to accurately estimate either the number or the types of viruses found in various soils. Given the fact that soil is an enormously complex environment, it is no surprise that plant growth and development is dramatically impacted by soil organisms, minerals, nutrients, physical properties as well as a wide range of chemicals.

1.3.2   Root Exudation

The concentration of microbes around the roots of plants is typically 10- to 1000-fold greater than the concentration of microbes found in the rest of the soil, which is sometimes called the bulk soil (Fig. 1.6). This is generally attributed to the fact that plants commonly exude from ~5% to ~30% of all photosynthetically fixed carbon through their roots. Both the quantity and the composition of root exudates are determined by a number of different factors including plant species, plant age, temperature, the amount of available light, plant nutrition, soil moisture, plant cultivar (a variety of a plant that has been created or selected intentionally and maintained through cultivation), and the presence of biotic and/or abiotic environmental stressors. Although root exudation represents a significant cost to plants in terms of energy and carbon resources, root exudates have been suggested to play important roles in the functioning of plants. In the top layers of the soil, root exudates provide microorganisms with easily degradable carbon rich substrates that drive

Fig. 1.6   Scanning electron micrograph of a PGPB bound to a canola root hair

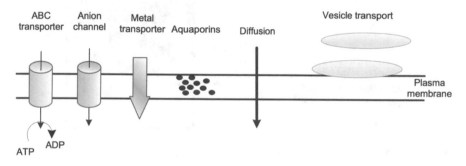

Fig. 1.7   Overview of the mechanisms which plant roots use to exude various compounds

microbial metabolic processes. Root exudates, depending upon their chemical composition, can shape microbial communities that are more or less specific to particular plant species. This occurs as a consequence of certain exudates stimulating or inhibiting particular microbial populations. Root exudates may regulate the local soil microbial community by secreting both chemical attractants and repellants, helping to cope with herbivores, changing the chemical and physical properties of the soil, and preventing the growth of potentially crop-inhibiting plant species. Root exudates are generally composed of low molecular weight compounds such as amino acids, sugars, organic acids and phenolics as well as some higher molecular weight compounds such as mucilage (viscous polysaccharides present in plant roots) and proteins. Sugars typically attract various microbes; flavonoids interact with mycorrhizae (a type of plant growth-promoting fungi), rhizobia (bacteria that form a symbiotic relationship with specific legume plants) and pathogenic fungi; aliphatic acids also attract certain bacteria.

The mechanism(s) by which plant roots secrete/exude compounds is considered to be a largely passive process (shown schematically in Fig. 1.7). This can occur through diffusion, ion channels, or vesicle transport. It is thought that ABC transporters, driven by ATP hydrolysis, are responsible for the exudation of lipids and flavonoids; anion channels for sugars and other carbohydrates; metal

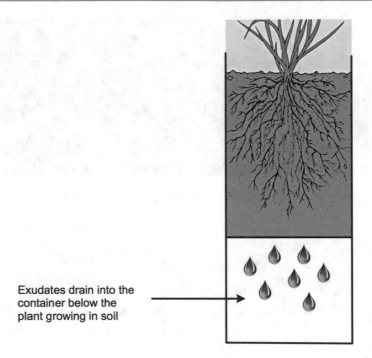

Fig. 1.8 A simple apparatus for collecting root exudates

transporters for various metals; aquaporins (proteins embedded in the cell membrane) for water and uncharged molecules; diffusion for a range of low molecular weight compounds; and vesicles for high molecular weight compounds like proteins.

Sometimes root exudates contain antibiotics that inhibit the growth of either bacteria or fungi, as well as numerous other bioactive compounds. For example, root exudates of the plant *Lithospermum erythrorhizon*, a Chinese herbal medicine, produce pigmented naphthoquinones that have biological activity against soilborne bacteria and fungi. Isolating and identifying some of these compounds is an important step in developing some of them for use in combatting either plant or animal diseases. One way to isolate, identify, and quantify unique bioactive compounds includes growing plants in special containers with porous bottoms and collecting the root exudates (Fig. 1.8), bioassaying the root exudates for various activities and then fractionating the root exudates until a pure compound is obtained.

1.3.3   Effect of PGPB on Plants

PGPB have been extensively documented to positively affect plants in a number of different ways. Thus, as a consequence of treating plants with PGPB any one or more of the following may be observed (Fig. 1.9): (1) increased plant biomass (Fig. 1.10); (2) increased plant nitrogen, phosphorus or iron increased content; (3) increased root

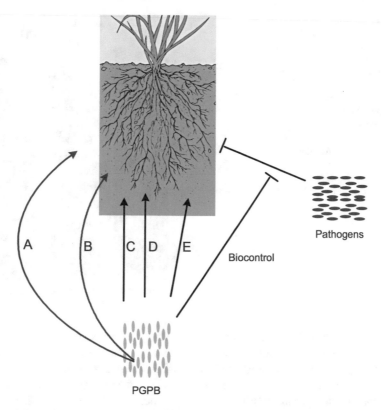

Fig. 1.9 Schematic representation of PGPB promoting plant growth. Some PGPB act as biocontrol agents that stimulate plant growth indirectly by inhibiting the growth and/or functioning of various phytopathogens. Other PGPB promote plant growth directly resulting in (**a**) increased plant biomass including increased root and shoot length; (**b**) increased plant assimilation of key metabolites such as fixed nitrogen, phosphorus, and iron; (**c**) enhanced seed germination; (**d**) increased plant resistance to a wide variety of abiotic stresses; and (**e**) increased production of various useful secondary metabolites

and/or shoot length; (4) enhanced seed germination; (5) increased plant disease resistance; (6) increased plant tolerance to various environmental stresses; (7) increased production of useful secondary metabolites (e.g., essential oils in *Origanum majorana* or sweet marjoram (Fig. 1.11a); and (8) better plant nutrition, especially the edible portions of the plant.

In the example shown in Fig. 1.10, the well-characterized endophytic PGPB *Paraburkholderia phytofirmans* PsJN (previously called *Burkholderia phytofirmans* PsJN) was found to significantly promote the growth (biomass production) of three different cultivars of switchgrass (*Panicum virgatum* cv. Alamo). Switchgrass is a non-crop plant that grows well on marginal land (generally considered to be unsuitable for agriculture) that may be useful as a so-called "biomass crop," in this case used for the production of ethanol or butanol, or as a component of phytoremediation protocols (see Chap. 10).

Fig. 1.10 Example of a
PGPB increasing plant
biomass. Increase of
switchgrass (*Panicum
virgatum*) cv. Alamo biomass
following 1 month of growth
and treatment with the
endophytic PGPB
*Paraburkholderia
phytofirmans* PsJN. Shawnee,
Nebraska, and Forestburg are
different cultivars of
switchgrass. The data is from
Biotechnol. Biofuels 5:37
(2012)

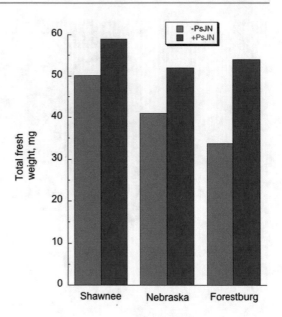

PGPB may stimulate plant growth in a variety of ways other than by merely increasing biomass. For example, several different PGPB strains (previously shown to increase plant biomass) were tested for the ability to promote the production of an essential oil in the plant *Origanum majorana* (sweet marjoram). In this case, the *Pseudomonas fluorescens* strain was slightly better than the other PGPB that were tested in facilitating plant growth (Fig. 1.11b). However, the *P. fluorescens* strain promoted the production of the essential oil (used to flavor foods and in fragrances) produced by this plant nearly 30 times compared to the untreated (control) plant (Fig. 1.11a). Importantly, the composition of the essential oil remained unchanged following bacterial treatment of the plant roots. In addition, another PGPB strain, also a strain of *Pseudomonas fluorescens*, was found to enhance both biomass yield and ajmalicine (an alkaloid that acts as an antihypertensive drug used in the treatment of high blood pressure) production in *Catharanthus roseus* (Madagascar periwinkle) plants that were subjected to water deficit stress.

Another example of the effectiveness and possible practical use of PGPB comes from the observation that it is possible to select bacterial strains that, in addition to increasing plant biomass, can significantly increase the yield of chili fruit (*Capsicum annuum* L.) under field conditions. In this work, scientists isolated a number of rhizosphere bacteria and then tested them for the ability to produce the plant hormone indoleacetic acid (IAA) and iron-binding siderophores; then 15 selected strains were tested for the ability to promote plant growth in pots in the greenhouse. The three most effective strains were subsequently tested in the field for the ability to increase the yield of chilies. As shown in Table 1.2, when two of the selected strains were applied together the fruit number and biomass increased dramatically. In the end, with the best strain combination it was possible to increase the number of chilies

Fig. 1.11 Example of a PGPB increasing plant oil production. Testing of four different putative PGPB for the ability to increase (**a**) essential oil production and (**b**) leaf number in *Origanum majorana* (sweet marjoram) plants treated with the indicated bacteria. The data is from Biochem. Systemat. Ecol. 36:766–771 (2008)

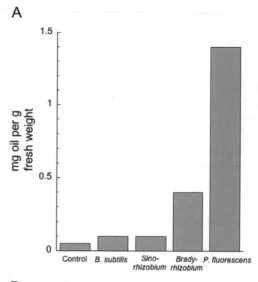

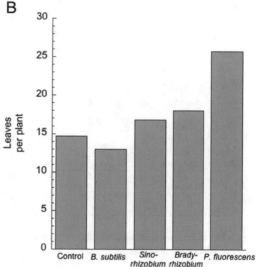

per plant by approximately threefold and the biomass of chilies per plant by sixfold compared to the control that was untreated by added PGPB.

PGPB may affect plant growth and development either directly or indirectly (Fig. 1.12). Direct promotion of plant growth occurs when a bacterium either facilitates the acquisition of essential nutrient resources from the environment or modulates the level of phytohormones within a plant. Nutrient acquisition facilitated by PGPB, or at least the best studied of the nutrients in this category, includes nitrogen, phosphorus, and iron. PGPB are known to directly affect plant growth by synthesizing and providing one or more of the phytohormones auxin, cytokinin, and

Table 1.2 Effect of bacterial isolates on the yield of chili (*Capsicum annuum* L.) under field conditions

Treatment	Plant height (cm)	Number of chili fruits per plant	Biomass of chili fruits per plant (g)
Growth medium	36.34	40.58	65
Strain C2	37.40	85.42	223.33
Strain C32	38.27	39.79	103.33
Strain C2 + strain C32	54.17	129.77	386.67

Strain C2 is a *Bacillus* sp.; strain C32 is a *Streptomyces* sp. The data are from Aust. J. Crop Sci. 4:531–536 (2011)

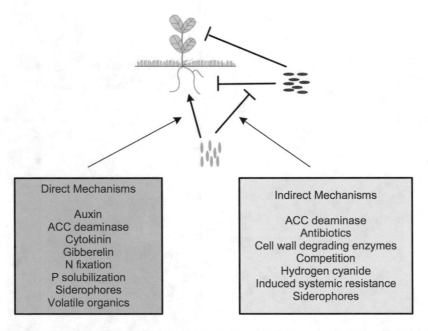

Fig. 1.12 Schematic representation of the main direct and indirect mechanisms used by PGPB to facilitate plant growth and development

gibberellin. In addition, some PGPB can lower levels of the phytohormone ethylene by synthesizing an enzyme, 1-aminocyclopropane-1-carboxylate (ACC) deaminase, that cleaves the compound ACC, the immediate precursor of ethylene in all higher plants. Indirect promotion of plant growth occurs when a PGPB (sometimes called a biocontrol PGPB or just a biocontrol bacterium) prevents, or at least decreases, the damage to a plant that might otherwise ensue as a consequence of infection of the plant with any one of a large number of different phytopathogens. These phytopathogens typically include various soil fungi and bacteria. In addition, biocontrol PGPB, by a range of different indirect mechanisms, may lessen the damage to plants from either insects or nematodes.

Throughout the detailed discussions of PGPB mechanisms that appear in subsequent chapters in this book, it is essential to keep several things in mind. (1) Not all PGPB utilize the same mechanisms. (2) Any particular PGPB strain may use any one or more of the many mechanisms that are discussed herein. (3) A PGPB strain may use either direct or indirect mechanisms, or both. (4) Not all PGPB strains of the same genus and species have the same PGPB activities. (5) Different plants (including different cultivars of the same plant) may respond differently to a particular PGPB strain, usually reflecting differences in plant physiology/biochemistry. These differences in response to a particular PGPB strain may be a consequence of plant age and growth conditions (including soil composition, plant growth temperature, and the presence or absence of stressful compounds and/or phytopathogens in the soil).

Since rhizobial strains, that usually form a highly specific relationship with their plant hosts, employ more or less the same set of mechanisms as other PGPB, they are sometimes able to bind to the roots of nonlegume plants and subsequently stimulate plant growth. Thus, it has been observed that *Rhizobium leguminosarum* strain PEPV16 was able to colonize the roots of lettuce and carrot plants (both nonlegumes) and subsequently promote the growth of both plants as well as increase the uptake of both N and P in the edible parts of both plants.

In addition to promoting the growth of crop plants in a traditional agricultural setting, PGPB may be used to stimulate the rooting of some vegetatively propagated plants (i.e., by asexual reproduction) such as olive plant seedlings. In Southern Spain, the world's largest olive oil producing region, there is significant motivation for production of new trees to be free of any chemical input including herbicides, pesticides, etc. (in this region, old trees are removed and replaced by new trees approximately every 12–15 years). In one set of experiments, researchers added various PGPB to cuttings of three different olive tree cultivars. Although the results were somewhat variable, depending on the cultivar and the mode of adding the PGPB, the *Pantoea* strain was the only bacterial strain that always produced a high rooting percentage (typically equivalent to or better than the addition of indolebutyric acid, an analog of indoleacetic acid or IAA which is often used to stimulate rooting). As this bacterial strain was not the best IAA producer of those strains that were tested, its effectiveness is thought to reflect a combination of IAA production and ACC deaminase activity (Table 1.3). This result was in agreement with other laboratory and greenhouse studies where PGPB that both synthesized IAA and contained ACC deaminase activity produced a large number of relatively long roots (which are more effective in helping the plant to become established) compared to IAA or indolebutyric acid that produced a large number of very short roots. In summary, these experiments suggest that PGPB with appropriate traits may become effective components of the organic production of olive trees used for producing olive oil.

In an effort to develop the single-celled green algae *Chlamydomonas reinhardtii* (a eukaryote that is considered to be a lower plant) as a commercial source of biodiesel (an environmentally friendly renewable and biodegradable fuel), one group of scientists cultured *C. reinhardtii* together with the soil bacterium

Table 1.3 PGPB properties of several PGPB strains and their ability to promote rooting of "Picual" olive tree cuttings

Bacterial strain	IAA production (mg/L)	Siderophore production	Phosphorus solubilization	ACC deaminase activity	% Rooted cuttings
Azospirillum	20	–	–	–	10
Pantoea	21	+	–	+	38
Chryseobacterium	7	–	–	–	7
Chryseobacterium2	13	++	–	–	30
Pseudomonas	45	+++	+	–	16
Water	n.a.	n.a.	n.a.	n.a.	4
IBA	n.a.	n.a.	n.a.	n.a.	40

Chryseobacterium and *Chryseobacterium2* are two different strains of this bacterium. IBA is indolebutyric acid. In this experiment, water acts as a baseline or negative control and IBA acts as a positive control. N.a. is not applicable

Azotobacter chroococcum. The aerobic nitrogen fixing bacterium, *A. chroococcum,* was chosen for this experiment based on earlier observations that both algal biomass and lipid accumulation decreased when there was insufficient nitrogen in the algal growth medium. Thus, it was thought that the addition of an efficient nitrogen fixing bacterium (that could provide nitrogen on a continuous basis) might increase the algal biomass and lipid accumulation. Since *A. chroococcum* grows well under aerobic conditions, a trait that is relatively uncommon for nitrogen fixing bacteria, it was considered to be an ideal organism to be cocultured with *C. reinhardtii* that, as a photosynthetic organism, produces copious amount of oxygen, a compound that is inhibitory to most other nitrogen fixing bacteria. As expected, when these two organisms were cocultured, both the algal biomass and the algal lipid production increased significantly (i.e., ~2.5- to 3.0-fold for algal biomass and ~3.0- to 4.0-fold for algal lipid production). A model depicting this cocultivation system is shown in Fig. 1.13. Of course, for this laboratory demonstration to become an industrial reality it will be necessary for scientists to develop large-scale and continuous methods for the cultivation and harvesting of these organisms.

1.4 PGPB Mechanisms

Historically, most of the interest in PGPB was based on the ability of either (1) some PGPB to act as biocontrol agents decreasing the damage to plants from phytopathogens or (2) some rhizobia strains to form symbiotic relationships with specific legumes and then providing those plants with fixed nitrogen. In fact, these are still major interests of a large number of researchers in this field. As well, the vast majority of commercialized PGPB strains have been developed to address these two issues. This focus reflects the generally held view that in the field it is easier to reproducibly demonstrate the efficacy of biocontrol plant growth-promoting bacteria

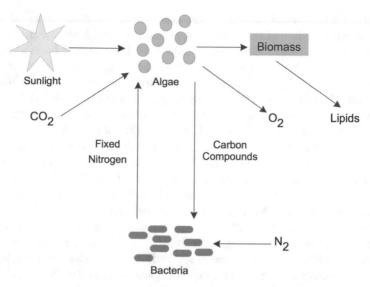

Fig. 1.13 Schematic representation of the cocultivation of the green algae *C. reinhardtii* and the aerobic nitrogen fixing soil bacterium *A. chroococcum*. Here, the algae, driven by the energy from sunlight, convert CO_2 into readily utilized low molecular weight carbon compounds while the bacteria fix atmospheric nitrogen. Thus, the algae provide a carbon source for the bacteria while the bacteria provide a nitrogen source for the algae. The increased algal biomass contains an increased amount of algal lipids. Molecular oxygen is a by-product of algal photosynthesis. Neither the algae nor the bacteria are drawn to scale

or nitrogen fixing rhizobia as opposed to PGPB that promote plant growth in other ways.

As mentioned briefly in Sect. 1.3.3, there are several ways in which different PGPB can directly facilitate the proliferation of their plant hosts (Fig. 1.12). PGPB may fix atmospheric nitrogen and supply it to plants; synthesize low molecular weight siderophores that can solubilize and sequester iron from the soil and provide it to their own as well as plant cells; synthesize plant hormones (phytohormones) such as auxins, cytokinins, and gibberellins which can enhance plant growth; provide mechanisms for the solubilization of phosphorus from the soil making it more readily available for plant growth; and synthesize the enzyme ACC deaminase which lowers potentially inhibitory plant ethylene levels. Any particular bacterium may affect plant growth and development using any one, or more, of these mechanisms. Moreover, a bacterium may utilize different traits at various times during the life cycle of the plant. For example, following seed germination a PGPB may lower the plant's ethylene concentration thereby decreasing the ethylene inhibition of seedling root length. Once the seedling has depleted the resources that are contained within the seed, the PGPB may help to provide the plant with a sufficient amount of iron and phosphorus from the soil. Through early plant development, PGPB may stimulate cell division by providing appropriate levels of phytohormones.

Some PGPB can act to thwart the proliferation or functioning of various phytopathogens including fungi, bacteria, nematodes, and insects and in this way indirectly promote plant growth. These biocontrol bacteria may be used commercially instead of various chemical fungicides, nematicides, insecticides, and other pesticides. Biocontrol PGPB employ a wide variety of mechanisms including the production of antibiotics, hydrogen cyanide, siderophores, cell wall-degrading enzymes, enzymes that lower plant ethylene levels, systemic resistance, quorum quenching mechanisms, and insecticidal and nematicidal proteins.

It is not unusual for a PGPB strain to contain several different plant beneficial properties. These properties include, but are not limited to, the following direct and indirect mechanisms: ACC deaminase, NO production, IAA production, phosphate solubilization (of both inorganic and organic phosphates), extracellular protease production, phenazine production, pyrrolnitrin production, HCN production, DAPG production, pyoluteorin production, cytokinin production, and chitinase production. In addition, some PGPB may fix nitrogen, induce systemic resistance to various pathogens, produce surfactants, produce gibberellins, and synthesize siderophores. Simplistically, it appears that PGPB that contain more beneficial properties are likely to be more effective at promoting plant growth. This is because it is assumed that a greater range of beneficial traits will protect the plant partner of the PGPB from a wide range of environmental pressures. In one study, the occurrence of numerous properties involved in facilitating plant nutrition, plant hormone modulation, or pathogen inhibition was analyzed in a collection of maize (corn) rhizosphere and bulk soil isolates of fluorescent pseudomonads. Surprisingly, and contrary to expectations, maize rhizospheres preferentially selected pseudomonads with a limited number (up to five) of plant beneficial traits. Interestingly, approximately 75–95% of the pseudomonads produced IAA, around 50–60% solubilized phosphate or produced extracellular proteases, about 40–50% produced NO or HCN, and around 15–25% synthesized ACC deaminase. All of the other tested biological properties (mostly mechanisms for limiting pathogen proliferation) were found at levels from 0 to 20% of the samples. This data suggests that at least for fluorescent pseudomonads, most strains have adapted to certain environments. That is, rather than being generalists and trying to do everything, they have become specialists performing optimally in specific niches. One possible reason for the limited number of plant beneficial traits in a single PGPB strain may have to do with the attempt by the bacterium to limit the number of nonessential genes that it carries and expresses, i.e., the bacterium is effectively limiting its metabolic load.

1.5   Screening for New PGPB

Until relatively recently, the isolation of new PGPB strains typically required that researchers use a long and quite tedious approach (Fig. 1.14). Briefly, this task typically entailed isolating a large number of bacterial strains (often hundreds, thousands, or even tens of thousands) from the soil around the roots of various

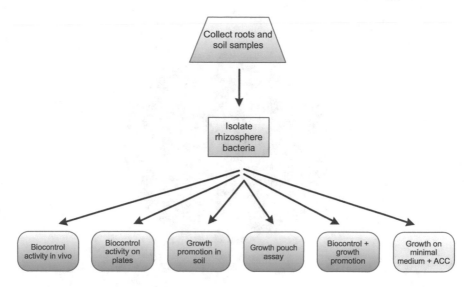

Fig. 1.14 Flowchart depicting various methods for selecting new PGPB. Older approaches are highlighted in blue

crop plants (i.e., the rhizosphere). These bacteria were then tested using any one of a number of different biological assays. The simplest of these screens entailed testing the isolated bacteria for biocontrol activity against one or more fungal pathogens. One way to do this is to assess the in vivo biocontrol activity of the bacterial isolates (adding these bacteria to plants or seeds prior to adding a phytopathogen). This approach tests the ability of isolated bacterial strains to protect plants in soil against damage by certain fungal or bacterial pathogens (by mechanisms unknown). Another way to test bacterial isolates for biocontrol activity is to test each bacterial strain for the ability to prevent the proliferation of one or more phytopathogens on agar plates (Fig. 1.15). In this assay, effective biocontrol strains secrete compounds that inhibit the growth of the phytopathogen as evidenced by a zone of clearance around the bacterial strain being tested. While these procedures for selecting new biocontrol bacteria are effective, they are both time consuming and limited in that they only select for certain types of biocontrol strains and are completely unable to select for bacteria that promote plant growth directly. Typically, bacteria that promote plant growth directly are selected from a large number of soil bacteria by testing each individual bacterial strain for its ability to promote plant growth, either in growth pouches (generally grown under controlled conditions in a growth chamber) or in soil, a process that typically requires growing a large number of plants (often 30–50, or more) for each strain that is being assessed. The large number of plants required to directly test for plant growth promotion is necessitated by the natural variation in plant seeds and their growth, especially in short term (i.e., a few

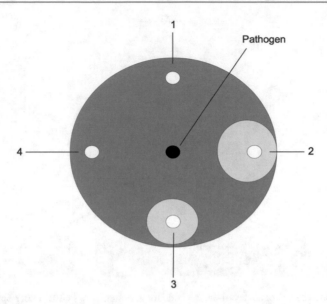

Fig. 1.15 Representation of method used to assess PGPB for biocontrol activity. The phytopatho-gen is spotted in the middle of the agar plate (shown in black) while various potential biocontrol PGPB strains are spotted around the edges of the plate (spots 1–4; shown in white). Following incubation, the phytopathogen proliferates over the entire plate except where it is prevented from growing by one of the strains being tested. The strains spotted in positions 1 and 4 do not exhibit any biocontrol activity while the strains spotted in positions 2 and 3 prevent the pathogen from proliferating and show a zone of clearance in their vicinity

days) assays. In addition, plant growth assays may be tested with several different plants, different soils, and different growth conditions. The growth of plants in soil is typically assessed by measuring the plant's aboveground plant biomass (wet or dry weight, although dry weight is generally preferred). Growth pouches (which do not include soil and typically follow plant growth for only a few days) are often used to measure relative root growth since measurement of plant root biomass or length is technically difficult to do when plants are grown in soil. This is because even the careful removal of soil form plant roots often results in the breakage of smaller plant roots. Nevertheless, with care it is sometimes possible to assess the root biomass of plants grown in soil (Fig. 1.16). The abovementioned procedures often take several months or more before a strain suitable for more extensive testing can be identified. As an alternative to these labor intensive methods for isolating new PGPB, a rapid and novel procedure for the isolation of ACC deaminase-containing bacteria that, by definition, are able to promote plant growth was developed (see Chap. 5). This procedure is both rapid and highly effective, yielding new PGPB, ready for further characterization and testing, within 1–2 weeks.

Fig. 1.16 Mung bean roots
from plants grown in soil
either in the absence (left) or
the presence (right) of an ACC
deaminase-containing PGPB
(see Chap. 4 for details)

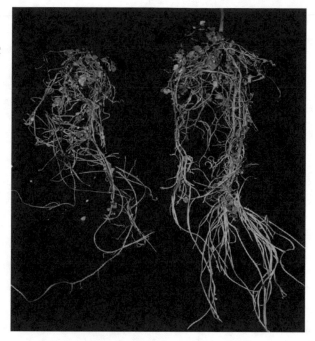

In addition to the abovementioned methods of identifying new and unique PGPB, some workers have tested some well -known bacteria that have been utilized for a variety of other purposes to see whether they might also promote plant growth. For example, a number of researchers have shown that many lactic acid bacteria can act as PGPB. Lactic acid bacteria are Gram-positive, acid-tolerant, generally nonsporulating, nonrespiring, bacteria that are typically found in decomposing plants and milk products and produce lactic acid as the major metabolic end product of carbohydrate fermentation. As a consequence, lactic acid bacteria have histori-cally been used in the production of fermented foods such as yogurt, pickled vegetables such as kimchi, and sourdough bread, as the resulting acidification inhibits the growth of spoilage agents. Lactic acid bacteria are considered generally regarded as safe (GRAS) organisms, due to their ubiquitous appearance in food and their contribution to the healthy microflora of human mucosal surfaces. While some lactic acid bacteria can directly promote plant growth, many more have been shown to be effective biocontrol agents.

Two very different groups of workers use PGPB in their studies. On the one hand, scientists whose efforts are motivated by the desire to increase plant crop yields primarily isolate, characterize, and primarily utilize rhizospheric PGPB. On the other hand, scientists whose focus is using PGPB to facilitate phytoremediation (environ-mental cleanup using plants; see Chap. 10) rarely use rhizospheric PGPB and instead for the most part use endophytic PGPB (see Chap. 2). More recently, workers focused on augmenting crop yields have increasingly turned their attention to the roles played by endophytic bacteria in promoting plant growth. Endophytic bacteria

Fig. 1.17 Isolation of new strains of endophytic PGPB. Soil samples are collected from the rhizospheres of various plants; seeds are planted in each soil sample; once the plants have reached a suitable size, the plant tissue, typically shoots and leaves, is first washed to remove any surface adhering bacteria and then macerated; the bacteria from the macerated tissue is isolated and characterized

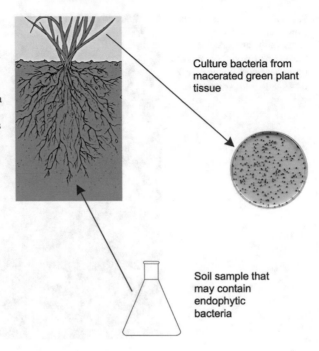

Culture bacteria from macerated green plant tissue

Soil sample that may contain endophytic bacteria

include (1) facultative endophytes that can live inside plants as well as in other habitats; (2) obligate endophytes that can only live inside of a plant; and (3) opportunistic endophytes that can occasionally enter plants and live inside their tissues. However, the fact that scientists can isolate and characterize specific endophytic strains means that they are most likely dealing exclusively with facultative endophytes. Facultative endophytes may be isolated by collecting rhizosphere soil samples, planting seeds in those samples, isolating and macerating plant tissue grown from those seeds, and then culturing the bacteria contained within that tissue (Fig. 1.17).

1.6   Commercial Inoculants

The earliest recorded instance of the commercialization of PGPB occurred in the United States and the United Kingdom in 1895. This commercialization included the inoculation of legumes with strains of *Rhizobia* spp. This early history notwithstanding, it was nearly 100 years before the commercial interest in PGPB of various types became more widespread.

In 2012, organic agriculture accounted for a small fraction, i.e., ~6% of the land used for agriculture within the European Union (EU). Moreover, the amount of land devoted to organic farming within the EU continues to increase as does the number of farmers using bacterial inoculants in an effort to increase crop yield.

Table 1.4 Items that
should be considered in
commercializing a PGPB
strain

1. Rhizospheric vs. endophytic
2. Appropriate biological activity
3. Nonpathogenic
4. Easy to prepare
5. Easy of ship and store
6. Stable
7. Compatible with chemical additives
8. Optimized for local conditions
9. Suitable application technology
10. Patent protected
11. Genetically engineered
12. Similarity to other strains in the marketplace

1.6.1   Commercial Considerations

What are the features of a PGPB strain that a company needs to consider when
developing that strain for the marketplace? In the first instance, most of the laboratory studies of PGPB have focused on rhizospheric bacteria (Table 1.4). However, it
may be advantageous to utilize endophytic strains (which use essentially the same
mechanisms to promote plant growth as rhizospheric bacteria, and are discussed in
some detail in Chap. 2) that are less subject to variations of soil type, soil pH,
weather, etc. [It has been suggested that shortly after PGPB are introduced into the
soil in the field (without carrier), the population of those bacteria declines rapidly.]
To date, many of the PGPB strains that have been commercialized have not been
characterized as being either rhizospheric or endophytic. Second, it is important to
understand what types of biological mechanisms are required to promote plant
growth under different conditions so that these activities are present when needed
(see Chaps. 4–7). Moreover, depending upon the particular end use (e.g., biocontrol
of certain fungal pathogens) it is essential to select PGPB with specific useful and
desirable traits. Third, it is absolutely essential that the selected strain be nonpathogenic to both animals (including humans) and plants. Fourth, it is advantageous if the
selected strain is easy to grow on inexpensive medium that does not require any
specialized or costly fermentation equipment. In this regard, Gram-negative bacteria
are often easier to grow and manipulate in a laboratory environment. Fifth, the
selected strain should be sufficiently stable that it will not appreciably lose activity
during storage or shipping. It is advantageous if the strain can be shipped in a
lyophilized form (preferably without a chemical or physical carrier) since
lyophilized bacteria weigh much less than bacteria in solution. In addition, when
Gram-positive bacteria are grown to a late stage of growth so that they significantly
decrease or even exhaust the supply of nutrients in the medium, they often form
spores that are very stable and extremely long lived (they can be stable for many
years) and may be reactivated into vegetatively growing bacteria when the nutrient
conditions change upon the addition of fresh growth medium. Sixth, it is best to
avoid using strains that are genetically unstable. Thus, numerous workers have

found that strains sometimes have a particular metabolic activity when they are originally isolated but then lose that activity upon repeated subculturing in the laboratory. Obviously, using genetically unstable strains may result in a manufacturer growing and shipping a PGPB strain that is no longer effective as a plant growth promoter. Seventh, since many seeds are chemically coated, it is imperative that the PGPB strain does not interact in any way with the chemicals that are planted along with the seed. Otherwise, the chemical seed coating may inhibit either the growth or the functioning of the PGPB strain. [For example, in one recent study, scientists selected PGPB strains that were resistant to the inhibitory effects of the widely used herbicide glyphosate and yet were still able to increase the availability of soluble soil phosphates to treated plants.] Eighth, even an endophytic strain will perform more efficiently if it has been selected for certain soil conditions. Thus for example, a bacterium from a cold environment may function in a hot and dry environment, but probably not optimally. Ninth, it is necessary to develop and optimize the plant inoculation procedure so that the bacterium is as stable and active, especially prior to planting, as possible (see the section below). Tenth, some companies have sought to patent protect their PGPB strains while others have kept the precise nature of the active ingredients of their inoculants a trade secret (see Chap. 11). Regardless of what approach is taken, it is necessary for companies to develop some strategies that protect their intellectual property. Eleventh, although there are a number of reports where genetically engineered PGPB strains are more effective under laboratory conditions, for the foreseeable future, only naturally occurring PGPB are likely to be acceptable worldwide to both regulatory agencies and farmers. Only a very small number of bacterial strains have been approved by regulatory agencies for deliberate release into the environment. Twelfth, when introducing a new PGPB strain into the market, it is important to consider what similar or comparable strains already exist in order to distinguish the newly introduced strain from all others. While not unique to the commercialization of PGPB, it is necessary to actively market any new product so that the potential user of that product is convinced that it is different from all existing products and there is a real need for it.

1.6.2   Inoculation Methods

Many procedures work simply and easily in the lab; however, the scale up of these procedures, often by 1000- to 10,000-fold, is not necessarily straightforward. Bacterial preparations suitable for inoculating crop plants on a commercial scale may take any one of several different forms. Those bacterial inoculant preparations should provide some measure of physical protection for the PGPB once it is introduced into the soil (Table 1.5). For example, wet/liquid inoculants of bacterial suspensions are often sold in 1–5 L plastic bags. Perhaps surprisingly, wet inoculants generally exhibit good stability and lifetime (generally having a shelf life of 6 months to a year). Liquid formulations typically are aqueous, oil-based, or polymer-based. Polysaccharides such as gums, carboxymethyl cellulose, and polyalcohol derivatives

Table 1.5 Some examples of carriers used to deliver PGPB

Carrier material	Bacterium
Sterilized oxalic acid industrial waste	*Bradyrhizobium japonicum*
Perlite	*Rhizobium leguminosarum, B. japonicum*
Sawdust	*B. japonicum, Sinorhizobium meliloti, Mesorhizobium loti*
Alginate beads	*Bacillus subitis, Pseudomonas corrugate, Azospirillum brasilense, Enterobacter* sp., *Chlorella vulgaris, Phyllobacterium myrsinacearum, P. fluorescens, B. polymyxa, P. putida, Streptomyces* sp., *Azospirillum lipoferum*
Vermiculite	*B. japonicum, S. meliloti, A. lipoferum, Azotobacter chroococcum, P. fluorescens*
Peat	*S. meliloti, Rhizobium* spp.
Coal	*R. leguminosarum*
Xanthan gum	*Rhizobium* spp., *Arthrobacter* spp.
Bentonite	*Rhizobium* spp.
Carrageenan	*A. brasilense*
Biochar	*P. putida, E. cloacae, B. japonicum, P. libanensis, B.* sp., *A.* sp., *R.* sp.

are often used to alter the fluid properties of liquid formulations. However, unless they are produced locally, cell suspensions (which are mostly water) need to be shipped (often long distances) and shipping costs can be expensive. On the other hand, various carrier materials to which the PGPB are added can influence inoculum success by providing protective pore spaces and also by modifying the soil structure. These carriers typically increase the survival rate and shelf life of PGPB by protecting them from desiccation and also provide an effective environment for the rapid growth of these bacteria after their release.

Peat moss-based inoculants, where PGPB are mixed with peat (partially decayed vegetation that typically forms in wetland conditions where the rate of decomposition is slowed because of the lack of oxygen) are very commonly used in the commercial application of PGPB. These inoculants are generally quite stable and may last up to several years. Peat slurry inoculants, made from finely milled and sterilized peat, are commonly used in Australia to inoculate legume crops with various Rhizobia strains. These inoculants, which typically contain 10^9–10^{10} bacterial cells per gram of peat, often include the use of adhesives and other polymers to ensure good contact between the Rhizobia strain and the legume seed coat. Seeds prepared in this way are generally sown within hours of the inoculant preparation. In addition, adsorption onto inert materials like talc (a soft light-colored mineral), kaolinite (a hydrous aluminum silicate), lignite (a soft brownish-black coal), bentonite (an aluminum silicate clay from volcanic ash), biochar (charcoal produced by thermal decomposition of biomass in the absence of oxygen), or vermiculite (a hydrated silicate mineral) may also be used as a simple method of delivering PGPB to plant roots. Finally, PGPB may be delivered to plants via encapsulation, generally either in alginate or in calcium carbonate. The latter procedure is used commonly in Japan.

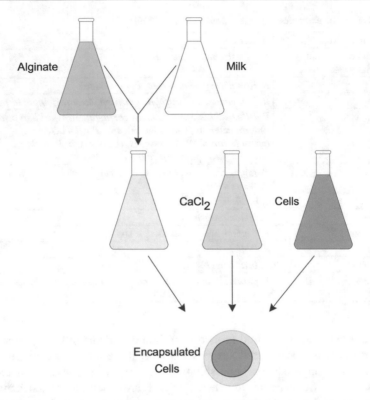

Fig. 1.18 A schematic overview of the encapsulation of PGPB cells in alginate. The dissolved alginate is mixed with skim milk before the mixture is combined with bacterial cells. The cell suspension is slowly added to a solution of $CaCl_2$ so that the alginate forms a gel around the bacterial cells. The encapsulated cells are shown much larger than their actual size relative to the flasks.

Bentonite is often used to prepare rhizobia inoculants. As a dry powder even after it is mixed with a suspension of the bacterial inoculant, it is easier to ship and store as it does not require refrigeration. Bentonite also buffers rhizobia against the possible effects of chemical pesticides that may be used to coat the seed.

Several groups of researchers have reported very promising results using alginate microbeads to encapsulate PGPB (Fig. 1.18). To prepare these microbeads, a concentrated suspension of PGPB was mixed with alginate and skim milk. The bacteria were then forced through a 0.2 nm pore and sprayed onto a 0.1 M solution of $CaCl_2$. The $CaCl_2$ causes the alginate to form microbeads which, when they are dried, can be stored with the bacteria within the alginate remaining viable for long periods of time. It has been argued that this method is much more reproducible than using peat as a carrier since different batches of peat may be quite variable from one another. However, while this approach works well in a laboratory setting, it remains to be seen whether it will be adopted for large-scale commercial use.

Table 1.6 Effect of different additives on the survival of populations of the PGPB *R. palustris* stored at 25 °C and 40 °C for 30 days. Each bacterial population was initially adjusted to 1×10^9 CFU/mL

Additive	Bacterial survival (log CFU/mL)	
	25 °C	40 °C
None	5.02	4.60
Alginate	5.90	3.72
Glucose	5.25	4.45
Glycerol	5.01	4.90
Horticultural oil	5.97	4.74
Polyethylene glycol	5.30	4.70
Polyvinyl pyrrolidone	5.80	4.38

Another group of researchers compared the effects of six different additives on the survival and efficacy of a strain of *Rhodopseudomonas palustris*, a Gram-negative, photosynthetic and nitrogen-fixing bacterium that is an effective PGPB in the laboratory but not in the field. This lack of effectiveness in the field has been attributed to inadequate formulation of this bacterium resulting in unstable bacterial preparations that have a short shelf life. Prior to this study, *R. palustris* was packaged as a liquid formulation; this reflects the fact that liquid formulations are considered easier to process and are somewhat less expensive than solid-based formulations. The additives that were compared in this study (each at a range of different concentrations) included alginate, glucose, glycerol, horticultural oil, polyethylene glycol, and polyvinylpyrrolidone (Table 1.6). When the bacterial survival rate was assessed at different temperatures and different additive concentrations, it was observed that the concentration of all of the additives had little effect on bacterial survival. However, the most effective additive was horticultural oil (0.5–1.0% w/v) that could maintain relatively high populations of the bacterium under a range of temperatures (from 4 °C to 40 °C). Moreover, horticultural oil is relatively inexpensive and has been shown to be environmentally safe. Horticultural oils are mineral oils that have been used for many years as a spray to control insect pests on fruit trees. Before horticultural oil is used as a preservative agent in PGPB formulations, it will be necessary to show that this work is highly reproducible when used in a field situation and is effective with a range of different PGPB.

Unfortunately, most of the research on microbial seed inoculation involves agrichemical and seed companies, and since the techniques and processes developed with particular crops and microorganisms can lead to a commercial advantage for the company involved, these approaches are hardly ever published, instead they are kept as "trade secrets." On the other hand, most of the published studies on microbial seed inoculation have not been scaled up for commercial use. Since growers and seed companies have to date indicated a strong preference for pre-inoculated seeds that can be used in the field in essentially the same way as conventionally treated seeds, optimizing methods of seed inoculation will, in the foreseeable future, likely remain a major concern in the commercialization of this technology.

In addition to applying PGPB to the roots of plants (as either seed treatment, root dip, or soil amendment) some PGPB are applied as a foliar spray. In this way, plants may be treated as needed throughout the growing season. This is important as a method of fighting plant pests and outbreaks of plant diseases. In this regard, by far the major bioinsecticidal spray includes various subspecies of *Bacillus thuringiensis* while more than 20 different strains of *Bacillus* spp. and *Pseudomonas* spp. have been commercialized as biocontrol agents for a variety of fungal and bacterial plant diseases.

1.6.3 Large-Scale Growth of PGPB

An essential aspect of the commercialization of PGPB is the need to develop strategies for the large-scale growth of these bacteria. Optimal growth conditions (for any microorganism) include determining, for any particular size bioreactor, the optimal temperature, pH, rate of mixing of the growing cells, and, the means of providing the growing cells with sufficient oxygen. Because oxygen is only sparingly soluble in water, it must be supplied continuously to growing (aerobic) bacterial cultures. However, the introduction of air (which is ~20% oxygen) into a bioreactor produces bubbles, and if the bubbles are too large, the rate of transfer of oxygen to the cells may be insufficient to support the optimal growth of the bacterial cells. Thus, fermenter design should include provision for monitoring the dissolved oxygen level of the culture, providing oxygen to the culture and adequately mixing the culture to efficiently disperse the bubbles. Adequate mixing of a microbial culture is essential for many aspects of a fermentation, including assurance of an adequate supply of nutrients to the cells and prevention of the accumulation of any toxic metabolic by-products in local, poorly mixed regions of the bioreactor. Effective mixing is easily attained with small-scale cultures, but it may become a major problem when the scale of fermentation is increased. Here, it is important to bear in mind that the optimal microbial growth conditions generally change with each tenfold increase in the volume of a bioreactor.

There are also other technical considerations. The design of the bioreactor is important. In this regard, it is necessary to ensure adequate sterility so that the culture does not become contaminated, and probes that permit the accurate and continuous on-line monitoring of as many critical reaction parameters as possible so that adjustments can easily be made throughout the course of the fermentation reaction (i.e., the growth of the microorganism). In addition, because high temperature sterilization may alter the composition of the microbial growth medium (e.g., by destroying vitamins or causing some components to precipitate), it is important to ascertain that the medium composition is still optimal for maximal microbial growth following sterilization.

Generally, in the scale-up of the growth of microorganisms from laboratory scale to production scale, fermentations are developed in a stepwise manner. A typical procedure (Fig. 1.19) begins with formulation and sterilization of the growth medium, and sterilization of the fermentation equipment. The cells are typically stored frozen in small aliquots (<0.5 mL), grown first as a stock culture (5–10 mL), then in a shake flask (200–1000 mL), and in a seed fermenter (10–100 L). Finally,

Fig. 1.19 Schematic representation of the large-scale growth of PGPB cells. A stored frozen aliquot (typically 50–500 µL) of bacterial cells are used to initiate the growth of a small liquid inoculum (5–10 mL) that in turn are used to inoculate 200–1000 mL of cellular growth medium. These cells, typically in mid-log phase of growth, are used to inoculate a small fermenter with a capacity of 10–100 L. Finally, the mid-log cells from this phase are used as an inoculum for 1000–10,000 L of growth medium in an industrial scale fermenter. Once the cells have reached the desired stage of growth, they are harvested, separated from the growth medium, by either continuous centrifugation or by filtration. None of the items are drawn to scale

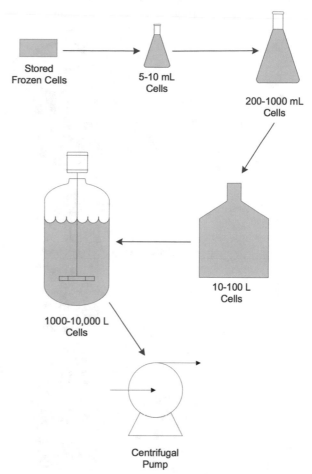

the production fermenter (1000–10,000 L) is inoculated. After the last fermentation step is completed, the cells are separated from the culture medium by either centrifugation or filtration.

Bacteria can be grown in batch, fed-batch, or continuous culture (Fig. 1.20). In batch fermentation (by far the most common means of growing bacterial cells), the sterile growth medium is inoculated with the appropriate bacterium, and the fermentation proceeds without the addition of fresh growth medium. In fed-batch fermentation, small volumes of concentrated nutrients are added at various times during the fermentation reaction (nutrients are typically added during the late logarithmic phase of bacterial cell growth); no growth medium is removed until the end of the process. In continuous fermentation, fresh growth medium is added continuously during fermentation, but there is also simultaneous removal of an equal volume of spent growth medium containing suspended bacteria. Notwithstanding its many perceived advantages, continuous fermentation is generally considered to be an experimental

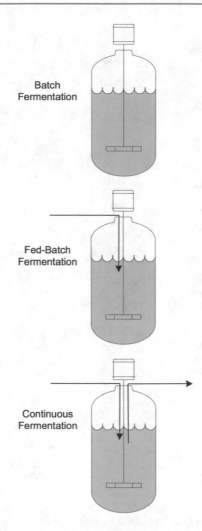

Fig. 1.20 A schematic representation of three different types of fermentation: batch, fed-batch, and continuous. In batch fermentation, all of the necessary ingredients are added at the start of the fermentation. In fed-batch fermentation, small amounts of concentrated growth medium may be added at one or more times during cell growth. This typically has the effect of maintaining the cells in mid-log phase for a longer period of time than under batch fermentation conditions thereby increasing the amount of cells produced, often from two to eightfold. In continuous fermentation, once the cells reach the mid-log phase of growth, growth medium is continuously added, at the same time that an equal volume of cells suspended in spent medium are removed. Continuous fermentations have the advantage of being able to produce a large amount of cells using a relatively small fermenter. However, from an industrial perspective, this is considered an experimental procedure and is not widely employed

procedure and is seldom used for industrial scale fermentations. For each type of fermentation, oxygen (usually in the form of sterile air), an antifoaming agent (especially at high cell densities), and, if required to maintain a constant pH, acid or base are injected into the bioreactor as needed.

During a batch fermentation, the composition of the culture medium, the concentration of bacteria (i.e., biomass concentration), and the internal chemical composition of the bacteria, all change as a consequence of the physiological state of the cells, reflecting growth, cellular metabolism, and availability of nutrients. Under these conditions, cells typically go through lag phase (where the inoculum cells adjust their metabolism to the new medium), acceleration phase, logarithmic or exponential phase, deceleration phase, stationary phase (caused by depletion of a critical growth substance where growth ceases and cellular metabolism changes dramatically), and death phase.

Nutrient additions to a batch fermentation, i.e., a fed-batch fermentation, prolongs both the log and stationary phases of bacterial growth, thereby significantly increasing the amount of cellular biomass. Because it is often difficult to measure the nutrient (substrate) concentration directly during the fermentation reaction, other indicators that are correlated with the consumption of the substrate, such as the production of organic acids, changes in the pH, or the production of CO_2, can be used to estimate when additional substrate is needed. Generally, fed-batch fermentations require more monitoring and greater control than batch fermentations and are therefore used to a lesser extent despite the fact that they may result in a significantly increased amount of bacterial biomass. Depending upon the particular microorganism, a fed-batch fermentation strategy can increase the yield by from 25% to more than 1000% compared with batch fermentation.

1.6.4 Commercialized Inoculant Examples

A limited but growing number of PGPB are available commercially (Table 1.7). The current biofertilizer market represents about 5% of the total chemical fertilizers market. The majority of the commercialized strains are either biocontrol agents that contribute indirectly to promoting the growth of various crops or a number of *Rhizobia* spp. that have been developed as inocula for various legume crops. In addition, a few of the commercially available bacterial strains include both rhizospheric and endophytic bacterial strains that directly promote plant growth.

More recently, several manufacturers have commercialized a mixture of a strain of *Rhizobia* together with either a rhizospheric or endophytic PGPB strain with the two types of bacteria working synergistically to promote plant growth. For example, a product sold under the name BioYield™ consists of a mixture of a strain of *Paenibacillus macerans* and a strain of *Bacillus amyloliquefaciens*. BioYield™ is directed toward crops such as tomato, cucumbers, cantaloupes, and peppers. Prior to treating seeds or the roots of plants, BioYield™ is combined with the other components of the planting mix, and then mixed thoroughly into a uniform composition including sufficient water to completely wet the mixture. When this procedure

Table 1.7 Some examples of the development of PGPB as commercial products

Bacterial ingredient	Product	Company	Intended crop
Agrobacterium radiobacter	Diegall, Galltrol-A, Nogall, Norbac 84C	Becker Underwood	Fruit, nut, ornamental nursery stock and trees
Azotobacter chroococcum, Bacillus megaterium, Bacillus mucilaginous	Biogreen	AgroPro	Field crops
Azospirillum brasilense	Azo-Green	Grama Karshaka Fertilizer Company	Rice, millet, cotton, wheat, maize
Bacillus firmus	BioNem L	Lidochem	Turf
Bacillus licheniformis SB3086	EcoGuard	Novozyme	Turf
Bacillus pumilus GB34	Yield Shield	Bayer CropScience	Soybean
Bacillus pumilus QST 2808	Sonata AS, Ballad	AgraQuest, Inc.	Food and field crops
Bacillus spp., *Streptomyces* spp., *Pseudomonas* spp.	Compete	Plant Health Care	Field crops
Bacillus subtilis var. *amyloliquifaciens* FZB24	Taegro	Earth BioSciences	Field crops
Bacillus subtilis GB03	Companion	Growth Products	Turf, greenhouse, nursery crops, ornamental, food and forage crops
Bacillus subtilis GB03	Kodiak	Bayer Crop Science	Cotton, vegetables, cereals
Bacillus subtilis	Epic	Gustafson	Beans
Bacillus subtilis (with *Bradyrhizobium japonicum*)	HiStick N/T, Turbo-N	Becker Underwood	Soybean
Bacillus subtilis (with *Bradyrhizobium japonicum*)	Patrol N/T	United Agri Products	Soybean
Bacillus subtilis MBI 600	Subtilex	Microbio	Field crops
Bacillus subtilis QST 713	Serenade	AgraQuest	Vegetables, fruit, nut and vine crops
Burkholderia cepacia type Wisconsin	Deny	Market VI LLC	Field crops
Delftia acidovorans	BioBoost	Brett-Young Seeds	Canola
Pantoea agglomerans E325	BlightBan C9-1	Nufarm Agricultural	Apple and pear
Pseudomonas aureofaciens Tx-1	Spot less	Eco Soil Systems	Turf grass
Pseudomonas chlororaphis 63–28	AtEze	Eco Soil Systems	Ornamentals and vegetables

(continued)

Table 1.7 (continued)

Bacterial ingredient	Product	Company	Intended crop
Pseudomonas fluorescens	Victus	Sylvan Spawn Laboratories	Mushrooms
Pseudomonas fluorescens	Pseumo Green	Grama Karshaka Fertilizer Company	Biocontrol for banana, vegetables, groundnut, cardamom, coffee, tea
Pseudomonas fluorescens A506	BlightBan A506	Nufarm Agricultural	Apple and pear
Pseudomonas syringae	Bio-Save	Jet Harvest Solutions	Citrus, pome fruit and potato
Streptomyces griseoviridis K61	Mycostop	AgBio Development	Field, ornamental and vegetable crops
Streptomyces lydicus WYEC 108	Actinovate	Natural Industries	Fruit, nut, vegetable and ornamental crops

was tested, plants that were treated with BioYield™ showed a significant growth response both in greenhouse and in field experiments, with fruit yield increases of these plants in different trials ranging from 5 to 35%. In these trials, plants treated with BioYield™ typically displayed earlier flowering and increased yield of earlier and larger fruit as well as significant reductions in disease severity and incidence. However, the manufacturer cautions the potential users that BioYield™ should not be used in place of commercial pest control measures as the reduction in disease resistance that is observed is likely to due to the increased health of the treated plants and not to any biocontrol activity in this combination of PGPB.

Another mixed bacterial inoculant called SoySuperb™ includes the bacterium *Delftia acidovorans* RAY209 and a strain of *Bradyrhizobium japonicum* in one package. *D. acidovorans* RAY209 is a sulfur-oxidizing Gram-negative bacterium that facilitates sulfur uptake, an important trait for crops grown in sulfur limited soils while *B. japonicum* is a nitrogen-fixing *Rhizobia* that specifically nodulates soybean roots. SoySuperb™ is sold as a liquid formulation with an estimated shelf life of 8 months. The company believes that the potential market for SoySuperb™ includes 75 million acres in the United States and ~3 million acres in Canada that are under soybean cultivation. In addition, an increasing amount of soybean is being grown in Latin America.

Questions

1. Why should PGPB be used in agriculture instead of chemicals?
2. What are the different types of PGPB?
3. How common are bacteria in typical soils? What other organisms are found in soil?
4. Why are bacteria concentrated in the rhizosphere?
5. What is a bacterial endophyte?
6. Briefly, how are compounds exuded from plant roots?

7. What are some of the ways in which PGPB positively impact plant growth?
8. What is the difference between direct and indirect mechanisms of plant growth promotion?
9. What are some of the mechanisms that PGPB use to directly promote plant growth? Indirectly promote plant growth?
10. How would you test whether or not a particular potential PGPB strain had biocontrol activity against a particular phytopathogen?
11. How would you isolate new strains of facultative endophytes?
12. How can specific soil bacteria facilitate the synthesis of lipids by green algae?
13. What items would you consider to be important in developing a laboratory PGPB strain for commercialization and use in the field?
14. Why do some companies prefer PGPB to be Gram-positive rather than Gram-negative?
15. What are some of the disadvantages of using peat-based bacterial inoculants?
16. What are the advantages of using alginate-encapsulated bacterial inoculants?
17. How can bacterial cells be encapsulated in alginate beads?
18. What are the differences between batch, fed-batch, and continuous fermentations?
19. What parameters should be monitored and controlled in an optimized fermentation process?
20. Briefly, outline the major step in the large scale production of PGPB cells in culture.

Bibliography

Agboola AA, Ekundayo FO, Ekundayo EA, Fasoro AA, Ayantola KJ, Kayode AJ (2018) Influence of glyphosphate on rhizosphere microorganisms and their ability to solubilize phosphate. J Microbiol Antimicrob Agents 4:15–21

Albareda M, Rodriguez-Navaroo DN, Camacho M, Temprano FJ (2008) Alternatives to peat as a carrier for rhizobia inoculants: solid and liquid formulations. Soil Biol Biochem 4:2771–2779

Backer R, Rokem JS, Ilangumaran G, Lamont J, Praslickova D, Ricci E, Subramanian S, Smith D (2018) Plant growth-promoting rhizobacteria: context, mechanisms of action, and roadmap to commercialization of biostimulants for sustainable agriculture. Front Plant Sci 9:1473

Bailey JE, Olis DF (1977) Biochemical engineering fundamentals. McGraw-Hill Book, New York, NY

Bais HP, Weir TL, Perry LG, Gilroy S, Vivanco JM (2006) The role of root exudates in rhizosphere interactions with plants and other organisms. Annu Rev Plant Biol 57:233–266

Bashan Y, Holguin G (1998) Proposal for the division of plant growth-promoting rhizobacteria into two classifications: biocontrol-PGPB (plant growth-promoting bacteria) and PGPB. Soil Biol Biochem 30:1225–1228

Bashan Y, Hernandez JP, Leyva LA, Bacillio M (2002) Alginate microbeads as inoculant carriers for plant growth-promoting bacteria. Biol Fertil Soils 35:359–368

Bashan Y, de-Bashan LE, Prabhu SR, Hernandez J-P (2014) Advances in plant growth-promoting bacterial inoculant technology: formulations and practical perspectives (1998–2013). Plant Soil 378:1–33

Brígido C, Glick BR, Oliveira S (2017) Survey of plant growth-promoting mechanisms in native Portuguese chickpea mesorhizobia. Microb Ecol 73:900–915

Calvo P, Nelson L, Kloepper JW (2014) Agricultural uses of plant biostimulants. Plant Soil. https://doi.org/10.1007/s11104-014-2131-8

Charles M (1985) Fermentation scale-up: problems and possibilities. Trends Biotechnol 3:134–139

Compant S, Clément C, Sessitsch A (2010) Plant growth-promoting bacteria in the rhizo- and endosphere of plants: their role, colonization, mechanisms involved and prospects for utilization. Soil Biol Biochem 42:669–678

de Vrieze J (2015) The littlest farmhands. Science 349:680–683

Deaker R, Roughley RJ, Kennedy IR (2004) Legume seed inoculation technology—a review. Soil Biol Biochem 36:1275–1288

Denton MD, Pearce DJ, Ballard RA, Hannah MC, Mutch LA, Norng S, Slattery JF (2009) A multisite field evaluation of granular inoculants for legume nodulation. Soil Biol Biochem 41:2508–2516

Dixon ROD, Wheeler CT (1986) Nitrogen fixation in plants. Blackie and Son, Glasgow

do Vale Barreto Figueiredo M, Seldin L, de Araujo FF, de Lima Ramos Mariano R (2010) Plant growth promoting rhizobacteria: fundamentals and applications. In: Maheshwari DK (ed) Plant growth and health promoting bacteria. Springer-Verlag, Berlin

Dykhuizen D (2005) Species numbers in bacteria. Proc Calif Acad Sci 56:62–71

Egamberdieva D, Hua M, Reckling M, Wirth S, Bellingrath-Kimura SD (2018) Potential effects of biochar-based microbial inoculants in agriculture. Environ Sustain 1:19–24

Flores-Félix JD, Menéndez E, Rivera LP, Marcos-Garcia M, Martinez-Hidalgo P, Mateos PF, Martinez-Molina E, Velazques ME, Garcia-Fraile P, Rivas P (2013) Use of *Rhizobium leguminosarum* as a potential fertilizer for *Lactuca sativa* and *Daucus carota* crops. J Plant Nutr Soil Sci 176:876–882

Gerland P, Rafferty AE, Sevcikova H, Li N, Gu D, Spoorenberg T, Alkema L, Fosdick BK, Chunn J, Lalic N, Bay G, Buettner T, Hweilig GK, Wilmoth J (2014) World population and stabilization unlikely this century. Science 346:234–237

Glick BR (1995) The enhancement of plant growth by free-living bacteria. Can J Microbiol 41:109–117

Glick BR (2004) Bacterial ACC deaminase and the alleviation of plant stress. Adv Appl Microbiol 56:291–312

Glick BR (2005) Modulation of plant cthylcnc levels by the enzyme ACC deaminase. FEMS Microbiol Lett 251:1–7

Glick BR, Bashan Y (1997) Genetic manipulation of plant growth-promoting bacteria to enhance biocontrol of fungal phytopathogens. Biotechnol Adv 15:353–378

Glick BR, Penrose DM, Li J (1998) A model for the lowering of plant ethylene concentrations by plant growth promoting bacteria. J Theor Biol 190:63–68

Glick BR, Cheng Z, Czarny J, Duan J (2007a) Promotion of plant growth by ACC deaminase-containing soil bacteria. Eur J Plant Pathol 119:329–339

Glick BR, Todorovic B, Czarny J, Cheng Z, Duan J, McConkey B (2007b) Promotion of plant growth by bacterial ACC deaminase. Crit Rev Plant Sci 26:227–242

Hale L, Luth M, Crowley D (2015) Biochar characteristics relate to its utility as an alternative soil inoculum carrier to peat and vermiculite. Soil Biol Biochem 81:228–235

Hartmann A, Schmid M, van Tuinen D, Berg G (2009) Plant-driven selection of microbes. Plant Soil 321:235–257

Herrmann L, Lesueur D (2013) Challenges of formulation and quality of biofertilizers for successful inoculation. Appl Microbiol Biotechnol 97:8859–8873

Jaleel MA, Sankar P, Kishorekumar B, Gopi A, Somasundaram R, Panneerselvam R (2007) *Pseudomonas fluorescens* enhances biomass yield and ajmalicine production in *Catharanthus roseus* under water deficit stress. Colloids Surf B Biointerfaces 60:7–11

Lamont JR, Wilkins O, Bywater-Ekegärd M, Smith DL (2017) From yogurt to yield: potential applications of lactic acid bacteria in plant production. Soil Biol Biochem 111:1–9

Lee S-K, Lur H-S, Lo K-J, Cheng K-C, Chuang C-C, Tang S-J, Liu C-T (2016) Evaluation of the effects of different liquid inoculant formulations on the survival and plant-growth-promoting

efficiency of *Rhodopseudomonas palustris* strain PS3. Appl Microbiol Biotechnol 100:7977–7987
Lugtenberg B (ed) (2015) Principles of plant-microbe interactions. Springer, Heidelberg
Lynch JM (ed) (1990) The rhizosphere. Wiley-Interscience, Chichester
Maheshwari K (ed) (2015) Bacterial metabolites in sustainable agroecosystem, sustainable development and biodiversity, vol 12. Springer International Publishing, Cham
Martin BC, Gleeson D, Statton J, Siebers AR, Grierson P, Ryan MH, Kendrick GA (2018) Low light availability alters root exudation and reduces putative beneficial microorganisms in seagrass roots. Front Microbiol 10:2667. https://doi.org/10.3389/fmicb.2017.02667
McClung CR (2014) Making hunger yield. Science 344:699–700
Montero-Calasanz MC, Santamaria C, Albareda M, Daza A, Duan J, Glick BR, Camacho M (2013) Alternative rooting induction of semi-hardwood olive cuttings by several auxin-producing bacteria for organic agriculture systems. Span J Agric Res 11:146–154
Nehra V, Choudhary M (2015) A review on plant growth promoting rhizobacteria acting as bioinoculants and their biological approach towards the production of sustainable agriculture. J Appl Nat Sci 7:540–556
O'Callaghan M (2016) Microbial inoculation of seed for improved crop performance: issues and opportunities. Appl Microbiol Biotechnol 100:5729–5746
Olanrewaju OS, Glick BR, Babalola OO (2017) Mechanisms of action of plant growth promoting bacteria. World J Microbiol Biotechnol 33:197
Pastor-Bueis R, Sánchez-Cañizares C, James EK, González-Andrés F (2019) Formulation of a highly effective inoculant for the common bean based on an autochthonous elite strain of *Rhizobium leguminosarum* bv. *phaseoli*, and genome-based insights into its agronomic performance. Front Microbiol 10:2724
Patten C, Glick BR (1996) Bacterial biosynthesis of indole-3-acetic acid. Can J Microbiol 42:207–220
Patten CL, Glick BR (2002) The role of bacterial indoleacetic acid in the development of the host plant root system. Appl Environ Microbiol 68:3795–3801
Pilon-Smits E (2005) Phytoremediation. Annu Rev Plant Biol 56:15–39
Preininger C, Sauer U, Bejarano A, Berninger T (2018) Concepts and applications of foliar spray for microbial inoculants. Appl Microbiol Biotechnol 102:7265–7282
Reed MLE, Glick BR (2004) Applications of free living plant growth-promoting rhizobacteria. Anton Van Leeuwen 86:1–25
Reed MLE, Glick BR (2013) Applications of plant growth-promoting bacteria for plant and soil systems. In: Gupta VK, Schmoll M, Maki M, Tuohy M, Mazutti MA (eds) Applications of microbial engineering. Taylor and Francis, Enfield, CT, pp 181–229
Riesenberg D, Guthke R (1999) High-cell-density cultivation of microorganisms. Appl Microbiol Biotechnol 51:422–430
Rocha I, Ma Y, Souza-Alonso P, Vosatka M, Freitas H, Oliveira RS (2019) Seed coating: a tool for delivering beneficial microbes to agricultural crops. Front Plant Sci 10:1357
Rosier A, Medeiros FHV, Bais HP (2018) Defining plant growth promoting rhizobacteria molecular and biochemical networks in beneficial plant-microbe interactions. Plant Soil 428:35–55
Schoebitz M, Lopez MD, Roldan A (2013) Bioencapsulation of microbial inoculants for better soil-plant fertilization. A review. Agron Sustain Dev 33:751–765
Shih PM (2019) Early cyanobacteria and the innovation of microbial sunscreens. mBio 10:e01262–e01219. https://doi.org/10.1128/mBio.01262-19
Strandberg L, Andersson L, Enfors S-O (1994) The use of fed batch cultivation for achieving high cell densities in the production of a recombinant protein in *Escherichia coli*. FEMS Microbiol Rev 14:53–56
Strobel SA, Allen K, Roberts C, Jiminez D, Scher HB, Jeoh T (2018) Industrially-scalable microencapsulation of plant beneficial bacteria in dry cross-linked alginate matrix. Ind Biotechnol 14:138–147

Timmusk S, Behers L, Muthoni J, Muraya A, Aronsson AC (2017) Perspectives and challenges of microbial application for crop improvement. Front Plant Sci 8:49. https://doi.org/10.3389/fpls.2017.00049

Vacheron J, Desbrosses G, Bouffaud M-L, Touraine B, Moënne-Loccoz Y, Muller D, Legendre L, Wisniewski-Dyé F, Prigent-Combaret C (2013) Plant growth-promoting rhizobacteria and root system functioning. Front Plant Sci 4:356

Vacheron J, Moënne-Loccoz Y, Dubost A, Goncalves-Martins M, Muller D, Prignet-Combaret C (2016) Fluorescent *Pseudomonas* strains with only a few plant-beneficial properties are favored in the maize rhizosphere. Front Plant Sci 7:1212

Van Brunt J (1985) Scale-up: the next hurdle. Biotechnology 3:419–424

Walker TS, Bais HP, Grotewold E, Vivanco JM (2003) Root exudation and rhizosphere biology. Plant Physiol 132:44–51

West PC, Gerber JS, Engstrom PM, Mueller ND, Brauman KA, Carlson KM, Cassidy ES, Johnston M, MacDonald GK, Ray DK, Siebert S (2014) Leverage points for improving global food security and the environment. Science 345:325–328

White MD, Glick BR, Robinson CW (1995) Bacterial, yeast and fungal cultures: the effect of microorganism type and culture characteristics on bioreactor design and operation. In: Asenjo JA, Merchuk J (eds) Bioreactor system design. Marcel Dekker, New York, pp 47–87

Wu CH, Bernard SM, Andersen GL, Chen W (2009) Developing microbe–plant interactions for applications in plant-growth promotion and disease control, production of useful compounds, remediation and carbon sequestration. Microb Biotechnol 2:428–440

Xu L, Cheng X, Wang Q (2018) Enhanced lipid production in *Chlamydomonas reinhardtii* by co-culturing with *Azotobacter chroococcum*. Front Plant Sci 10:741. https://doi.org/10.3389/fpls.2018.00741

Zaidi A, Khan MS (eds) (2017) Microbial strategies for vegetable production. Springer Nature, Cham

Zaidi SS, Vanderschuren H, Qaim M, Mahfouz MM, Kohli A, Mansoor A, Tester M (2019) New plant breeding technologies for food security. Science 363:1390–1391

Zvinavashe AT, Lim E, Sun H, Marelli B (2019) A bioinspired approach to engineer seed microenvironment to boost germination and mitigate soil salinity. Proc Natl Acad Sci USA 116:25555–25561

Microbiomes and Endophytes

2

Abstract

The nature of microbiomes and endophytic plant growth-promoting bacteria is elaborated. For a start, the plant rhizosphere (bacteria around the roots) and the plant endosphere (bacteria inside of the plant) are colonized by different subsets of the bacteria that are found in the bulk soil. In the development of these interactions, to some extent, the plant (and its metabolites) and the soil composition help to select the composition of bacteria that inhabit its tissues (its microbiomes). Root, seed, and synthetic microbiomes are each discussed in some detail. In addition, since the vast majority of higher plants are colonized by endophytic bacteria, the nature of these bacteria and what makes them both similar and unique to rhizospheric bacteria is discussed.

2.1 Microbiomes

There are an enormous number of bacteria in soil, sometimes including as many as 10^6 different taxa (taxonomic groups such as bacteria of different genera and species) in a single gram of soil. To classify bacteria into a specific genus and species, scientists often determine, using next-generation sequencing, the DNA sequence of all or part of the bacterial genomic DNA that encodes the 16S ribosomal RNA genes (a key component of the 30S small subunit of all prokaryotic ribosomes). In one large study, a detailed analysis of soil bacteria 16S rRNA genes from 237 locations worldwide, including all six continents, found that only 2% of soil bacterial phylotypes (bacterial taxa with an overall similarity of $\geq 97\%$ across the sequenced region) accounted for around half of the soil bacterial communities. This observation suggests that while there are an enormous number of different bacteria in soil, a relatively small number of different types of bacteria predominate in soils throughout the world (Fig. 2.1).

© Springer Nature Switzerland AG 2020
B. R. Glick, *Beneficial Plant-Bacterial Interactions*,
https://doi.org/10.1007/978-3-030-44368-9_2

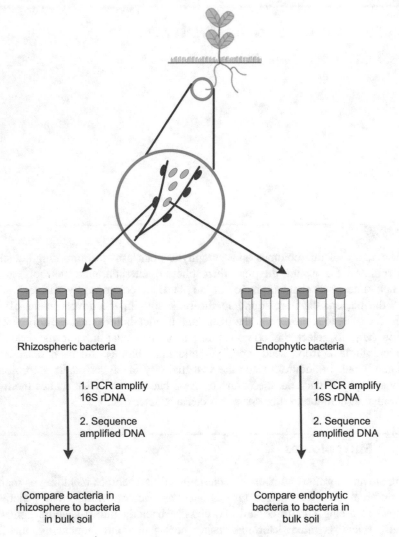

Fig. 2.1 Schematic representation of bacteria found in the soil, the plant rhizosphere (the area immediately surrounding plant roots) and the endosphere (within plant roots). Both the soil and the rhizosphere contain a large number of different types of bacteria with the root exudates providing a food source for the rhizosphere bacteria. The endosphere selects for a limited number of bacteria that colonize the space between root cells, and in a few cases, the interiors of the root cells. In this representation, different bacteria are depicted as different colors and shapes

Given the diversity of soil bacteria, a question that arises is whether there are any particular features that facilitate the colonization of plant rhizospheres by certain bacteria and not by others. To address this question, two groups of scientists separately studied the DNA sequences of all of the 16S rRNA genes in soil, rhizospheric and endophytic bacteria using different strains of the plant *Arabidopsis*

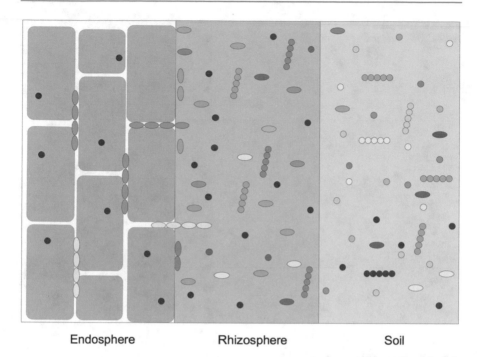

Endosphere   Rhizosphere   Soil

Fig. 2.2 Schematic representation of the identification of the bacteria that comprise the microbiome of the plant *Arabidopsis thaliana*. Red-colored shapes represent rhizospheric bacteria bound to a root surface while blue-colored shapes represent endophytic bacteria located within the root

thaliana (Fig. 2.2). *Arabidopsis thaliana* (thale cress) is a small flowering plant (Fig. 2.3) related to cabbage and canola that is commonly used as a model organism in plant biology, in part as a consequence of its very small genome (compared to nearly all other plants) that facilitates both genetic and biochemical studies. These two groups of researchers took soil, rhizospheric and endophytic samples from several different *Arabidopsis* strains, each grown in different soils and observed that the bacterial composition of both the rhizospheric and the endophytic microbiomes were defined primarily by the soil type. In addition, the researchers reported that *A. thaliana* plants contain a typical "core" microbiome, ~60% of which is "recruited" from common soil bacteria. This core microbiome is selected by the ability of individual bacteria to grow well in *A. thaliana* root exudates. The remaining bacteria that were identified appeared to be responding and subsequently binding to the plant root's lignocellulosic surface. These results were also independent of plant age. A limited number of experiments have demonstrated that several other plants (e.g., oat, wheat, and pea) behave similarly to *A. thaliana* in selecting each of their microbiomes from the soil; thus, this is likely to turn out to generally be the case. As one researcher has suggested, "root-derived carbon sources fuel a very active and specialized (microbial) community."

Fig. 2.3 The plant
Arabidopsis thaliana

2.1.1   Root Microbiome

Roots are the major site where plants with microorganisms typically interact. Thus, for example, it has been estimated that of the approximately 10^{29} bacterial cells found in all of the earth's soils, around 10^{26}–10^{27} are associated with plant roots. The root microbiome (rhizospheric and endophytic) may develop based on both positive and negative selection pressures that occur as a result of the composition of root exudates. For a start, soil bacteria are largely dependent upon exogenous carbon sources for both energy, and growth and development. The release of certain specific compounds by plant roots therefore select for the limited number of bacteria that are able to use those compounds and proliferate. Moreover, root exudates are generally plant specific and the use of these specific compounds often act as signals in establishing a plant's affinity for particular bacteria. For example, some rhizosphere bacteria can use indole acetic acid (IAA, auxin) produced by the plant as a carbon source for growth. Similarly, only a limited number of soil bacteria can utilize sulfur-containing compounds such as methionine for their growth. Thus, those bacteria that can utilize the sulfur-containing compounds that are typically produced by crucifer-ous plants such as canola and cauliflower will therefore be found commonly in the microbiomes of those plants.

Rhizosphere soil is readily washed off of the roots and the microbes in that fraction of the soil may be analyzed, either following cultivation or by amplification of 16S rRNA genes (a technique whereby portions of the genomes of both cultivat-able and non-cultivatable bacteria are amplified). Since removal of rhizosphere soil leaves a significant number of microorganisms attached to the root surface (the rhizoplane), the simplifying assumption that is sometimes made is that rhizosphere and rhizoplane organisms are essentially the same. However, in practice, the rhizosphere is colonized by a portion (or subset) of the organisms that are found in the bulk soil while the rhizoplane is colonized by a portion of the organisms that are present in the rhizosphere. Similarly, only a small fraction of the organisms found in the rhizosphere are able to enter the plant tissues and permanently colonize the plant endosphere. In one study using rice plants, the researchers were able to distinguish three separate root-associated compartments each with its own unique microbiome, i.e., the endosphere (root interior), the rhizoplane, and the rhizosphere (soil close to the root surface). These generalizations notwithstanding, some plants have root microbiomes that are very different from the microorganisms in the bulk soil while other plants have root microbiomes that are similar to the collection of microorganisms in the bulk soil. Generally speaking, the composition of the root microbiome is a function of the plant and its developmental stage.

Interestingly, different parts of the plant root are found associated with different bacteria. Thus, the bacteria that are associated with mature root zones are different from those associated with root tips which are in turn different from those associated with lateral roots. These differences may reflect differences in the types of molecules exuded by different root parts.

Plant root tips also contain loosely associated border cells that separate from the root tip and disperse into the soil producing mucilage and numerous other compounds, where mucilage is a thick glue-like substance containing glycoproteins and exopolysaccharides.

Following the discovery that many plants can regulate both the amount and type of molecules that they exude and, based on the knowledge that different root exudates select for different types of bacteria, it was suggested that some plants could thereby modulate their microbiomes. This behavior might, for example, enable plants subjected to a particular abiotic stress to adjust their exudates to select for a microbiome that decreased the level of stress and therefore facilitated the proliferation of the plant. For example, one group of researchers observed (Fig. 2.4) that when pepper plants were infected with aphids (small sap-sucking insects that feed on plant juices), the plants increased the amount of root exudates that they produced. In turn, the increased level of root exudates (the chemical identity of the causative root exudate was not determined) resulted in an increased growth rate for some strains of *Paenibacillus* spp. (and not for Gram-negative bacteria), and those bacteria stimulated the plant's innate defenses against aphid infestation. Thus, through the intervention of bacteria in the root microbiome, aphid infestation and damage could be limited.

Plants have the ability to synthesize a large number of different specialized metabolites, a capability that reflects different plants' adaptation to diverse environments. In the case of *A. thaliana*, the plant is able to synthesize and subsequently exude from its roots, several triterpenes. Recently, scientists observed that these compounds promote the growth of certain bacteria while inhibiting the growth of others. Thus, by synthesizing these triterpenes, *A. thaliana* directs the assembly of an *A. thaliana*-specific microbiome. Whether this phenomenon is true for other plants and their microbiomes remains to be demonstrated.

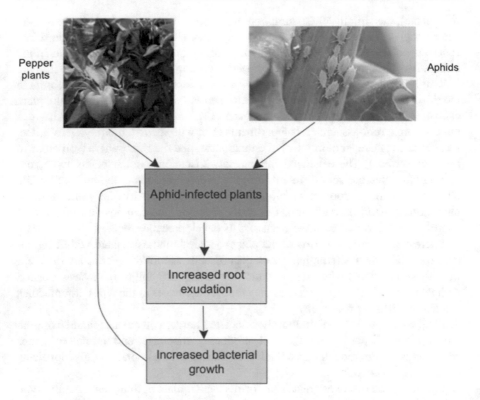

Fig. 2.4 Aphid infection increases pepper plant root exudation which in turn increases the growth of selected soil bacteria that are able to inhibit the infection of pepper plants by aphids

The common grape vine (*Vitis vinifera*) is native to the Mediterranean region including parts of Europe, Africa, and Asia. In winemaking, to protect the fruit from drought and certain pathogen-caused diseases, the fruit-producing part of the plant is often grafted onto a rootstock that has a well-developed root system. The rootstock provides water and minerals to the plant while the portion that is grafted on (called the scion) dictates the type of grape that will be produced. In one set of experiments a *Vitis vinifera* cultivar Lambrusco was used as the scion and grafted onto two different rootstocks where one of the rootstocks demonstrated potassium absorption problems (Fig. 2.5). At the time of the experiment, it had recently been demonstrated that rootstocks had the ability to select for particular microbes at the root–soil interface from the soil. Thus, by examining the microbiomes (both the root endophytes and the rhizosphere microbes) of the two grafted plants it was possible to ascertain whether either the rootstock or a microbiome selected by the rootstock was the key determinant of the plant's poor absorption of potassium. In fact, when the four microbiomes were characterized by 16S rRNA sequencing it became apparent that the rootstock with poor potassium absorption had selected for rhizosphere and root endophytic microbiomes with reduced diversity (compared to the more effective rootstock). Moreover, the microbiomes from the rootstock with poor

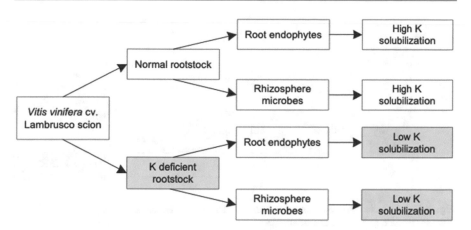

Fig. 2.5 Schematic representation of how the same *V. vinifera* scion is grafted onto two different *V. vinifera* rootstocks, one with normal potassium absorption and one that poorly absorbs potassium from the soil. Isolation of root endophytic and rhizospheric microbiomes from the mature plants reveals that the microbiomes from the plants containing the rootstock that poorly absorbs potassium from the soil contain a decreased number of potassium absorbing microbes

potassium absorbance were specifically low on microbes known to be efficient potassium solubilizers. Thus, in this case the poor ability of one rootstock to take up potassium could be attributed to its microbiome.

Carbon dioxide (CO_2) is a trace gas in our atmosphere that is used by plants (and a few photosynthetic microbes) to produce carbohydrates where all other living organisms in nature utilize these carbohydrates as a source of carbon and energy. As a consequence of human activity (particularly the burning of fossil fuels), it is estimated that the global concentration of CO_2 in the atmosphere has increased from around 280 ppm (parts per million) before the industrial revolution (the mid-eighteenth century) to approximately 410 ppm (as of late-2018). While this increase in atmospheric CO_2 levels may not seem especially large, it has been estimated that this is the highest value for these levels in the last 1–20 million years. Approximately 30–40% of the CO_2 that is released by human activity into the atmosphere dissolves into oceans, rivers, and lakes, leading to their acidification. Another large fraction of the atmospheric CO_2 is absorbed by soil, and the CO_2 concentration in soil can become much higher than the levels in the atmosphere thereby significantly impacting the growth of plants. In a recent series of experiments, researchers examined the effect of artificially increased CO_2 levels in a field experiment (where CO_2 was literally pumped into the field) in order to ascertain the effect of the CO_2 on both different cultivars of rice and their microbiomes (Fig. 2.6). When the rice plant rhizosphere microbiomes were characterized by sequencing the 16S rRNA genes, it was found that the presence of CO_2 had decreased the richness and diversity of the microbes in the microbiome. In the first instance, increased atmospheric CO_2 led to increases in soil CO_2 that in turn led to an increase in the carbon content of the root exudates. The altered root

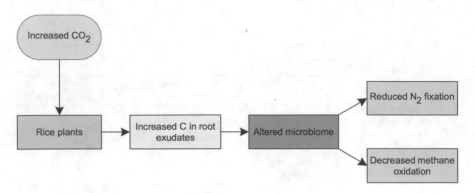

Fig. 2.6 Schematic representation of the effects of increased atmospheric CO_2 on rice plant root nitrogen fixation and methane oxidation

exudates selected for different microbes in the microbiome. These changes occurred for several different rice cultivars. The altered microbiome fixed a decreased amount of N_2 and oxidized a lower level of methane. Thus, increased atmospheric CO_2 suppresses methane oxidation and promotes methanogenesis in rice roots, resulting in a higher level of methane and altering the carbon cycle in rice paddy fields.

Sorghum bicolor (sorghum) is the world's fifth most important cereal crop (after rice, wheat, corn, and barley). It is cultivated widely for its grain in tropical and subtropical countries (including Nigeria, India, Mexico, and the United States) and is used for food for humans and animals, as well as for ethanol production. Sorghum cultivation is particularly suited to warm and dry climates as it employs the more efficient C4 carbon fixation and therefore requires much less water than plants employing (the much more common but less efficient) C3 carbon fixation. In one series of recent experiments, a group of researchers examined the detailed changes that occurred in the sorghum root microbiome during periods of drought (Fig. 2.7). These researchers found that when sorghum plants were watered normally the preponderance of the endophytic and rhizospheric bacteria were diderms (i.e., they had two cell membranes; Gram-negative). However, when the plants were subjected to drought conditions, the majority of the endophytic and rhizospheric bacteria were found to be monderms (i.e., they had a single cell membrane; Gram-positive). To explain these results, it was suggested that when the plants were well watered, the Gram-negative bacteria grew more rapidly and were superior root colonizers compared to the Gram-positive bacteria. In contrast, it was speculated that the much thicker peptidoglycan cell wall of the Gram-positive bacteria (compared to Gram-negative bacteria) facilitated their survival during drought stress. Moreover, it was shown that (even in the absence of any bacteria) in an effort to ameliorate the drought stress (which includes elements of oxidative stress), the plant significantly increased the exudation of glycerol-3-phosphate, where glycerol-3-phosphate is a major precursor of the peptidoglycan cell wall of Gram-positive bacteria. In fact, when the Gram-positive bacteria were isolated and used to inoculate drought stressed plants, it was found that these microbes helped the plants to overcome the drought stress.

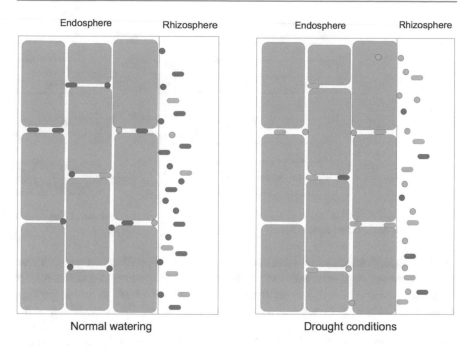

Fig. 2.7 Schematic view of a cross-section of a plant root and its associated rhizosphere following either normal watering or drought conditions. The orange-colored shapes represent Gram-negative bacteria while the blue-colored shapes represent Gram-positive bacteria

Scientists generally agree that the plant species and soil type are the major determinants of the bacteria that inhabit the plant and its roots. This is presumably largely a consequence of the fact that various plant species produce different root exudates that attract only certain types of bacteria from the soil (which is considered to be a reservoir for rhizosphere bacteria). However, superimposed on the role of plants and soil in selecting rhizosphere bacteria are a host of environmental conditions that may alter the composition of root exudates and affect which bacteria are selected. These environmental conditions may include (but are not limited to) the amount of water in the soil, the soil pH, and the climate in which the plant is propagated. For example, in one set of experiments, researchers examined the microbiomes of two different cold climate plants, mountain sorrel (*Oxyria digyna*) and purple saxifrage (*Saxifraga oppositifolia*) growing in three Arcto-alpine regions (i.e., alpine, low Arctic, and high Arctic). These microbiomes were then compared with bacteria from the bulk soil in those three regions as well as with one another (Fig. 2.8). In the first instance, for both of the plants tested, the bacterial communities from the three Arcto-alpine regions were significantly different from one another. Second, the relative abundance of Proteobacteria (Gram-negative bacteria), in the bulk soil, decreased progressively from the alpine to the high Arctic soils. Moreover, the endospheres (inside the root) of plants from the alpine region contained many more Proteobacteria while the endospheres of plants from the high Arctic had a

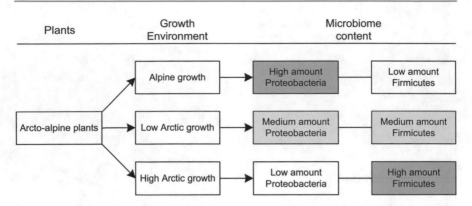

Fig. 2.8 Schematic representation of the selection of specific bacteria from common core taxa in three different Arcto-alpine climate zones

much higher percentage of Firmicutes (mostly Gram-positive bacteria). These observations are consistent with the possibility that more severe cold environments (i.e., high Arctic versus alpine) are more stressful to these plants so that in the high Arctic environment as compared to the alpine environment the composition of the plant root exudates contains very different molecules thereby selecting for a different set of bacteria in their microbiome. Whether this response is specific for cold and drought stress (see the example in the previous paragraph) or is common to a wide range of abiotic stresses, remains to be determined.

In addition to the effect of plant species, soil type and environmental conditions on determining the composition of the root microbiome, scientists are beginning to realize that the deliberate addition of PGPB to plant roots can have a significant effect on the microbiome composition. Thus, in some instances it has been shown that inoculation of plants with a particular PGPB induces variations in the fungal and bacterial root microbiome and these changes in the relative abundance of various microbiota may persist over several growing seasons. Thus, despite the fact that an added PGPB may not persist for a prolonged period of time, the changes to the soil microbiome that this addition causes may persist for a longer period of time.

2.1.2 Seed Microbiome

In nature, seeds are essential for initiating the plant life cycle and reproducing the species. In addition, seeds also facilitate a plant's dispersal and its adaptation to a new environment. Both seeds and seedlings are especially vulnerable to a wide range of abiotic and biotic stress, compared to older and more established plants. Microbes, both those that are growth promoting and those that are pathogenic, interact with seeds throughout plant growth and development. Seed microbiomes include both endophytic microbes that reside within seed tissues (this region is sometimes called the endospermosphere) and epiphytic microbes that are typically localized on seed surfaces. The endophytic microbes are presumably able to survive desiccation as they are derived from the previous generation of a plant. The bacteria that are commonly found associated with seeds include a limited range of species, generally Proteobacteria, Actinobacteria, Firmicutes and Bacteroidetes, reflecting the dominance of these phyla in most soils. Nevertheless, it has been observed that seed endophytic microbiomes are often distinct from the soil where the plants have been grown. This observation is consistent with the possibility that seed endophytic microbiomes are derived from the mother plant. In fact, it has been suggested that the preexisting seed microbiome provides the basis for successful plant growth and the subsequent development of the microbiomes that are associated with many plant organs.

It would be advantageous if it were possible to introduce new, growth-promoting, microbes into seeds so that seeds or seedlings no longer needed to be treated in traditional ways to deliver the microbes to growing plants. Initially, this might be achieved by modifying the seed microbiome with one or two particularly effective endophytic plant growth-promoting bacteria. Following the successful development of this approach, it may eventually be possible to design an artificial seed microbiome that facilitates optimal plant growth and development. In a recent series of experiments, researchers reported introducing the plant growth-promoting endophyte *Paraburkholderia* (formerly *Burkholderia*) *phytofirmans* PsJN into the seed microbiomes of several different plants (Fig. 2.9). First, the chromosomal DNA of *P. phytofirmans* PsJN was labeled with an *E. coli* β-glucuronidase gene. The labeled bacteria were grown for 2 days at 28 °C to stationary phase and then (after the cell suspension was adjusted to a high cell density) they were used to spray flowering plants (corn, pepper, or soy prior to continued growth in the greenhouse, or wheat growing in the field). The plants were then allowed to continue growing until maturity when they produced seeds that were then harvested. Following storage of the new seeds for 2 or 7 months, they were tested in various ways for the presence of *P. phytofirmans* PsJN. These tests included, plate counts of bacteria associated with seed extracts (identified by DNA hybridization), staining of seed tissues for the presence of a functional β-glucuronidase enzyme (that turns the tissue blue in the presence of a specific substrate) that is encoded by the *gusA* gene, qPCR to detect *P. phytofirmans* PsJN genes, and measurements of the altered physiology of mature plants produced from seeds with an altered microbiome. The efficiency of introducing strain PsJN into various seeds was as high as 90%, although this varied with different plants and the precise conditions employed. While this procedure seems quite attractive, if it is going to be used on a wide scale, it will need to be optimized for each different type of plant and perhaps even for each different cultivar. Interestingly, the researchers who performed this study noted that strain PsJN was not found in the next generation of seeds so that this procedure would need to be repeated for each separate crop. Finally, these researchers point out that this technique may be used along with conventional chemical treatment of seeds prior to planting. This is because the chemicals that are used to treat seeds will be on the outside of the seeds while the bacteria are in the inside of the seeds.

While root and seed microbiomes have received quite a lot of attention, recently scientists have begun to examine the microbiomes of various fruits. In particular, one

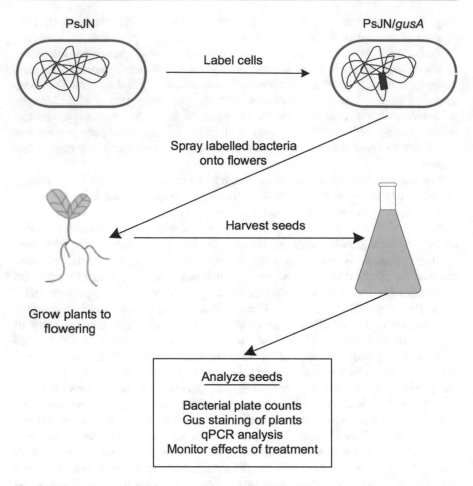

PsJN   PsJN/*gusA*

Label cells

Spray labelled bacteria
onto flowers

Harvest seeds

Grow plants to
flowering

Analyze seeds

Bacterial plate counts
Gus staining of plants
qPCR analysis
Monitor effects of treatment

Fig. 2.9 Overview of the engineering of plant seed microbiomes

study analyzed the apple (cv. Arlet) microbiome and asked whether there was any significant difference between the microbiomes of apples raised conventionally or organically. Moreover, as each apple has several different tissues including stem, peel, fruit pulp, seeds, and calyx (sepals of the apple flower that remains on the bottom of the apple), the question arises as to whether these tissues are colonized by different microorganisms. The presence of bacteria in the abovementioned apple tissues was measured by qPCR of bacterial 16S DNA. Interestingly, not only were there significant differences in the numbers of bacteria that were present in each of these tissues, there was also a significant difference in the numbers of bacteria when comparing conventionally and organically grown apples with the conventionally grown apples generally having lower levels of bacteria (Table 2.1). Interestingly, the bacterial composition was significantly different in conventionally and organically produced apples with bacteria known for health-affecting potential being enhanced

Table 2.1 Abundance of bacteria found in various apple fruit tissues from either organically or conventionally grown apples

Apple tissue	Organic, 16S rRNA gene copies per g apple tissue	Conventional, 16S rRNA gene copies per g apple tissue
Stem	8×10^7	2×10^8
Stem end	3×10^6	3×10^5
Peel	8×10^4	1.5×10^4
Fruit pulp	3×10^5	8×10^4
Seeds	1×10^8	3×10^7
Calyx end	5×10^6	4×10^6

in conventionally grown apples. Whether this data turns out to be the case for other apple cultivars and for other fruit remains to be determined. However, as it stands, these experiments indicate that with each apple we simultaneously consume about 1×10^8 bacterial cells.

2.1.3 Synthetic Microbiomes

Without question, most of the current microbiome research includes elaborating what organisms are present in different soils and climates, and when examining different plants. However, a few researchers have endeavored to develop reproducible synthetic microbiomes. This approach is still in its infancy as this approach is so far easier said than done. In the first instance, it is argued that it is not necessary to inoculate the whole soil microbiome to target plants. This is because microbial communities typically contain many functional redundancies so that only the so-called core microbes are required. Further, "microbial networks can be compartmentalized into several network modules within which microbial species are highly connected with each other" and the hub species from each module can be identified and serve as the core microbes. Since around 99% of soil microbes cannot be cultured in the lab, it is suggested that culturing hub species might be possible using Web-based platforms such as KOMODO (Known Media Database) to predict media components for culturing the core microbiome. Of course, once core microbes have been identified and cultured, they need to be tested, first in the greenhouse including variations in soil properties, climatic factors, and crop traits. Following successful greenhouse testing, it is necessary to ensure that the synthetic microbiome is also effective in the field over several seasons and with what range of crops it is effective. Simultaneously with field trials, it is necessary to assess the ecological impact of the developed synthetic microbiome. In these endeavors, there is still a lot to be done and at this point success is far from guaranteed. Nevertheless, if researchers can overcome the difficulties in growing many of the hub species in a reproducible manner and then combining these microorganisms into stable synthetic microbiomes, there is every reason to believe that employing these consortia will eventually be superior to using individual microorganisms to facilitate plant growth. However, it is important to bear in mind that it is economically disadvantageous to commercially produce a PGPB consortium compared to the production of a single strain.

2.2   Endophytes

Plant bacterial endophytes ubiquitously colonize a plant's various tissues without causing any apparent inhibition of plant growth. The term endophyte comes from "endo" meaning inside and "phyte" meaning plant. These bacteria can enter the plant in a variety of ways including through tissue wounds, stomata (leaf pores used for gas exchange), lenticels (raised pores in the stems of woody plants), germinating radicles (the first part of the seedling to emerge from the seed), and roots cracks (which is generally thought to be the main portal of entry for these bacteria). Various estimates have suggested that healthy plant roots typically contain around 10^5–10^7 bacterial cells per gram of tissue while leaves and stems contain about 10^6–10^8 bacterial cells per gram of plant tissue (the larger area of leaves and stems may allow greater numbers of endophytes to colonize these tissues compared to the roots). However, these numbers may be skewed one way or the other by nutrients that are present in the soil and by plant metabolism. Bacterial endophytes grow within plant tissues in a mutualistic relationship both intercellularly and intracellularly. It is estimated that this relationship evolved around 60 million years ago with endophytes probably originating from rhizospheric organisms. Presumably, this is why endophytes and rhizospheric bacteria utilize essentially the same mechanisms to promote plant growth. For example, various bacterial endophytes reportedly produce numerous plant polymer hydrolytic enzymes (facilitating entry of bacteria into the plant), the enzyme ACC deaminase (which lowers plant ethylene levels thereby decreasing the inhibitory effects of abiotic stress), IAA (which promotes plant cell elongation and proliferation), siderophores (that sequester iron and provide it to plants), gibberelins (hormones that regulate various plant developmental processes), molecules involved in inorganic phosphate solubilization, trehalose (which can lower salt and drought stress), nitrogen fixation, enzymes that can degrade a range of the complex organic molecules produced by some plant and antibiotics (that inhibit pathogen proliferation). Although, it is not discussed in this chapter, it is nevertheless important to point out that there are numerous endophytic plant growth-promoting fungi. Moreover, for the most part these organisms use the same mechanisms as endophytic PGPB to facilitate the growth of plants.

2.2.1   Endophytic Genes

What specialized genes or mechanisms confer on some bacteria the ability to exist as endophytes? In other words, what are the differences between endophytes and

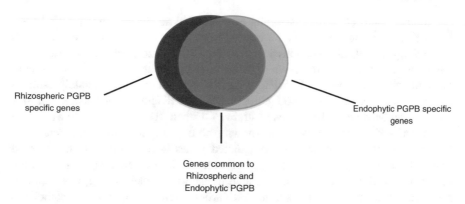

Rhizospheric PGPB
specific genes

Endophytic PGPB specific
genes

Genes common to
Rhizospheric and
Endophytic PGPB

Fig. 2.10 Venn diagram schematically showing the subtraction of the genome of a rhizospheric plant growth-promoting bacterium (PGPB) from the genome of an endophytic PGPB of the same genus. The genes of interest, the endophytic PGPB specific genes are shown in light green

rhizospheric plant growth-promoting bacteria? In one approach to addressing these questions, the complete genome sequences of several plant growth-promoting bacteria were compared. In the first instance, the genomic sequence of a rhizospheric plant growth-promoting strain of *Burkholderia xenovorans* was compared with the genomic sequence of an endophytic strain of *Paraburkholderia phytofirmans* (Fig. 2.10). Then, the set of genes encoded by the rhizospheric strain was subtracted from the set of genes encoded by the endophytic strain, thereby removing all of the genes that were common to the two bacterial strains. Subsequently, the set of genes that were common to both strains were removed from the gene complement of the endophytic strain, leaving a small number of genes that might be involved in endophytic behavior. Finally, these putative endophytic genes were compared with the genomes of eight different plant growth-promoting endophytic bacterial strains with the idea that genes that were present in all or most of these eight strains might be involved in endophytic behavior. In this last step, the assumption that was made was that most endophytes use similar, if not identical, strategies to colonize plants. Following these bioinformatics manipulations, a set of ~40 putative endophytic genes were identified including genes encoding transporter proteins (involved in transporting different molecules across membranes), secretion proteins, plant polymer degradation/modification enzymes, transcriptional regulator proteins, detoxification enzymes, redox potential maintenance proteins and a few general purpose enzymes such as dehydrogenases, synthases, and hydratases. Many, but not all of the identified genes/proteins, had previously been suggested in other studies to be a component of the endophytic behavior of some bacteria, lending credence to the genome comparison approach taken here. However, additional biochemical and genetic evidence is required to prove the involvement of all of the identified proteins in endophytic behavior. Moreover, since this particular study was limited to proteobacterial endophytes, it is not known whether the same set of genes that were found to be present in proteobacterial endophytes are also employed by other types of bacteria.

Another approach to understanding how endophytic and non-endophytic PGPB differ from one another included a transcriptomic study of two strains of *Bacillus mycoides*; strain EC18 is endophytic and strain SB8 is rhizospheric. Strain EC18 has a genome that is 5.75×10^6 bp while the genome of SB8 is 5.98×10^6 bp, and both of them have a GC content of 35.1%. When the two strains were separately exposed to potato plant root exudates, 408 genes in strain EC18 and 261 genes in strain SB8 were upregulated while 307 genes in strain EC18 and 219 genes in strain SB8 were downregulated. The majority of the known (the function of many of the genes in a bacterial genome is not known) upregulated genes were involved in amino acid metabolism and signal transduction. The bulk of the downregulated genes were involved in cell cycle control, amino acid transport and metabolism, cell wall and membrane biogenesis, transcription, and carbohydrate transport and metabolism. However, there were significant differences in the gene expression changes observed between EC18 and SB8 in response to the potato root exudates. In total 53 genes were upregulated in EC18 and downregulated in SB8; these genes were mainly coding for membrane proteins, transcriptional regulators, or proteins involved in amino acid metabolism and biosynthesis. On the other hand, 52 genes that were downregulated in EC18 were upregulated in SB8; these genes mainly coded for proteins involved in sugar transport. It remains for a comprehensive and detailed model of endophytic behavior to be developed. Nevertheless, it is clear that endophytic behavior involves a large set of genes that enable endophytes to bind and enter plant roots and then alter their metabolism in response to their new environment inside of the plant.

2.2.2   Isolating Endophytes

Biologically, there are three different types of endophytes. Opportunistic endophytes only occasionally enter into plant tissues and then colonize those tissues. Obligate endophytes can only live inside of a plant tissue. Facultative endophytes can live inside of plant tissue as well as in other habitats such as soil. Thus, for the most part, all of the endophytes that have been isolated and studied in the lab and are being developed as bacterial inoculants are facultative endophytes. In addition, while most bacterial endophytes colonize plant roots, not all endophytes can colonize a wide range of plant tissues including roots, stems, leaves, and flowers.

The simplest and most direct way to isolate new bacterial endophytes is to first keep in mind that endophytes are typically localized inside of plant tissues and not on the outer surface of the plant (Fig. 2.11a). Thus, following the isolation of plant tissue (typically roots, although other tissues may also act as source material), the outer surface of the plant tissue is cleaned/sterilized in order to ensure that no bacteria remain on the surface; in the case of roots this means eliminating the rhizospheric bacteria (Fig. 2.11b). The clean plant tissue is then macerated so that the bacteria that are present inside are released. Once endophytic bacteria have been isolated, they are characterized to ascertain which plant growth-promoting traits they possess. In addition, the genus and species of the newly isolated bacterial strains are

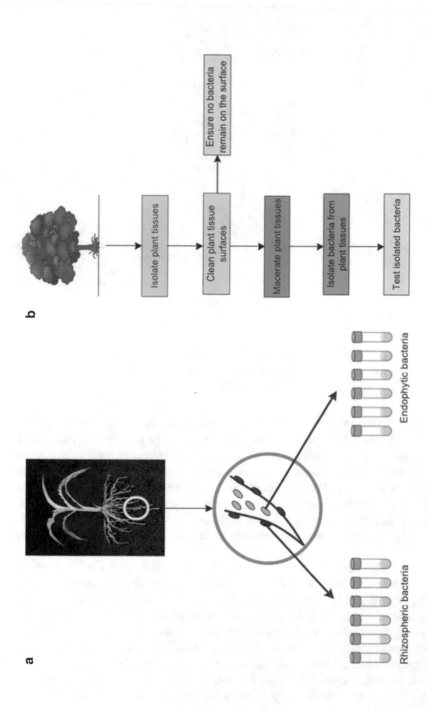

Fig. 2.11 (a) Schematic representation of a cross section of a plant root, that shows the localization of individual rhizospheric (red) and endophytic (blue) bacteria. (b) Flow chart showing how endophytic bacteria may be isolated from various plant tissues including leaves, stems, roots, bark and flowers, eliminating rhizospheric bacteria

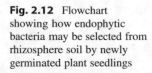

Fig. 2.12 Flowchart showing how endophytic bacteria may be selected from rhizosphere soil by newly germinated plant seedlings

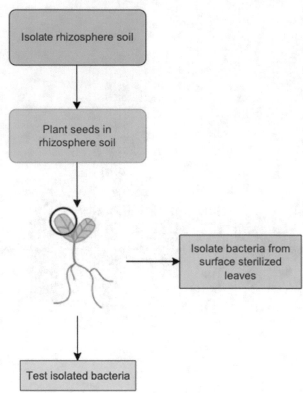

generally determined at this point, by sequencing all or a large portion of the bacterial gene encoding the 16S ribosomal RNA. Since the isolated endophytic bacteria may contain opportunistic and obligate endophytes as well as (the more desirable) facultative endophytes, it is necessary to determine whether these strains can be grown in culture. The selected strains (usually chosen on the basis of superior plant growth-promoting traits) are then typically grown to the mid-log phase of growth and then stored in small aliquots in glycerol at very cold temperatures (such as −80 °C), a condition that will generally ensure the long-term survival of these bacteria.

Another procedure, for the isolation of endophytes is sometimes called bio-prospecting and involves collecting a number of different soil samples (often from rhizospheric soils). Each soil sample is put into one or more pots into which a number of seeds (of the investigator's choice) are sown (Fig. 2.12). In this procedure, the growing plants select endophytes from the soil samples. Following several weeks of plant growth (generally under controlled conditions) the plants are harvested and plant tissue that has not been in contact with the soil (such as leaf tissue) is surface sterilized, macerated, and then used as source material for the isolation of new bacterial endophytes. Similar to the case with endophytes isolated directly from plant tissues, the new bacterial endophytes from this procedure are thoroughly characterized. However, in this case, since the starting point was soil that contained bacteria (and not plant tissue), it is likely that this procedure will yield a preponderance of facultative endophytes that can readily be grown in culture.

2.2.3 Endophytes and Flowers

The worldwide market for cut flowers is enormous, with some estimates suggesting that in 2018 this market was valued at more than US$500 billion. Unfortunately, many common flowers such as roses and carnations are highly sensitive to ethylene-caused senescence resulting in a decreased lifetime for these flowers. Flower senescence may include petal in-rolling, loss of color, petal wilting, shedding of flower parts, and gradual fading of the blossom. To decrease the rate of flower senescence, flower sellers often add one of several different chemical compounds that effectively increase the lifetime of the flowers. Unfortunately, many of these chemicals are potentially environmentally hazardous so that a biological means of extending the lifetime of cut flowers would be extremely beneficial. Some plant growth-promoting bacteria contain the enzyme 1-aminocyclopropane-1-carboxylate (ACC) deaminase that can lower plant ethylene levels by cleaving the compound ACC which is the immediate precursor of ethylene in all higher plants. In one series of experiments, the ability of an endophytic strain and a rhizospheric strain of ACC deaminase-containing plant growth-promoting pseudomonads to affect the senescence of fresh cut mini-carnations was compared. In addition, the effects of an ACC deaminase minus mutant of the endophytic strain, the compound ACC, the chemical ethylene inhibitor L-α-(aminoethoxyvinyl)-glycine (AVG) and no addition at all, on flower senescence, were studied. Mini-carnations were cut from growing plants exactly when the flowers were at full bloom, and then monitored visually for 10 days with the extent of flower senescence recorded on a scale of zero to four, where zero is a freshly cut flower and four is a completely senesced flower. In the first instance, addition of ACC, the ethylene precursor, significantly sped up the rate of flower senescence while the addition of AVG, an ethylene inhibitor, significantly slowed the rate of flower senescence compared to the control where nothing was added to the flowers (Table 2.2). Second, addition of the wild-type endophytic bacterium, but not the wild-type rhizospheric bacterium or the ACC deaminase minus mutant of the endophytic bacterium, significantly slowed flower senescence. In fact, the wild-type endophytic bacterium was significantly better at preventing flower senescence than AVG, the chemical ethylene inhibitor. The ineffectiveness of the ACC deaminase minus mutant of the endophytic bacterium provided strong evidence that ACC deaminase was responsible for the observed activity. Since all of the flowers contained approximately 8 cm of stem material, the data suggested that the wild-type endophyte but not the wild-type rhizosphere bacterium was able to travel up the stem and thereby sequester and cleave much of the ACC released by the flowers before it could be converted to ethylene. In fact, by using the technique of transmission electron microscopy, it was possible to visualize the presence of the endophyte but not the rhizosphere bacterium all the way up the stem (but not in the flower). And, although the ACC deaminase minus mutant was ineffective in preventing flower senescence, it was also able to travel up the flower stem. Based on this experiment, the use of ACC deaminase-containing endophytes, instead of chemical ethylene inhibitors, as a means of delaying the senescence of certain cut flowers may be an attractive prospect for the flower industry.

2.2.4 Endophytic Fungi

It is important to point out that numerous fungi, as well as bacteria, are able to form an endophytic relationship with plants. Initially, endophytic fungi might attach to the root surface. Subsequently, the attached fungi penetrate the outer layers of the plant root and then migrate into and colonize the plant's internal tissues. Among other mechanisms, endophytic fungi can activate the plant host immune system affecting host genetics and phenotype expression to resist both pathogens and herbivores (Table 2.3).

Table 2.2 Senescence of mini-carnations following various treatments

Day	Level of senescence					
---	No addition	+ AVG	+ ACC	+ wild-type endophyte	+ mutant endophyte	+ wild-type rhizospheric bacterium
0	0	0	0	0	0	0
2	0.1	0.1	0.5	0	0.1	0.2
4	1.1	0.8	2.2	0.1	1.1	1.0
6	2.2	1.7	3.5	1.2	2.3	2.4
8	3.5	3.1	4.0	2.6	3.5	3.5
10	4.0	3.8	4.0	3.7	4.0	4.0

Each number is the mean of 100–110 flower observations. The standard error on all measurements is typically around 5% or less of the values shown. A level of senescence of 0 indicates no senescence while a level of 4.0 indicates a completely senesced flower

Table 2.3 Endophytic fungi that protect plants against pathogens

Fungi	Host plant
Rhexocercosporidium sp.	*Sophora tonkinensis*, a Chinese medicinal herb
Aspergillus spp., *Penicillium* spp., *Fusarium* spp. and *Phoma* spp.	*Eleusine coracana*, finger millet
Phialocephala sphaeroides	*Picea abies*, Norway spruce
Paraconiothyrium SSM001	*Taxus* spp., Yew trees
Trichoderma hamatum UoM	*Pennisetum glaucum*, pearl millet
Colletotrichum tropicale	*Theobroma cacao*, Cacao tree
Beauveria bassiana	*Solanum lycopersicum* and *Gossypium* spp., tomato and cotton
Neotyphodium coenophialum	*Festuca arundinacea*, tall fescue
Epichloë occultans	*Lolium multiflorum*, annual ryegrass
Exophiala pisciphila	*Zea mays*, maize or corn

Questions

1. Why are large numbers of bacteria found around the roots of plants?
2. What are the main types of bacteria found around plant roots?
3. Why is the plant *Arabidopsis thaliana* often used in genetic and biochemical plant research?
4. How might aphid infection of plants affect plant growth-promoting bacteria? What effect might this have on the continued aphid infection of the same plants?
5. When the same *Vitis vinifera* scion is grafted onto two different rootstocks, one that is normal and one that is potassium deficient, how does this affect the root endophytes and rhizospheric bacteria?
6. How might increasing levels of atmospheric carbon dioxide affect microbiome nitrogen fixation and methane oxidation?
7. How might drought conditions affect the distribution of Gram-negative-vs. Gram-positive bacteria in the rhizosphere and the endosphere?
8. How does the bacterial composition of the microbiome of Arcto-alpine plants change when the plants are grown under Alpine, Low Arctic, and High Arctic conditions, respectively?
9. How can fruit microbiomes be used to distinguish between organic and conventionally grown apples?
10. What is a synthetic microbiome? How might it be developed?
11. Starting with a common Gram-negative endophytic bacterium, how would you engineer a plant seed microbiome to contain a significant amount of that bacterium? How would you prove the presence and demonstrate the effectiveness of the bacterium?
12. Based on the knowledge of the genomic sequence of an endophytic bacterium and a rhizospheric bacterium of the same genus, how would you estimate what genes might be involved in the endophytic activity of the first bacterium?
13. What are the steps that one might follow to isolate new endophytic bacterial strains from specific tissues of a target plant?
14. Starting from several rhizosphere soil samples, how would you isolate new endophytic strains?
15. Why are bacterial endophytes and not rhizospheric bacteria that contain the same panel of plant growth-promoting genes able to prolong the shelf life of cut flowers? What bacterial trait is necessary for an endophytic bacterial strain to be effective at prolonging the shelf life of cut flowers?

Bibliography

Afzal I, Iqrar I, Shinwari ZK, Yasmin A (2017) Plant growth-promoting potential of endophytic bacteria isolated from roots of wild *Dodonaea viscosa* L. Plant Growth Regul 81:399–408

Ali S, Charles TC, Glick BR (2012) Delay of flower senescence by bacterial endophytes expressing 1-aminocyclopropane-1-carboxylate deaminase. J Appl Microbiol 113:1139–1144

Ali S, Duan J, Charles TC, Glick BR (2014) A bioinformatics approach to the determination of genes involved in endophytic behavior in *Burkholderia* spp. J Theor Biol 343:193–198

Ali S, Charles TC, Glick BR (2017) Endophytic phytohormones and their role in plant growth promotion. In: Doty SL (ed) Functional importance of the plant microbiome: implications for agriculture, forestry and bioenergy. Springer, Berlin, pp 89–105

Bashir M, Asif M, Naveed M, Qadri RWK, Faried N, Anjum F (2019) Postharvest exogenous application of various bacterial strains improves the longevity of cut 'Royal Virgin' tulip flowers. Pak J Agric Sci 56:71–76

Berg G, Smalla K (2009) Plant species and soil type cooperatively shape the structure and function of microbial communities in the rhizosphere. FEMS Microbiol Ecol 68:1–13

Brígido C, Singh S, Menéndez E, Tavares MJ, Glick BR, Félix MR, Oliveira S, Carvalho M (2019) Diversity and functionality of culturable endophytic bacterial communities in chickpea plants. Plan Theory 8:42

Bulgarelli D, Rott M, Schlaeppi K, van Themaat EVL, Ahmadinejad N, Assenza F, Rauff P, Huttel B, Schmelzer E, Peplies J, Gloeckner FO, Amann R, Eickhorst T, Schulze-Lefert P (2012) Revealing structure and assembly cues for *Arabidopsis* root-inhabiting microbiota. Nature 488:91–95

Bulgarelli D, Schlaeppi K, van Themaat EVL, Spaepen S, Schulze-Lefert P (2013) Structure and function of the bacterial microbiota of plants. Annu Rev Plant Biol 64:807–838

Compant S, Samad A, Faist H, Sessitsch A (2019) A review on the plant microbiome: ecology, functions, and emerging trends in microbial application. J Adv Res. https://doi.org/10.1016/j.jare.2019.03.004

D'Amico F, Candela M, Turroni S, Biagi E, Brigidi P, Bega A, Vancini D, Rampelli S (2018) The rootstock regulates microbiome diversity in root and rhizosphere compartments of *Vitis vinifera* cultivar Lamrusco. Front Microbiol 9:2240. https://doi.org/10.3389/fmicb.2018.02240

Delgado-Baquerizo M, Oliverio AM, Brewer TE, Benavent-González A, Eldridge DJ, Bardgett RD, Mestre FT, Singh BJ, Fierer N (2018) A global atlas of the dominant bacteria found in soil. Science 359:320–325

Dudeja SS, Giri R, Saini R, Suneja-Madan P, Kothe E (2012) Interaction of endophytic microbes with legumes. J Basic Microbiol 52:248–260

Edwards J, Johnson C, Santos-Medellin C, Lurie E, Podishetty NK, Bhatnagar S, Eisen JA, Sundaresan V (2015) Structure, variation, and assembly of the root-associated microbiomes of rice. Proc Natl Acad Sci U S A 112:E911–E920

Esmaeel Q, Miotto L, Rondeau M, Leclère V, Clément C, Jacquard C, Sanchez L, Barka EA (2018) Paraburkholderia phytofirmans PsJN-plants interaction: from perception to the induced mechanisms. Front Microbiol 9:2093. https://doi.org/10.3389/fmicb.2018.02093

Haney CH, Ausubel FM (2015) Plant microbiome blueprints. Science 349:788–789

Harodim PR, van Overbeek LS, van Elsas JD (2008) Properties of bacterial endophytes and their proposed role in plant growth. Trends Microbiol 16:463–471

Harodim PR, van Overbeek LS, Berg G, Pirttilä AM, Compant S, Campisano A, Döring M, Sessitsch A (2015) The hidden world with plants: ecological and evolutionary considerations for defining functioning of microbial endophytes. Micribiol Molec Biol Rev 79:293–320

Hartmann A, Schmid M, van Tuinen BG (2009) Plant-driven selection of microbes. Plant Soil 321:235–257

Hidayati U, Chniago IA, Munif A, Siswanto SA (2014) Potency of plant growth promoting endophytic bacteria from rubber plants (*Hevea brasiliensis* Müll Arg.). J Agron 13:147–152

Huang AC, Jiang T, Liu Y-X, Bai Y-C, Reed J, Qu B, Goossens A, Nützmann H-W, Bai Y, Osbourn A (2019) A specialized metabolic network selectively modulates *Arabidopsis* root microbiota. Science 364:eaau6839

Khare E, Mishra J, Arora NK (2018) Multifaceted interactions between endophytes and plant: developments and prospects. Front Microbiol 9:2732. https://doi.org/10.3389/fmicb.2018.02732

Kim S, Lowman S, Hou G, Nowak J, Flinn B, Mei C (2012) Growth promotion and colonization of switchgrass (*Panicum virgatum*) cv. Alamo by bacterial endophyte *Burkholderia phytofirmans* strain PsJN. Biotechnol Biofuels 5:37

Kim B, Song GC, Ryu CM (2016) Root exudation by aphid leaf infestation recruits root-associated Paenibacillus spp. to lead plant insect susceptibility. J Microbiol Biotechnol 26:549–557

Kong Z, Hart M, Liu H (2018) Paving the way from the lab to the field: using synthetic microbial consortia to produce high-quality crops. Front Plant Sci 9:1467

Kumar M, Brader G, Sessitsch A, Mäki A, van Elsas JD, Nissinen R (2017) Plants assemble species specific bacterial communities from common core taxa in three acto-alpineclimate zones. Front Microbiol 8:12. https://doi.org/10.3389/fmicb.2017.00012

Lacava PT, Azevedo JL (2013) Endophytic bacteria: a biotechnological potential in agrobiology system. In: Maheshwari DK, Saraf M, Aeron A (eds) Bacteria in agrobiology: crop productivity. Springer, Berlin, pp 1–44

Li Y, Wu X, Chen T, Wang W, Liu G, Zhang W, Li S, Wang M, Zhao C, Zhou H, Zhang G (2018) Plant phenotypic traits eventually shape its microbiota: a common garden test. Front Microbiol 9:2479

Lin C, Xiuli X, Jiabao Z, Redmille-Gordon M, Nie G, Wand Q (2019) Soil characteristics overwhelm cultivar effects on the structure and assembly of root-associated microbiomes of modern maize. Pedosphere 29:360–373

Lunberg DS, Lebeis SL, Paredes SH, Yourstone S, Gehring J, Malfatti S, Tremblay J, Engelbrektson A, Kunin V, Glavina del Rio T, Edgar RC, Eickhorst T, Ley RE, Hugenholtz P, Tringe SG, Dangl JL (2012) Defining the core *Arabidopsis thaliana* root microbiome. Nature 488:86–90

Mitter B, Petric A, Shin MW, Chain PS, Hauberg-Lotte L, Reinhold-Hurek B, Nowak J, Sessitsch A (2013) Comparative genome analysis of *Burkholderia phytofirmans* PsJN reveals a wide spectrum of endophytic lifestyles based on interaction strategies with host plants. Front Plant Sci 4:120. https://doi.org/10.3389/fpls.2013.00120

Mitter B, Pfaffenbichler N, Flavell R, Company S, Antonielli L, Petric A, Berninger T, Naveed M, Sheibani-Tezerji R, von Maltzahn G, Sessitsch A (2017) A new approach to modify plant microbiomes and traits by introducing beneficial bacteria at flowering into progeny seeds. Front Microbiol 8:11. https://doi.org/10.3389/fmicb.2017.00011

Nelson EB (2018) The seed microbiome: origins, interactions, and impacts. Plant Soil 422:7–34

Okubo T, Liu D, Tsurumaru H, Ikeda S, Asakawa S, Tokida T, Tago K, Hayatsu M, Aoki N, Ishimaru K, Ujiie K, Usui Y, Nakamura H, Sakai H, Hayashi K, Hasegawa T, Minamisawa K (2015) Elevated atmospheric x levels affect community structure of rice root-associated bacteria. Front Microbiol 6:136

Pinski A, Betekhtin A, Hupert-Kocurek K, Mur LAJ, Hasterok R (2019) Defining the genetic basis of plant-endophytic bacteria interactions. Int J Mol Sci 20:1947

Rashid S, Charles TC, Glick BR (2012) Isolation and characterization of new plant growth-promoting bacterial endophytes. Appl Soil Ecol 61:217–224

Reinhold-Hurek B, Bünger W, Burbano CF, Sabale M, Hurek T (2015) Roots shaping their microbiome: global hotspots for microbial activity. Annu Rev Phytopathol 53:403–424

Ripa FA, Cao W-D, Tong S, Sun J-G (2019) Assessment of plant growth promoting and abiotic stress tolerance properties of wheat endophyhtic fungi. BioMed Res Int 2019:6105865

Rodriguez CE, Mitter B, Barret M, Sessitsch A, Compant S (2018) Commentary: seed bacterial inhabitants and their routes of colonization. Plant Soil 422:129–134

Santoyo G, Moreno-Hagelsieb G, Orozco-Mosqueda MC, Glick BR (2016) Plant growth-promoting bacterial endophytes. Microbiol Res 183:92–99

Sasse J, Martinoia E, Northen T (2018) Feed your friends: do plant exudates shape the root microbiome? Trends Plant Sci 23:25–41

Stone BWG, Weingarten EA, Jackson CR (2018) The role of the phyllosphere microbiome in plant health and function. Annu Plant Rev 1:1–24

Voges MJEEE, Bai Y, Schulze-Lefert P, Sattely ES (2019) Plant-derived coumarins shape the composition of an *Arabidopsis* synthetic root microbiome. Proc Natl Acad Sci U S A 116:12558–12565

Wassermann B, Muller H, Berg G (2019) An apple a day: which bacteria do we eat with organic and conventional apples? Front Microbiol 10:1629

Wilpiszeski RL, Aufrecht JA, Retterer ST, Sullivan MB, Graham DE, Pierce EM, Zablocki OD, Palumbo AV, Elias DA (2019) Soil aggregate microbial communities: towards understanding microbiome interactions at biologically relevant scales. Appl Environ Microbiol 85(14):e00324-19

Xu L, Naylor D, Dong Z, Simmons T, Pierroz G, Hixson KK, Kim Y-M, Zink EM, Engbrecht KM, Wang Y, Gao C, DeGraaf S, Madera MA, Sievert JA, Hollingsworth BD, Scheller HV, Hutmacher R, Dahlberg J, Jansson C, Taylor JW, Lemaux PG, Coleman-Derr D (2018a) Drought delays development of the sorghum root microbiome and enriches for monderm bacteria. Proc Natl Acad Sci U S A 115:E4284–E4293

Xu J, Zhang Y, Zhang P, Trivedi P, Riera N, Wang Y, Liu X, Fan G, Tang J, Coletta-Filho HD, Cubero J, Deng X, Ancona V, Lu Z, Zhong B, Roper MC, Capote N, Catara V, Pietersen G, Vernière C, Al-Sadi AM, Li L, Yang F, Xu X, Wang J, Yang H, Jin T, Wang N (2018b) The structure and function of the global citrus rhizosphere microbiome. Nat Commun 9:4894

Yaish MW, Antony I, Glick BR (2015) Isolation and characterization of endophytic plant growth-promoting bacteria from date palm tree (*Phoenix dactylifera* L.) and their potential role in salinity tolerance. Antonie Van Leeuwenhoek 107:1519–1532

Yan L, Zhu J, Zhao X, Shi J, Jiang C, Shao D (2019) Beneficial effects of endophytic fungi colonization on plants. Appl Microbiol Biotechnol 103:3327–3340

Yi Y, de Jong A, Frenzel E, Kuipers OP (2017) Comparative transcriptomics of *Bacillus mucoides* strains in response to potato-root exudates reveals different genetic adaptation of endophytic and soil isolates. Front Microbiol 8:1487

Zhang Y, Gao X, Shen Z, Zhu C, Jiao Z, Li R, Shen Q (2019a) Pre-colonization of PGPR triggers rhizosphere microbiota succession associated with crop yield enhancement. Plant Soil 439:553–567

Zhang J, Liu Y-X, Zhang N, Hu B, Jin T, Xu H, Qin Y, Yan P, Zhang X, Guo X, Hui J, Cao S, Wang X, Wang C, Wang H, Qu B, Fan G, Yuan L, Garrido-Oter R, Chu C, Bai Y (2019b) NRT1.1B is associated with root microbiota composition and nitrogen use in field-grown rice. Nat Biotechnol 37:676

Some Techniques Used to Elaborate Plant–Microbe Interactions

3

Abstract

This chapter addresses some of the major biochemical and genetic tools that have been used to study plant–bacterial interactions. Most of these techniques are already a mainstay of modern biology. However, they are discussed here so that the reader will comprehend the multiplicity of approaches that scientists use to understand the many mechanisms that bacteria employ to facilitate plant growth and development. These techniques include DNA sequencing (including the sequencing of entire bacterial genomes), use of the polymerase chain reaction (to amplify and analyze minute amounts of plant or bacterial DNA), transcriptomics (the study of the mRNAs synthesized by plants or bacteria following various treatments), proteomics (the study of the proteins synthesized by plants or bacteria following various treatments), metabolomics (the study of the metabolites synthesized by plants or bacteria following various treatments), labeling and imaging of bacteria interacting with plants, and microencapsulation of bacteria prior to their use in the environment.

With the advent of recombinant DNA technology in the early 1970s and the subsequent development of techniques for DNA sequencing, the polymerase chain reaction (PCR), DNA synthesis, and Clustered Regularly Interspaced Short Palindromic Repeats (CRISPR), it is possible to isolate, characterize, amplify, modify, sequence, and synthesize virtually any gene of interest. Moreover, the techniques of transcriptomics, proteomics, and metabolomics may be employed to understand in considerable detail changes in plants that occur as a consequence of an interaction with PGPB as well as changes in PGPB that result from interactions with plants. The use of these techniques is enabling researchers to develop a more profound understanding of the detailed nature of plant–bacterial interactions, an important step in the development of new and more efficacious PGPB strains. Here, a number of those techniques are briefly examined and discussed.

© Springer Nature Switzerland AG 2020
B. R. Glick, *Beneficial Plant-Bacterial Interactions*,
https://doi.org/10.1007/978-3-030-44368-9_3

3.1   Next-Generation DNA Sequencing

The function of a gene may often be deduced from its nucleotide sequence. For example, the amino acid sequence of a protein, determined from the nucleotide sequence of the DNA that encodes that protein, can be compared with amino acid sequences of known proteins, and significant sequence similarity generally indicates a protein with an equivalent function. In addition, distinctive regions of chromosomal DNA such as promoters and other regulatory regions can be determined. Nucleotide sequence information may also facilitate gene isolation and characterization.

For several decades, the procedures originally developed (separately) by Fred Sanger and Walter Gilbert were used for DNA sequencing. Sanger and Gilbert shared the 1980 Nobel Prize in Chemistry. Notwithstanding the enormous advances brought about by this work, with time it became clear that to determine the DNA sequence of entire genomes, from bacteria to humans, these approaches were much too slow and expensive.

A variety of DNA sequencing techniques are currently being utilized. Generally, these methods involve repeated cycles of enzymatic addition of nucleotides to a primer based on complementarity to a template DNA fragment and detection and identification of the nucleotide(s) added. The techniques differ in the method by which the nucleotides are extended, employing either DNA polymerase to catalyze the addition of a single nucleotide or DNA ligase to add a short, complementary oligonucleotide, and in the method by which the addition is detected.

In the more than 40 years since Sanger and Gilbert shared the Nobel Prize in Chemistry, both the cost of DNA sequencing and the amount of time need to acquire DNA sequences has decreased by several orders of magnitude. These dramatic changes are a direct result of the introduction of automated DNA sequencers that can generate several million bases of DNA sequence information per day.

3.1.1   DNA Sequences of Complete Bacterial Genomes

With the advances in our ability to rapidly determine the sequence of large amounts of DNA, it is currently possible to sequence entire bacterial and plant genomes. By late 2019, the complete genome sequence of more than 200,000 bacterial and several hundred plant genomes (including both higher/vascular plants and lower plants including algae, ferns and bryophytes) had been determined. Moreover, the completely characterized bacterial genomes include more than a thousand different bacterial strains that specifically interact with plants, including both PGPB and phytopathogens. A representation of the genome sequence of the PGPB *Pseudomonas* sp. UW4 is shown in Fig. 3.1.

At a first glance, the amount of data contained in a complete genome sequence may seem overwhelming. Thus, many bacteria genomes contain approximately 1–10 million base pairs (bp) of DNA. However, this data may be readily analyzed using widely available computer software. For example, it was determined that the

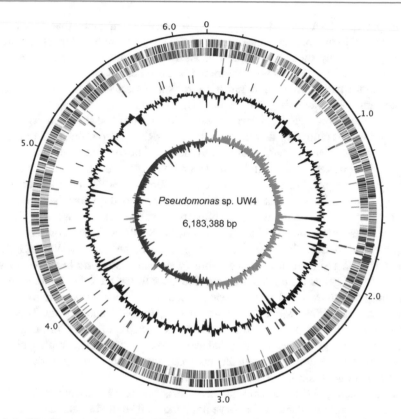

Fig. 3.1 Circular genome map of the PGPB *Pseudomonas* sp. UW4. The outer black circle on the outside shows the scale line in Mbps; circles 2 and 3 represent the coding region with the colors of the Clusters of Orthologous Groups (COG) categories (where circle 2 represents the plus and circle 3 the minus strand); circles 4 and 5 show tRNA (green) and rRNA (red), respectively; circle 6 displays the Insertion Sequence (IS) elements (blue); circle 7 represents mean centered G+C content (bars facing outside-above mean, bars facing inside-below mean); circle 9 shows the GC skew of different portions of the genome. COG represents standard (and generally agreed upon) categories of proteins with similar functions

entire genome of *Pseudomonas* sp. UW4 contains 6,183,388 base pairs (bp) of DNA, a G + C content of 60.05%, 5570 predicted protein coding genes, and 22 ribosomal RNA and 72 tRNA genes in seven separate clusters. More detailed analysis of the encoded proteins reveals that this bacterium contains genes for a number of traits that are often associated with PGPB activity. For example, *Pseudomonas* sp. UW4 encodes seven enzymes that appear to be involved in the biosynthesis of the phytohormone indoleacetic acid (IAA). These seven enzymes appear to synthesize IAA via two different (but overlapping) biosynthetic pathways. This information provides a starting point for experimentally dissecting, and therefore better understanding the functioning of this pathway(s). For example, knowledge of the genomic sequence of this PGPB has enabled investigators to use the technique of PCR (discussed in Sect. 3.2) and recombinant DNA technology to isolate any or all of the putative IAA biosynthesis genes, to construct mutants that block these pathways and to overproduce one or more of the enzymes involved in a biosynthetic pathway and study them biochemically. As the genomic DNA sequences of an increasing number of PGPB are determined, it becomes possible for researchers to use this information to gain a much more sophisticated understanding of the mechanisms and functioning of these bacteria.

As mentioned previously, one way of assigning a genus and species to a particular bacterium is by comparing that bacterium's 16S rRNA gene sequences to the 16S rRNA gene sequences from a large number of known and well-characterized bacteria. Unfortunately, the number of 16S rRNA genes in different bacteria can vary from 1 to as many as 15, and these gene sequences are often not identical to one another, despite the fact that they are encoded within the same bacterium. For example, strain UW4 contains seven 16S rRNA gene sequences. When a phylogenetic tree was constructed of all the unique 16S rRNA gene sequences of all completely sequenced *Pseudomonas* spp. genomes, the strain UW4 16S rRNA genes were most similar to the genes from *Pseudomonas putida* but also not very different to the genes from *Pseudomonas fluorescens*. On the other hand, when the DNA from the entire UW4 genome was compared to other *Pseudomonas* genome sequences, strain UW4 was most similar to *Pseudomonas fluorescens*. This result serves as a cautionary tale that it is not always possible to make an unequivocal assignment of the genus and species of a bacterium even when the complete genomic DNA of that bacterium is known. In this case, given the conflicting data, it was decided to assign strain UW4 as *Pseudomonas* sp.

By sequencing entire bacterial genomes, it is possible to discern quite a lot of information regarding bacteria that have not been well studied in the laboratory. For example, *Frankia*, a genus of nitrogen-fixing, filamentous bacteria that lives in symbiosis in root nodules with actinorhizal plants, has not been studied in any great detail although several *Frankia* genomes have been sequenced. Recently, a group of researchers compared the genomes of 21 different strains of *Frankia*. They found that *Frankia* genomes encoded (1) enzymes that were associated with the synthesis of both IAA and nitrogenase, (2) genes that were known to modulate the effects of biotic and abiotic environmental stress, and (3) genes for the biosynthesis of siderophores, several lytic enzymes and cytokinin.

3.2   The Polymerase Chain Reaction

The polymerase chain reaction, commonly called PCR, is a simple and highly effective means of generating large quantities of a specific fragment of DNA. With this procedure, developed in 1983 by Kary Mullis, who was subsequently awarded the Nobel Prize in Chemistry in 1993 for this work, DNA fragments from ~100 to ~35,000 bp in length may be amplified several million times. Since its early development, PCR has become a mainstay of modern molecular biology. It is commonly used as a component of DNA cloning and sequencing, in directed mutagenesis protocols and as a key element in a wide range of DNA diagnostic

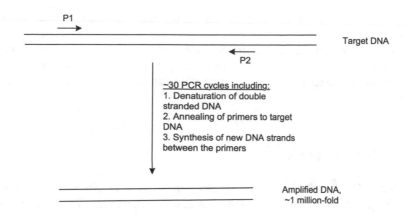

Fig. 3.2 Schematic overview of the polymerase chain reaction. P1 and P2 are oligonucleotide primers 1 and 2

procedures. PCR may be used to amplify whole genes or gene fragments, to chemically synthesize genes, to detect the presence of very low levels of target DNA and to produce genes with specific changes in their DNA.

A PCR amplification reaction (Fig. 3.2) typically includes the DNA that is targeted for amplification, two synthetic oligonucleotide primers (~20 nucleotides each) that are complementary to the ends of the target DNA sequence, a DNA polymerase that can withstand being repeatedly heated to approximately 95 °C and copies the DNA template with high fidelity, and a vast molar excess of the four deoxyribonucleotides (DNA building blocks) over the target DNA. In each PCR cycle the quantity of target DNA is doubled. Each cycle of DNA synthesis is catalyzed by the enzyme DNA polymerase, that is added at the very beginning of the procedure. All of the steps of a PCR cycle are carried out in a programmable block heater. In the first step, the double-stranded DNA in the sample is denatured (i.e., made single stranded) as a consequence of raising the temperature to 95 °C. This step typically lasts about 1 min. In the second step which also lasts about 1 min, the reaction mixture is cooled to ~55 °C, allowing the primers to bind (anneal) to complementary sites on the target DNA. In the third step, lasting about 2 min, the temperature is raised to ~75 °C which is near the optimal temperature of the thermostable DNA polymerase. DNA synthesis is initiated at the 3′ hydroxyl end of each primer and both strands of the template DNA are copied.

If DNA that is amplified in a PCR reaction is labeled with a fluorescent dye, it is possible to monitor the production of the PCR products in real time. In addition, the technique of real-time PCR allows a researcher to quantitate the amount of a specific DNA fragment in the starting material.

PCR may be used in a variety of ways in the study of plant–microbe interactions. For example, when all or a portion of the genomic DNA sequence of a bacterium is known, PCR may be used to amplify and then isolate genes of interest within the characterized region of the genome; PCR may also be used to aid in strategies to express the isolated gene in a new host strain. PCR may be employed as a tool to construct mutants of specific target genes and/or their regulatory regions thereby facilitating a better understanding of the role of that gene in the functioning of the bacterium. PCR may be utilized in the synthesis of whole genes that are being introduced into a test bacterium. And, of course, PCR may also be used to facilitate the isolation, characterization and genetic manipulation of plant, as well as bacterial, genes.

3.2.1   Real-Time PCR

Real-time PCR (RT-PCR; also called quantitative PCR or qPCR) is a technique where target DNA is simultaneously amplified and quantified. With this method, DNA that is amplified is labeled with the fluorescent dye SYBR green as the double-stranded DNA forms, and it is possible to monitor the fluorescence that results when the dye that is bound to double-stranded DNA (the product of the reaction) is irradiated with light. The SYBR green dye is excited using blue light ($\lambda_{max} = 488$ nm) and it emits green light ($\lambda_{max} = 522$ nm). This approach allows researchers to quantify the amount of a specific DNA fragment in any particular sample. In the simplest approach to labeling DNA, researchers add dyes that bind to double-stranded DNA and emit fluorescence, and the fluorescence intensity increases in proportion to the concentration of double-stranded DNA (Fig. 3.3).

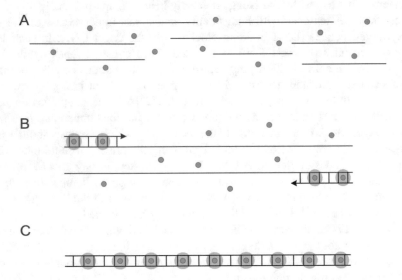

Fig. 3.3 The principle behind real-time PCR. The fluorescent dye SYBR green does not bind to single-stranded DNA and does not fluoresce when it is free in solution (**a**). As the DNA is amplified in the PCR reaction, the dye binds to the double-stranded regions and fluoresces (**b**). All of the completely amplified double-stranded DNA contains bound and fluorescing dye (**c**). The amount of fluorescence is proportional to the amount of bound dye which is directly related to the amount of double-stranded DNA

Compared to other techniques, real-time PCR offers a dramatic improvement in both the sensitivity of detection and the time that it takes to complete the analysis. Real-time PCR occurs in four phases. First, for ~10–15 cycles (linear phase), fluorescence emission at each cycle does not rise above the background level. Then, in the early exponential phase, the fluorescence reaches a threshold at which it is significantly higher than the background. The cycle at which this occurs is known as the threshold cycle and is inversely correlated with the amount of target DNA in the original sample. During the next (exponential) phase, the amount of product doubles in each cycle, while in the final (plateau) stage, the reaction components become limited and measurements of the fluorescence intensity are no longer useful. To quantitate the amount of target DNA in a sample, a standard curve is first generated by serially diluting a sample with a known number of copies of the target DNA, and the threshold values for each dilution are plotted against the starting amount of sample. The amount of a target DNA in a sample is estimated from the threshold value for the sample and extrapolating the starting amount from the standard curve.

3.3 Transcriptomics

A key question in developing a deeper understanding of the interaction between plants and various microbes is, what is the nature of the changes in the physiology and biochemistry of the plant that occur as a consequence of its interaction with a particular microorganism (Fig. 3.4). And, of course, the corresponding question is, what are the nature of the changes in the physiology and biochemistry of the microorganism that occur as a consequence of its interaction with a particular plant (Fig. 3.4). These two questions may be addressed by monitoring gene expression either by transcriptomics, proteomics, or metabolomics (all of which are briefly described below). Often, the results of monitoring changes in gene expression will provide scientists with information about the activation or inhibition of a particular metabolic pathway as a consequence of a specific treatment. Unlike most traditional biological experiments that are hypothesis driven, transcriptomic, proteomic, and metabolomics experiments are said to be "hypothesis generating." A less flattering way of characterizing these types of experiments is to say that they are sometimes considered to be fishing expeditions where you never know ahead of time exactly what you are going to find, or if you will find anything meaningful.

Transcriptomics, which is also called functional genomics, is concerned with determining which genes are expressed under a particular set of conditions and the levels of expression (i.e., the amount of transcription) of those genes. Transcription, within the context of plant–microbe interactions, is typically assessed as the response of plants to the presence of either phytopathogens, plant growth-promoting bacteria, abiotic stressors or some combination of these factors. In principle, this approach could also be used to assess the effect of plants, plant extracts, or plant exudates on the transcription of plant associated bacteria; however, this technique is only rarely used for this purpose. A transcriptomics approach aims to discover the genes that are

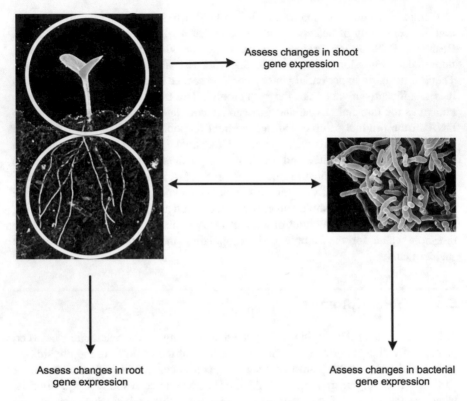

Fig. 3.4 Schematic view of monitoring the physiological and biochemical effects of plants and soil microorganisms on one another

either up- or downregulated under specific conditions. In the past, the transcription of only one or a few genes could be followed at a time. The approach outlined below permits researchers to track the simultaneous transcription of hundreds or even thousands of genes. However, it is important to keep in mind, in addition to monitoring newly synthesized transcripts, DNA microarray technology also examines DNA transcripts that are rapidly turned over.

3.3.1   DNA Microarray Technology

In one type of commercially available microarray system, approximately 28,000 different oligonucleotides (each 60 nucleotides long) that are complementary to portions of the DNA sequence of the model plant *Arabidopsis thaliana* are synthesized directly on a solid surface. Tens of thousands of copies of an oligonucleotide with the same specific nucleotide sequence are synthesized in a predefined position on the array surface. About 10,000 different probes can be arrayed in a 1 cm² area. Because *A. thaliana* has a quite small genome (135×10^6 base pairs, or about 20 times the size of a typical soil pseudomonad genome), the 28,000 oligonucleotides correspond to approximately 80% of the entire genome of this plant. However, for most other plants, a complete whole-genome oligonucleotide array would require more than 500,000 oligonucleotide probes.

Generally, computer programs determine probe sequences that are specific for their target sequences, are least likely to hybridize with nontarget sequences, have no secondary structure that would prevent hybridization with the target sequence, and have similar melting (annealing) temperatures, so that all target sequences can bind to their complementary probe sequences under essentially the same conditions. Repetitive sequences are not included in genomic DNA microarrays. Thus, to design a highly effective DNA microarray it is essential that the DNA sequences of a large number of plant genes, and preferably the entire genome sequence of the target plant, are known.

Typically, for most gene expression profiling experiments that utilize microarrays, mRNA is extracted from plant tissues and purified, and cDNA is synthesized using the enzyme reverse transcriptase, with the extracted mRNA as a template. Often in these experiments, mRNA is extracted from two or more sources whose expression profiles are compared; for example, from untreated plants versus plants treated with either a phytopathogen or a PGPB. The cDNA from each plant source is labeled with a different fluorophore by incorporating fluorescently labeled nucleotides during cDNA synthesis. For example, a green-emitting fluorescent dye (Cy3) may be used for the normal (untreated) sample and a red-emitting fluorescent dye (Cy5) for the sample treated with a PGPB (Fig. 3.5). After being labeled, the

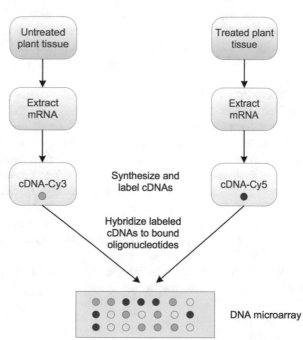

Fig. 3.5 Schematic overview of a DNA microarray experiment. Tissues are isolated from treated and untreated plants; the mRNA is extracted from those tissues; using the mRNAs as templates, cDNA is synthesized and then labeled with a dye that will fluoresce when irradiated with a specific laser; the labeled cDNAs are hybridized to oligonucleotides (or DNA fragments) bound to a microscope slide; finally the dye fluoresces after being irradiated with a laser, with the fluorescence signal being visualized captured and stored on computer

cDNA samples are mixed and hybridized to the same microarray. A laser scanner determines the intensities of bound Cy5 and Cy3 for each oligonucleotide probe. Each spot may have a different color depending on the relative expression levels of the gene in that spot in the two samples (Fig. 3.5). To avoid any experimental variation due to differences in labeling efficiencies between Cy3 and Cy5, the samples are usually reverse labeled and hybridized to another microarray. In this example, reverse labeling (also called dye swapping) would entail labeling the untreated sample with Cy5 and PGPB-treated sample with Cy3.

Notwithstanding the fact that there are many variations on the details of this experimental technique, using microarrays permits researchers to monitor changes to the abundance of a large number of different plant mRNAs (hundreds to thousands) that occur as a consequence of a range of different treatments.

Because of the large amount of data generated by microarray experiments, specialized software has been developed to maximize the output of information. Each spot (probe cell) of a two-dye hybridized microarray is scanned using a confocal scanning microscope. Following laser excitation of the dye, fluorescence emitted from each probe cell, detected at both 532 and 635 nm for Cy3 and Cy5, respectively, is collected and converted to an electrical signal. The intensities of the fluorescence emitted by both dyes for each probe cell, along with background readings for the microarray, are recorded and stored. To minimize errors, multiple probe cells for each gene are typically included on a single microarray, and replicate samples are independently prepared under the same conditions and hybridized to different microarrays.

One of the major purposes of a microarray experiment is to identify those genes whose expression changes in response to a particular biological condition. The response to a biological condition is determined by comparing the fluorescence intensity for each gene (as measured in each probe cell), averaged among replicates. The log ratios for all probe cells are compiled into a table called an expression matrix.

Microarray analysis is also used to identify genes that are co-expressed under different conditions or over a period of time, with the goal of determining which gene products function in a given pathway. A number of computational strategies are available that organize the data into related groups (clusters). For a clear presentation of the clustered data, ranges of log ratio values are assigned arbitrary colors. Usually, black is designated for a log ratio of zero, dark to bright red for increasing positive log ratios, and dark to bright green for decreasing negative log ratios. In other words, red is frequently used to denote gene over-expression and green to denote under-expression. A visualized representation of a clustered microarray is called a gene expression profile, where the rows represent the reordered genes and the columns represent either conditions or samples.

Very recently, ultra-high-throughput RNA sequencing has emerged as an alternative to the use of microarrays.

3.4   Proteomics

In the mid-1970s Patrick O'Farrell, who was at that time a graduate student at the University of Colorado, developed a high-resolution two-dimensional gel electrophoresis system that was able to separate and visualize approximately 1100 *E. coli* proteins on a single slab gel. This pioneering work, that enabled scientists to monitor changes in the cellular concentrations of hundreds of proteins at the same time, has opened the door to a wide range of possible applications and together with other technical developments has led to the development of the field that is called proteomics.

With the O'Farrell procedure, a protein mixture is first separated into individual protein components in a thin tube gel using isoelectric focusing (which separates proteins on the basis of their overall charge). After the gel has been run, the fragile gel is carefully removed from the glass tube and placed on the top of a standard sodium dodecyl sulfate (SDS) polyacrylamide slab gel. The isoelectric focusing separated proteins (which are now localized in multiple sites along the length of the tube gel) are then electro-transferred onto the SDS polyacrylamide slab gel and they are subsequently separated when an electric current is applied (separating proteins on the basis of their size). After the slab gel has been run, it is stained to visualize the proteins that were present in the original sample (Fig. 3.6). The gel may be stained with the dye Coomassie Blue, for greater sensitivity silver staining may be used, or

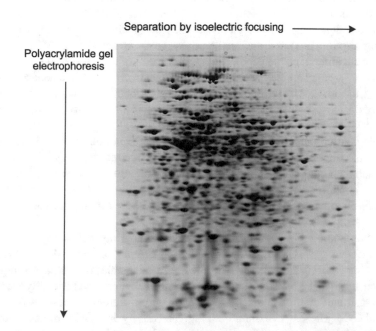

Fig. 3.6 Example of a two-dimensional gel stained with Coomassie Blue. The proteins were separated by isoelectric focusing in the first dimension and by SDS polyacrylamide gel electrophoresis in the second dimension

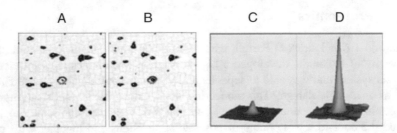

Fig. 3.7 Example of the quantitation of the amount of an individual bacterial protein (grown in the absence of any stress) and the change in that protein's concentration when the bacterium was grown in the presence of stress. (**a**) The portion of the two-dimensional gel highlighting the protein of interest following growth of the bacterium in the absence of stress. (**b**) The portion of the two-dimensional gel highlighting the protein of interest following growth of the bacterium in the presence of stress. (**c**) Quantitation of the amount of target protein from gel (**a**). (**d**) Quantitation of the amount of target protein from gel (**b**)

for much greater sensitivity (i.e., the ability to visualize much lower levels of protein) with the fluorescent dyes Cy3 and/or Cy5 (see Sect. 3.3.1). One technical problem with two-dimensional gels is the difficulty in detecting proteins whose abundance is relatively low, such as regulatory proteins, signal transduction proteins, and receptor proteins.

Once the gel has been run, it is necessary to identify individual proteins from the many spots that can be visualized. In one method, individual proteins are eluted from the gel and digested with a proteolytic enzyme (such as trypsin or papain), the resulting peptide mixtures are analyzed by mass spectrometry (MS) with each protein generating a unique peptide mass fingerprint (PMF) which can be compared to previously characterized proteins found in existing protein databases thereby enabling the identification of the target protein. Identification of any individual protein is achieved when there is a match between the PMF of the target protein and a specific protein in one of the databases.

To compare protein expression levels under different conditions it is possible to quantify the amount of protein produced in one set of conditions, e.g., the absence of stress, and then compare this to the amount of production of the same protein under different condition, e.g., the presence of stress (Fig. 3.7). Unfortunately, this approach often causes dramatic variability as a result of gel-to-gel variation when each test sample is separated on a different gel.

To avoid these problems, a common way of comparing protein expression levels under different conditions employs the method called "Differential In-Gel Electrophoresis" or DIGE (Fig. 3.8). The DIGE technique separates two or more samples on the same gel so that real and meaningful biological changes can be readily identified. Briefly, test and control samples are labeled with different fluorescent dyes (e.g., Cy3 and Cy5). These dyes exhibit both high sensitivity and a wide dynamic range in detection. Equal amounts of the differently labeled protein samples are then mixed and subjected to two-dimensional electrophoresis separation on the *same* gel, ensuring that identical proteins from separate samples co-migrate and that their

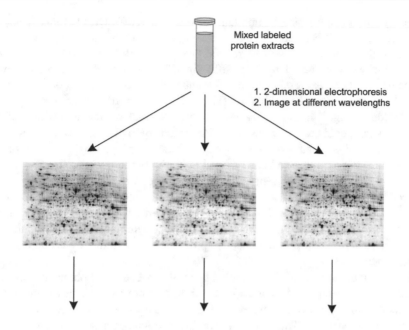

Mixed labeled
protein extracts

1. 2-dimensional electrophoresis
2. Image at different wavelengths

1. Quantitation of each protein spot at each wavelength
2. Comparison of the protein amounts at each wavelength
3. MS identification of all proteins of interest

Fig. 3.8 Differential in-gel electrophoresis (DIGE). Protein extracts (of the same bacteria or plant tissue, subjected to different treatments) and internal standards are typically labeled with different dyes and then mixed together. The mixed protein extract is separated by two-dimensional electrophoresis before the gel is imaged at wavelengths unique to each of the fluorescent label dyes. A computer comparison of the fluorescent images at each wavelength is used to indicate which proteins have increased expression and which have decreased expression. Protein spots may be identified by mass spectrometry

fluorescence images are super-imposed, thereby enabling more accurate differential expression analysis. The protein expression ratios between the samples are defined in the subsequent image analysis by comparison of the normalized intensities of each protein spot from the Cy3 and Cy5 channels. Overall, the DIGE technique is exceptionally reproducible, accurate, and sensitive.

In one set of experiments, canola plants were osmotically stressed by the addition of polyethylene glycol (which prevents the plants from taking up sufficient water) so that both the fresh and dry weights of roots and leaves decreased significantly. However, when a PGPB strain of *Pseudomonas fluorescens* was added to the hydroponic system in which the plants were growing, the osmotic stress was apparently lowered as the fresh and dry weights of the roots and leaves partially recovered. Following each of these treatments, plant leaf tissue was isolated and subjected to proteomic analysis. This analysis indicated that the osmotic stress increased the synthesis of ACC oxidase (an enzyme involved in ethylene synthesis) and D-2-hydroxyglutarate dehydrogenase (an enzyme involved in photorespiration, an energy-producing process important for plant survival). On the other hand, the presence of the PGPB resulted in a decrease in the synthesis of these two proteins. In addition, the expression levels of several proteins involved in metabolism/energy generation were increased by the PGPB (in the presence of the osmotic stress) compared to their expression in the presence of the osmotic stress without the PGPB. In summary, the addition of the PGPB reversed a number of protein expression changes in canola leaves that were caused by the imposition of the osmotic stress, changes that were otherwise inhibitory to plant growth. Importantly, proteomic analysis enabled researchers to precisely pinpoint the nature of the biochemical changes that had been caused by the imposed osmotic stress.

In one study of two different strains of *Sinorhizobium fredii* that both nodulate soybean plants, researchers found that the levels of protein that were expressed by genes that were encoded in the chromosomal genome were much higher than the levels of protein expressed that were encoded by the symbiotic plasmid. In the latter instance, the symbiotic plasmid encoded proteins were specifically accumulated in bacteroids.

Finally, it is important to bear in mind that the results that are obtained from gene expression studies (transcriptomics) and protein expression studies (proteomics) may show some significant differences in terms of assessing how various treatments affect gene expression. This is because on the one hand, transcriptomics monitors changes in transcripts (mRNAs) that are both newly synthesized and rapidly turned over. On the other hand, proteomics monitors the presence of stable proteins and does not generally distinguish between existing and newly synthesized proteins. Therefore, the results of these studies are expected to be complementary rather than identical.

3.5   Metabolomics

With recent developments in mass spectrometry, it is possible to rapidly and simultaneously measure thousands of different metabolites from relatively small samples. With the right tools, it is possible to assess the metabolome (the complete collection of metabolites, or small molecules) of a plant tissue or group of bacterial cells under any specific set of conditions. Moreover, as with transcriptomics and proteomics, it is possible to readily identify changes in an organism's metabolome as a consequence of changing environmental conditions (regardless of whether those conditions are biotic or abiotic).

Scientists have defined two different types of metabolomics experiments, targeted and non-targeted. In targeted metabolomics the experiment is designed to measure a limited number of specific metabolites, typically from one or more related biochemical pathways. These types of studies are often driven by efforts to study a specific pathway and are dependent upon having extensive information about the system being investigated prior to the start of the experiment. On the other hand, untargeted metabolic studies are intended to simultaneously measure the concentrations of as many metabolites as possible including metabolites that may be unknown or poorly

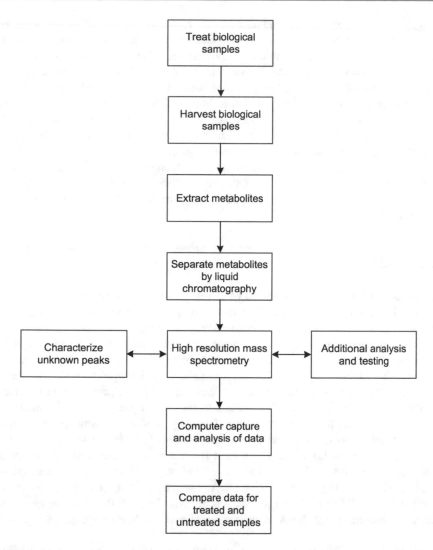

Fig. 3.9 Flowchart of a metabolomics experiment

characterized (Fig. 3.9). Untargeted metabolomics may be quite complex in that it has the potential to detect and quantify several thousand compounds at once and in a highly reproducible manner. Moreover, changes in the concentrations of various metabolites that occur in response to environmental stimuli can provide a strong indication of how the test organism copes with different environments.

In untargeted metabolomics, metabolites are extracted from biological samples, separated by liquid chromatography, analyzed by high-resolution mass spectrometry and finally the resultant data is quantified by computer and may then be compared to similar samples that are treated differently (Fig. 3.9). Notwithstanding the ability of this system to resolve a large number of different compounds, all of the peaks of interest need to be individually identified (structurally characterized). From the mass spec data, the mass to charge ratio of individual peaks is identified and compared to databases of known compounds. For uncommon or unknown compounds, a number of additional (and often complex) procedures may be required to identify them.

Metabolomics is a relatively recent technique that requires quite sophisticated equipment to resolve, separate, and measure all of the metabolites in an organism. Thus, at the present time there are a relatively limited number of reports of this technique being used to study plant–microbe interactions. However, because of the potential of this approach, it is likely that this technique will be employed with increasing frequency because the data that can be obtained from metabolomics studies is a logical complement to similar data from transcriptomic and proteomic experiments.

3.6 CRISPR

In the past 10 years, researchers have developed strategies to insert, replace, or disrupt DNA sequences at specific targeted sites in intact genomes in vivo. This procedure, known as CRISPR (for *c*lustered *r*egularly *i*nterspaced *s*hort *p*alindromic *r*epeats), is based on a naturally occurring system that protects bacteria against invasion by foreign DNA including bacteriophage DNA and plasmids. In nature, bacteriophage DNA fragments (called protospacers) are incorporated into a bacterial genome between repeat DNA sequences. Subsequently, the protospacers are transcribed to produce CRISPR RNA molecules that act to guide a bacterial nuclease (called Cas) to base pair with a sequence of that is part of an invading bacteriophage. For this system to work properly, the recognition of the target sequence on the invading DNA must be adjacent to a short DNA sequence known as a protospacer adjacent motif (PAM). The presence of the PAM sequence prevents the cleavage of the bacterium's endogenous genomic DNA at sequences that are complementary to the CRISPR RNA (since this site in the genome lacks a PAM sequence). The invading bacteriophage DNA is then cleaved and the bacteriophage is rendered harmless.

Following the development of an understanding of how the natural CRISPR system worked, scientists subsequently adapted this natural system as a genetic engineering tool in which specific bacterial genomic DNA sequences could easily be altered, deleted, or inserted (Fig. 3.10). To make CRISPR manipulations easy to perform, a single RNA molecule of 80–100 nucleotides was created; this RNA molecule contains a 20-nucleotide long guide sequence that is exactly complementary to the target site while the rest of the RNA molecule binds directly to the Cas nuclease. The guide sequence positions the entire complex by its complementarity to the target sequence (which is located adjacent to a PAM sequence). Once the complex has bound to the target sequence, the nuclease makes a double-stranded cleavage of the target DNA. During the subsequent repair of the double-stranded break in the genomic DNA, nucleotide sequences may be replaced, inserted, or

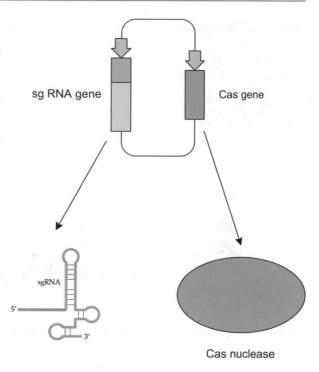

Fig. 3.10 A plasmid vector carrying genes for an sgRNA and a Cas nuclease. Following the introduction of the plasmid into target cells and the expression of these genes, the transcribed sgRNA guides the Cas nuclease to the target genomic DNA. The orange portion of the sgRNA contains the 20-nucleotide guide sequence while the remainder of the molecule binds to the Cas nuclease. The Cas nuclease makes a double-stranded break in the target DNA. The green arrows denote promoters for the two genes on the plasmid. The repair systems generate DNA alterations, deletions, or additions

deleted. This may occur in several different ways, e.g., nonhomologous end joining or homology-directed repair. In addition, the Cas nuclease can be engineered to create staggered double-stranded breaks.

Since its inception as a means of genome modification, CRISPR has taken the biological world by storm. There are already multiple variations on the original CRISPR methodology, each claiming to be easier, better at creating specific types of DNA alterations, or more specific than the earlier protocols. Moreover, in addition to employing this technique as a means to modify bacterial genomes, researchers have actively pursued the use of CRISPR to modify plant and animal genomes as well.

3.7 Imaging

Scientists have developed a number of different techniques that enable them to either directly visualize the interaction of (labeled) PGPB with plant tissues (generally roots) or to quantify the effect that various PGPB have on the growth and development of plant tissues. Such techniques provide much more detailed information regarding the effect of PGPB on plant growth and development than simply measuring plant tissue length or fresh and dry weights.

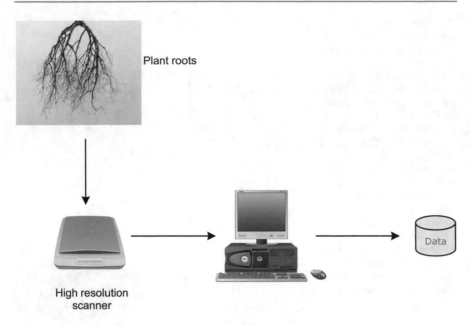

Plant roots

High resolution
scanner

Data

Fig. 3.11 Using a high-resolution scanner to capture data regarding the growth of plant roots

3.7.1 High-Resolution Scanning

The measurement of the length of plant tissues (e.g., roots or shoots) is generally done manually and as such it is potentially quite inaccurate as well as being tedious to perform. However, as described below, it is possible to automate this process so that many more plants may be analyzed and with much greater accuracy. In one example of the use of this technique, plant tissues are placed onto a special (commercially available) high-resolution scanner that can resolve up to 4800 dots per in. (which is equivalent to a pixel size of ~0.005 mm). Most standard flatbed scanners, which are used to scan documents and photographs, can only resolve 600 or 1200 dots per in. In one version of such a high-resolution scanner, the size of the area that is scanned is 21.6 × 28 cm (approximately 8.5 × 11 in.). The scanner is directly connected to a computer with specific computer software so that following a scan of plant tissue (root, stem, or leaf), the computer program is able to calculate total and average leaf area, total root length, total root surface area, the number of root tips, and the degree of root branching (Fig. 3.11). It should be emphasized that while it is possible to manually measure the length of a plant's main or tap root, it is nearly impossible to manually measure the length of all of a plant's roots, a task that is relatively straightforward for the computer program. Similarly, this approach enables researchers to readily calculate the surface area of leaves following various treatments (Fig. 3.12). In one reported experiment using this methodology, scientists compared the effect of adding either a PGPB or a strain of arbuscular mycorrhizae (AM) or both on the growth of cucumber plant tissues (Table 3.1). This data showed

Fig. 3.12 Example of
cucumber leaves from a single
plant that were placed in a
high-resolution scanner so
that the total leaf surface area
might be calculated

Table 3.1 The effect of adding a PGPB or an AM or both on the growth of cucumber plants

Addition	Shoot length, cm	Number of leaves	Total root length, cm	Total root surface area, cm	Number of root tips
None	5.942	7.813	1411.2	265.08	1405.0
PGPB	6.761	8.202	1622.4	288.82	1605.4
AM	6.902	9.021	1562.3	288.43	1875.8
PGPB + AM	8.312	9.204	1801.5	343.14	2119.8

that both the PGPB and the AM stimulated plant growth in every parameter
measured; these increases in plant growth were all statistically significant. Moreover,
in every instance, the addition of both the PGPB *and* the AM caused the greatest
(statistically significant) increase in all of the parameters measured. These results
were taken as a strong indication of a positive synergistic interaction between the
PGPB and the AM fungus, resulting in increased plant growth. Finally, when
examining the data that resulted from this experiment, it is interesting to note that
by using a high-resolution scanner connected to a computer it is possible to collect
data up to five significant figures where manually collected data is usually limited to
three significant figures.

3.7.2 Labeling PGPB

Although the majority of the studies with PGPB have been performed in controlled
situations including growth chambers and greenhouses, it ultimately is essential to
understand the functioning of PGPB in the field. And, in order to monitor the
functioning of PGPB in the field, it is necessary to label these bacteria in a manner

that does not interfere with their ability to promote plant growth. While a number of different techniques have been used to label PGPB (and other bacteria), a few more common approaches to labeling PGPB are discussed here.

Green fluorescent protein (GFP) is a protein originally isolated from the jellyfish *Aequorea victoria* that consists of 238 amino acids and exhibits green fluorescence when it is exposed to ultraviolet light. This fluorescence occurs at room temperature and does not require the addition of any cofactors; therefore, the cDNA for this protein can be expressed in a variety of different organisms. While many fluorescent dyes are phototoxic, the incorporation of green fluorescent protein into cells allows intact living cells to be monitored in real time. The use of this reporter molecule has revolutionized fluorescence microscopy. In addition to GFP, one group of researchers isolated a gene for a red fluorescent protein (RFP) from *Discosoma* coral and, by random mutation, generated a mutant that existed exclusively as a monomer instead of as a tetramer. To expand the repertoire of available fluorescent proteins, DNA encoding seven amino acids from the N terminus of GFP was added to the gene for RFP. Then, DNA for the six amino acids from the GFP C terminus was added to the gene for RFP. This construct then became the starting point for additional random mutagenesis. Eventually, seven different monomeric-colored fluorescent proteins were produced (Fig. 3.13). In this way, it is possible to label several different types of bacteria with different colors and monitor all of them at the same time. In order to label the target PGPB with GFP (or another fluorescent protein), The *gfp* gene (on a broad-host-range plasmid) is transferred from a donor strain of *E. coli* by triparental conjugation enlisting the assistance of another strain of *E. coli* that carries a helper plasmid (Fig. 3.14). The GFP-labeled PGPB may be selected from the mixture of bacterial cells by its resistance to the antibiotic whose resistance gene is present on the transferred plasmid carrying the *gfp* gene (Fig. 3.13). Once a PGPB has been labeled with GFP (or any other fluorescent protein), it is essential that it be checked to ascertain that all of its plant growth-promoting activities remain intact. In one set of experiments, it was found that the nickel-resistant PGPB exhibited a slightly decreased (but not statistically significant) biological activity (Table 3.2). This is not surprising as the labeled PGPB has the burden of expressing another major protein, i.e., GFP. Despite this small decrease in biological activity, the labeled PGPB is considered to be suitable for use in field experiments to monitor the behavior of this bacterium. It is also worth noting that this data provides strong evidence that the addition of the selected PGPB enables the canola plants to partially (but not completely) overcome the inhibitory effects on plant growth of the presence of nickel in the soil.

It is technically difficult to study roots in situ in a live plant because the roots are generally underground so that it is difficult to observe them without altering the plant. To get around this problem, researchers have developed a number of imaging techniques that allow roots to be visualized, typically at one point in time. For example, plants can be grown between thinly separated glass sheets, known as rhizotrons (or root windows), which contains a viewing chamber to monitor and visualize the roots within the soil. Unfortunately, to study root–bacterial interactions, rhizotrons are not especially well suited to simultaneously observing roots and

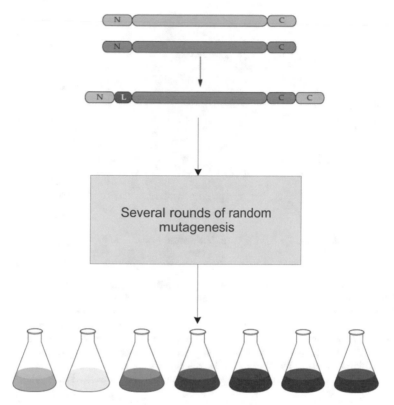

Fig. 3.13 Combining the genes (cDNAs) of green fluorescent protein (GFP) and red fluorescent protein (RFP), mutagenizing the construct and eventually deriving seven new fluorescent proteins. GFP and RFP are shown in green and red, respectively. The DNA encoding the N and C termini of GFP were spliced onto the gene for RFP following the removal of DNA encoding the N terminus of RFP and the addition of a short linker peptide (L). Following several rounds of random mutagenesis, seven new fluorescent proteins with colors designated (from left to right) as honeydew, banana, orange, tomato, tangerine, strawberry, and cherry, respectively

individual bacteria. To get around this problem, one group of scientists reported developing a technique that they called TRIS (tracking root interactions system). In this system, the roots are grown into a narrow root chamber (160 μM high) contained in a microfluidics device into which bacteria can subsequently be introduced. The microfluidics chamber was designed with nine separate chambers, each of which (in these experiments) contains a single *Arabidopsis* root. The bacteria that are interacting with the roots are fluorescently labeled so that they may be readily visualized by confocal microscopy. When a strain of *B. subtilis* was used as the test bacterium in this system, the bacteria were found to accumulate in the root elongation zone within 20 min of the bacteria being added (Fig. 3.15). In this particular experiment, to increase the sensitivity of the detection system, both the bacterium and the plant roots were fluorescently labeled. When both labeled *E. coli* and labeled *B. subtilis* (using different labels) were added to the roots at the same

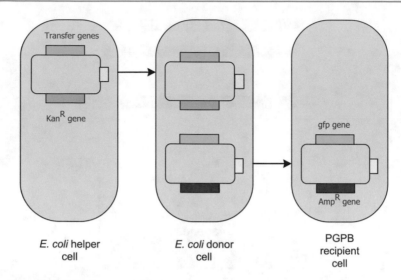

Fig. 3.14 Transferring a plasmid that carries a *gfp* gene from a strain of *E. coli* to the target PGPB. The helper plasmid is first transferred to the donor bacterium (which contains a plasmid with a *gfp* gene) where the proteins that mediate transfer (shown in blue) are expressed. For the plasmid carrying the *gfp* gene (shown in green) to be transferred it must have an appropriate origin of transfer (shown in yellow). The helper plasmid also contains a kanamycin resistance gene (shown in orange) while the donor plasmid contains an ampicillin resistance gene (shown in red). Following triparental plasmid transfer, the labeled PGPB is selected for the presence of the donor plasmid (that confers ampicillin resistance) and the absence of the helper plasmid (i.e., for kanamycin sensitivity)

Table 3.2 The effect of a PGPB (*Kluyvera ascorbata*) on the growth of 25-day-old canola plants in soil containing 3 mM nickel

Addition to soil	Plant fresh weight, g	Plant dry weight, g	Chlorophyll, mg/g leaf	Protein, mg/g leaf
Buffer	2.55	0.162	2.27	2.83
Buffer + 3 mM Ni	0.83	0.066	1.39	2.40
PGPB + buffer + 3 mM Ni	1.36	0.098	2.21	2.68
Labeled PGPB + buffer + 3 mM Ni	1.27	0.096	2.10	2.71

time, the *B. subtilis* excluded the *E. coli* from binding to the root, demonstrating that there is some specificity in this interaction. In the future, as a means of better understanding what exudates are attracting the bacteria, this system might be used to test different mutants of the *B. subtilis* strain that are defective in taking up certain metabolites. In addition, plants other than Arabidopsis might be used in this system. Finally, it should be possible, with this system, to simultaneously monitor the binding, in real time of two or more PGPB or phytopathogens to the roots of target plants.

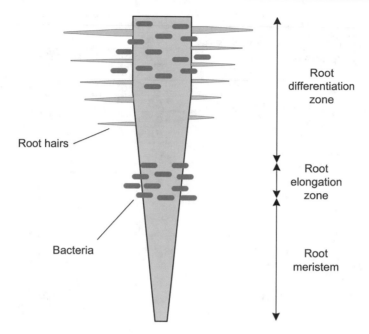

Fig. 3.15 Schematic representation of a plant root showing the pattern of bacteria binding. Bacteria first bind to the root elongation zone that is just behind the root tip. This binding likely reflects the root elongation zone as being the major site of root exudation. As the root grows, bacteria are bound to the root differentiation zone

3.8 Microencapsulation

As discussed in Chap. 1, inoculating crop plants with PGPB is not a trivial task. Thus, scientists are continually trying to improve the ways in which PGPB are delivered to plants. One significant drawback of both liquid and solid PGPB inoculants is that with both approaches, the PGPB have a relatively short shelf life. Although it is still largely a laboratory and experimental procedure, a number of researchers have been actively trying to develop the technique of encapsulating PGPB in alginate microbeads as an alternative delivery system. In one set of experiments, scientists were using a polycyclic aromatic hydrocarbon (see Chap. 10) resistant strain of *Pseudomonas asplenii* to promote the growth of plants that were growing in the presence of these contaminants. As part of that study, the researchers sought to compare the effectiveness of *P. asplenii* cells in liquid suspension to the same bacteria that had been encapsulated in alginate beads. This information would be important in commercializing this PGPB if was found to be effective at promoting plant growth in the presence of these contaminants. Thus, *P. asplenii* cells were encapsulated in alginate beads and tested for their ability to survive storage at different temperatures. As can be seen in Table 3.3, survival of the bacteria is greatest at 4 °C storage and least at 30 °C. The PGPB strain did not

Table 3.3 Viability of a strain of *Pseudomonas asplenii* immobilized in alginate beads at different temperatures

Days of storage	Cell count		
	4 °C	22 °C	30 °C
0	1×10^8	1×10^8	1×10^8
30	1×10^8	1×10^7	7×10^4
70	1.1×10^8	1×10^7	1×10^4
130	0.5×10^8	3×10^6	0.9×10^4
180	0.7×10^8	2×10^6	0.3×10^4
240	0.9×10^8	1×10^6	0.1×10^4

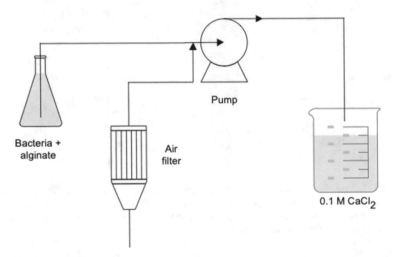

Fig. 3.16 Schematic representation of a laboratory scale apparatus that is used to encapsulate bacteria in alginate. A bacterial suspension mixed with alginate is pumped through a capillary tube at low pressure to form a mist that is dropped onto a $CaCl_2$ solution where the alginate forms microdrops

completely lose bacterial viability at any of the storage temperature treatments. Moreover, after 250 days of storage at 4 °C, bacterial viability of the PGPB strain was maintained at a level nearly equivalent to the starting amount. In practice, this suggests that alginate encapsulated cells might be shipped from the producer to the end user, arrive within 2–3 days (maintaining >90% of their viability), and then be stored by the end user for at least 250 additional days without any further loss of cell viability. In fact, when monitoring the growth of canola plants in the presence of polycyclic aromatic hydrocarbons, alginate encapsulated PGPB cells were significantly better than nonencapsulated cells at promoting both root and shoot growth. Although it is still used exclusively in laboratory studies, alginate encapsulated PGPB have been used to increase plant performance in a variety of plants including wheat, lettuce, sugar beet, tomato, cotton, and sorghum.

To encapsulate bacteria within an alginate matrix, the following steps are followed (Fig. 3.16). A bacterial suspension is mixed together with 2% sodium alginate. This mixture is passed through a capillary tube under low pressure to create a fine spray of tiny droplets. The mist is sprayed onto a gently rotating solution of 0.1 M $CaCl_2$ and allowed to stand for 30 min before the beads were removed from the $CaCl_2$ solution. The final size of the microbeads was 100–220 μm (although some workers, using a slightly different apparatus, have reported an average size of ~10 μm for these microbeads) and the bacterial concentration was ~2 × 10^{11} CFU g^{-1}. The wet beads are rinsed several times in saline solution before they are transferred to complete bacterial growth media for ~12 h. The microbeads were again rinsed in a saline solution. The beads could either be used immediately or dried at ~37 °C for 48 h, or by lyophilization. While some researchers have reported using encapsulated bacteria by directly adding the microbeads to plant tissues, others have reported mixing the microbeads with adhesives such as lecithin.

Questions
1. What sort of information can one glean from the complete genomic DNA sequence of a PGPB?
2. What is PCR and how does it work?
3. What is real-time PCR, how does it work, and why might it be useful?
4. What is a DNA microarray?
5. How would you use a DNA microarray to elaborate the set of genes that are induced when a PGPB is added to the roots of the plant *Arabidopsis thaliana*?
6. How can proteins be effectively separated on two-dimensional gels?
7. How would you identify an individual protein on a two-dimensional gel?
8. How would you quantify the amount of an individual protein on a two-dimensional gel?
9. What is metabolomics? How would you use this technique to assess the effect of a particular stress or the addition of a PGPB on a plant's growth and development?
10. How does the CRISPR system work? What does it do? How can it be used to modify a bacterial genome?
11. How can high-resolution scanning be used to monitor plant growth and development?
12. How can various PGPB strains be labeled so that they are readily visualized in their interaction with plant tissues?
13. How can the binding to a plant root of several different types of bacteria that compete with one another be visualized?
14. How can fluorescently labeled bacteria be used to provide information regarding the colonization patterns of the bacteria to plant roots?
15. How can fluorescently labeled bacteria be used to pinpoint specific plant exudate components that attract the bacteria to the plant root?
16. How can PGPB cells be encapsulated in alginate? Why might this be useful?

Bibliography

Alphey L (1997) DNA sequencing: from experimental methods to bioinformatics. BIOS Scientific Publishers, New York

Bashan Y (1986) Alginate beads as synthetic inoculant carriers for slow release of bacteria that affect plant-growth. Appl Environ Microbiol 51:1089–1098

Bashan Y, Hernandez J-P, Leyva LA, Bacillio M (2002) Alginate microbeads as inoculant carriers for plant growth-promoting bacteria. Biol Fertil Soils 35:359–368

Bondy-Denomy J, Davidson AR (2014) To acquire or resist: the complex biological effects of CRISPR-Cas systems. Trends Microbiol 22:218–225

Chan EY (2005) Advances in sequencing technology. Mutat Res 573:13–40

Cheng Z, Duan J, Hao Y, McConkey BJ, Glick BR (2009a) Identification of bacterial proteins mediating the interaction between the plant growth-promoting bacterium *Pseudomonas putida* UW4 and *Brassica napus* (canola). Mol Plant-Microbe Interact 22:686–694

Cheng Z, Wei Y-YC, Sung WWL, Glick BR, McConkey BJ (2009b) Proteomic analysis of the response of the plant growth-promoting bacterium *Pseudomonas putida* UW4 to nickel stress. Proteome Sci 7:18

Cheng Z, McConkey BJ, Glick BR (2010a) Proteomic studies of plant-bacterial interactions. Soil Biol Biochem 42:1673–1684

Cheng Z, Woody OZ, Glick BR, McConkey BJ (2010b) Characterization of plant-bacterial interactions using proteomic approaches. Curr Proteomics 7:244–257

Doudna JA, Charpentier E (2014) The new frontier of genome engineering with CRISPR-Cas9. Science 346:1258096-1–1258096-9

Downie HF, Adu MO, Schmidt S, Otten W, Dupuy LX, White PJ, Valentine TA (2015) Challenges and opportunities for quantifying roots and rhizosphere interactions through imaging and image analysis. Plant Cell Environ 38:1213–1232

Duan J, Heikkila JJ, Glick BR (2010) Sequencing a bacterial genome: an overview. In: Vilas AM (ed) Current research, technology and education topics in applied microbiology and microbial biotechnology. Formatex Research Center, Badajoz, pp 1443–1451

Duan J, Jiang W, Cheng Z, Heikkila JJ, Glick BR (2013) The complete genome sequence of the plant growth-promoting bacterium *Pseudomonas putida* UW4. PLoS One 8(3):e58640

Erlich HA, Gelfand D, Sninsky JJ (1991) Recent advances in the polymerase chain reaction. Science 252:1643–1651

Gamalero E, Berta G, Massa N, Glick BR, Lingua G (2008) Synergistic interactions between the ACC deaminase-producing bacterium *Pseudomonas putida* UW4 and the AM fungus *Gigaspora rosea* positively affect cucumber plant growth. FEMS Microbiol Ecol 64:459–467

Gharelo RS, Bandehagh A, Toorchi M, Farajzadeh D (2016) Canola 2-dimensional proteome profiles under osmotic stress and inoculation with *Pseudomonas fluorescens* FY32. Plant Cell Biotechnol Mol Biol 17:257–266

Hoheisel JD (2006) Microarray technology: beyond transcript profiling and genotype analysis. Nat Rev Genet 7:200–210

Huber W, von Heydebreck A, Vingron M (2005) An introduction to low-level analysis methods of DNA microarray data. Bioconductor Project Working Papers, Working Paper 9. http://www.bepress.com/bioconductor/paper9

Jiang W, Bikard D, Cox D, Zhang F, Marraffini LA (2013) RNA-guided editing of bacterial genomes using CRISPR-Cas systems. Nat Biotechnol 31:233–241

Jinek M, Chylinski K, Fonfara I, Hauer M, Doudna JA, Charpentier E (2012) A programmable dual-RNA-guided DNA endonuclease in adaptive bacterial immunity. Science 337:816–821

Ju J, Kim DH, Bi L, Meng Q, Bai X, Li Z, Li X, Marma MS, Shi S, Wu J, Edwards JR, Romu A, Turro NJ (2006) Four-color DNA sequencing by synthesis using cleavable fluorescent nucleotide reversible terminators. Proc Natl Acad Sci U S A 103:19635–19640

Ma W, Zalec K, Glick BR (2001) Biological activity and colonization pattern of the bioluminescence-labeled plant growth-promoting bacterium *Kluyvera ascorbata* SUD165/26. FEMS Microbiol Ecol 35:137–144

Massalha H, Korenblum E, Malitsky S, Shapiro OH, Aharoni A (2017) Live imaging of root-bacteria interactions in a microfluidics setup. Proc Natl Acad Sci U S A 114:4549–4554

Maxam AM, Gilbert W (1977) A new method for sequencing DNA. Proc Natl Acad Sci U S A 74:560–564

Meena KK, Sorty AM, Bitla UM, Choudhary K, Gupta P, Pareek A, Singh DP, Prabha R, Sahu PK, Gupta VK, Singh HB, Krishanani KK, Minhas PS (2017) Abiotic stress responses and microbe-mediated mitigation in plants: the omics strategies. Front Plant Sci 8:172

Mullis KB, Ferré F, Gibbs RA (eds) (1994) The polymerase chain reaction. Birkhäuser, Boston

Nouioui I, Cortés-Albayay C, Carro L, Castro JF, Gtari M, Ghodhbane-Gtari F, Klenk H-P, Tisa LS, Sangal V, Goodfellow M (2019) Genomic insights into plant-growth-promoting potentialities of the genus Frankia. Front Microbiol 10:1457

O'Farrell PH (1975) High resolution two-dimensional electrophoresis of proteins. J Biol Chem 250:4007–4021

Patti GJ, Yanes O, Siuzdak G (2012) Metabolomics: the apogee of the omics trilogy. Nat Rev Mol Cell Biol 13:263–269

Poole P (2017) Shining a light on the dark world of plant-microbe interactions. Proc Natl Acad Sci U S A 114:4281–4283

Reed MLE, Glick BR (2005) Growth of canola (Brassica napus) in the presence of plant growth-promoting bacteria and either copper or polycyclic aromatic hydrocarbons. Can J Microbiol 51:1061–1069

Rehman HM, Cheung W-L, Wong K-S, Xie M, Luk CY, Wong F-L, M-W L, Tsai S-N, To W-T, Chan L-Y, Lam H-M (2019) High-throughput mass spectrometric analysis of the whole proteome and secretome from Sinorhizobium fredii strains CCBAU25509 and CCBAU45436. Front Microbiol 10:2569

Saiki RK, Gelfand DH, Stoffel S, Scharf S, Higuchi R, Horn GT, Mullis KB, Erlich HA (1988) Primer-directed enzymatic amplification of DNA with a thermostable DNA polymerase. Science 239:487–491

Sanger F, Nicklen S, Coulson AR (1977) DNA sequencing with chain-terminating inhibitors. Proc Natl Acad Sci U S A 74:5463–5467

Schena M, Shalon D, Davis RW, Brown PO (1995) Quantitative monitoring of gene expression patterns with a complementary DNA microarray. Science 270:467–470

Shaner NC, Campbell RE, Steinbach PA, Giepmans BNG, Palmer AE, Tsien RY (2004) Improved monomeric red, orange and yellow fluorescent proteins derived from Discosoma sp. red fluorescent protein. Nat Biotechnol 22:1567–1572

Shendure J, Porreca GJ, Reppas NB, Lin X, McCutcheon JP, Rosenbaum AM, Wang MD, Zhang K, Mitra RD, Church GM (2005) Accurate multiplex polony sequencing of an evolved bacterial genome. Science 309:1728–1732

Shendure J, Balasubramanian S, Church GM, Gilbert W, Rogers J, Schloss JA, Waterston RH (2017) DNA sequencing at 40: past, present and future. Nature 550:345–353

Singh-Gasson S, Green RD, Yue Y, Nelson C, Blattner F, Sussman MR, Cerrina F (1999) Maskless fabrication of light-directed oligonucleotide microarrays using a digital micromirror array. Nat Biotechnol 17:974–978

Sørenson J, Nicolaisen MH, Ron E, Simonet P (2009) Molecular tools in rhizosphere microbiology—from single-cell to whole-community analysis. Plant Soil 321:483–512

Stobel SA, Allen K, Roberts C, Jimenez D, Scher HB, Jeoh T (2018) Industrially-scalable microencapsulation of plant beneficial bacteria in dry cross-linked alginate matrix. Ind Biotechnol 14:138–147

van Vliet AHM (2010) Next generating sequencing of microbial transcriptomes: challenges and opportunities. FEMS Microbiol Lett 302:1–7

Wilton R, Ahrendt AJ, Shinde S, Sholto-Douglas DJ, Jojnson JL, Brennan MB, Kemmer KM (2018) A new suite of plasmid vectors for fluorescence-based imaging of root colonizing pseudomonads. Front Plant Sci 8:2242. https://doi.org/10.3389/fpls.2017.02242

Resource Acquisition

4

Abstract

This chapter addresses the issue of how plant growth-promoting bacteria help plants to acquire resources from the environment (the soil and the air). The process of bacterial nitrogen fixation is examined and discussed (briefly for cyanobacteria and in some detail for Rhizobia) at the physiological and genetic level. The functioning of the proteins/enzymes involved in nodulation and hydrogen recycling as well as nitrogen fixation per se are considered. Subsequently, the structure and functioning of bacterial siderophores in providing minute amounts of iron from the soil to the plant is examined along with a brief discussion of the regulation of this process. Finally, how plant growth-promoting bacteria solubilize both inorganic and organic forms of phosphorus and provide it to the plant are considered.

To grow, both plants and bacteria require nutrients that are mainly acquired from the soil environment. In addition to small amounts of a number of different metals, both plants and bacteria require fixed nitrogen, iron, and phosphorus. In this chapter, the mechanisms and genes involved in PGPB and plant resource acquisition is discussed in some detail. In fact, one of the major benefits that PGPB provide to their plant partners is that they facilitate the plant's acquisition of various nutrient resources from the soil.

4.1 Nitrogen Fixation

One of the major nutrients necessary for the growth of all living organisms including plants and bacteria is nitrogen. Despite nitrogen's abundance in the earth's atmosphere, i.e., ~80%, it must first be reduced to ammonia before it can be metabolized by plants to become an integral component of proteins, nucleic acids, and other

© Springer Nature Switzerland AG 2020
B. R. Glick, *Beneficial Plant-Bacterial Interactions*,
https://doi.org/10.1007/978-3-030-44368-9_4

biological molecules. This conversion requires a large input of energy because the triple bond that holds the two molecules of nitrogen together to form nitrogen gas (N_2) is extremely stable. Consequently, contemporary agriculture, especially in more developed countries, relies heavily on the use of nitrogen fertilizers that are produced by the Haber–Bosch process which was first developed in 1909 and then on an industrial scale in 1913. In this process, atmospheric nitrogen gas is converted to ammonia by a reaction with hydrogen that requires a metal catalyst as well as high pressure (15–25 MPa) and temperature (400–500 °C); the pressure and temperature are derived at the expense of petroleum while the major source of hydrogen is methane gas. In fact, ~1.3 tons of oil, or an equivalent amount of energy, are needed to fix 1 ton of ammonia, and more than 100 million tons of fixed nitrogen are currently needed annually to sustain global food production. Besides being costly, the production of chemical nitrogen fertilizers significantly depletes nonrenewable energy resources and also causes human and environmental hazards. On the other hand, nature has ingeniously solved the problem of converting nitrogen gas to ammonia through the process of biological nitrogen fixation. Nitrogen fixing bacteria, or diazotrophs, annually produce about 60% of the earth's newly fixed nitrogen. With the objective of reducing the world's dependency on energy intensive chemical fertilizers, over the past 30–40 years there has been a large body of research directed toward understanding, and eventually utilizing biologically fixed nitrogen in agriculture.

A wide range of bacteria can fix nitrogen, and a number of these bacteria have potential as crop fertilizers; eukaryotes do not fix nitrogen. Diazotrophic bacteria include, but are not limited to, several genera of cyanobacteria; the symbiotic genera *Rhizobia*, *Sinorhizobia*, *Bradyrhizobia*, *Mesorhizobia*, and *Frankia*; rhizospheric and endophytic genera *Azospirillum*, *Azotobacter*, *Klebsiella*, *Bacillus*, and *Pseudomonas*.

4.1.1 Cyanobacteria

Cyanobacteria, previously called blue-green algae, share some characteristics of both bacteria and plants. Although they share many characteristics with Gram-negative bacteria, cyanobacteria contain both chlorophyll a and phycobiliproteins (proteins that capture light energy which is passed on to chlorophylls during photosynthesis). Cyanobacteria are capable of both photosynthesis (fixing atmospheric CO_2 into sugars) and nitrogen fixation; however, while all cyanobacteria are photosynthetic, only some can fix nitrogen. A number of different studies have been reported on the use of dried cyanobacteria as a fertilizer to inoculate soils and aid fertility. Moreover, there are a large number of studies in which cyanobacteria have been added to rice paddies with the result that the nitrogen that they fix and release may be taken up and used by the rice plants. This strategy is typically quite effective as sufficient fixed nitrogen is often considered to be rice's major growth limiting factor. Following seed germination and a short growth phase in a greenhouse, rice seedlings are generally transplanted to paddies where the roots and a portion of the stem of the plant are continuously submerged. It is estimated that globally, in 1980,

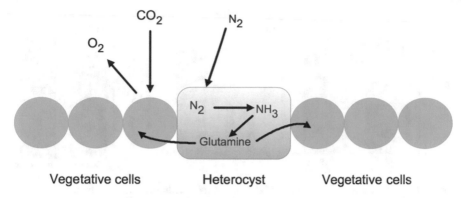

Fig. 4.1 Schematic representation of a portion of a chain of *Anabaena* cells showing CO_2 being taken up into photosynthetic vegetative cells and N_2 being taken up by nitrogen fixing heterocyst cells. The vegetative cells produce oxygen. The heterocysts transfer the newly fixed NH3 to glutamic acid to produce glutamine which is then transferred to nearby vegetative cells

there were more than 1.7×10^6 km^2 of land devoted to the cultivation of rice. Diazotrophic cyanobacteria may be added to the water in the paddy where they proliferate and release some of their fixed nitrogen into the water to be taken up by the rice plants. In addition to rice, in many tropical and subtropical environments, other crops including vegetables, wheat, sorghum, corn, cotton, and sugarcane are also grown using cyanobacteria as a biofertilizer. Moreover, in some regions the heterocystous cyanobacterium—*Anabaena azollae* is found within the leaf cavity of the fern Azolla with this partnership acting as a traditional biofertilizer.

Some cyanobacteria such as *Nostoc* spp. and *Anabaena* spp. consist of long chains of two different types of cells, vegetative cells and heterocysts (Fig. 4.1). The vegetative cells, typically comprising about 90–95% of the total number of cyanobacterial cells, are photosynthetic while the larger and thick-walled heterocysts are nitrogen fixing. The heterocysts protect the oxygen-sensitive nitrogen-fixing enzymes from inactivation that would otherwise occur if these enzymes were located in the oxygen-rich vegetative cells. Following photosynthesis in the vegetative cells and nitrogen fixation in the heterocysts, the vegetative cells and the heterocysts exchange nutrients so that all of the cyanobacterial cells have sufficient levels of both nitrogen and newly fixed carbon necessary to support the growth of the organism. Nitrogen is typically transported as the amino acid glutamine while carbon may be transported as the disaccharide sucrose.

In one interesting experiment, scientists were able to genetically manipulate *Anabaena* sp. strain PCC7120 so that it contained an increased number of heterocysts and therefore fixed more nitrogen than usual. Recently, the *hetR* gene, which encodes a serine-type protease, has been identified as a master regulator of heterocyst differentiation. Thus, one group of researchers cloned the *Anabaena hetR* gene, expressed the cloned gene under the transcriptional control of a strong light-inducible *Anabaena* promoter, and integrated the entire construct into an intergenic region within the *Anabaena* genome; with the final construct containing two copies

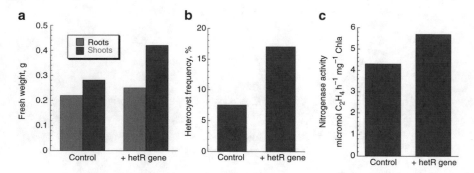

Fig. 4.2 Improvement of (**a**) fresh weight of root and shoot cells, (**b**) heterocyst frequency (i.e., the number of heterocysts divided by the total number of cells), and (**c**) the nitrogenase activity of control (wild-type) versus recombinant cells that overexpress the *hetR* gene in *Anabaena* sp. The data is from Appl. Environ. Microbiol. 77:395–399 (2011)

of the *hetR* gene, the original gene and the recombinant gene. Integrating the *hetR* gene into an intergenic region ensures that no *Anabaena* genes will be disrupted. The resulting high level of expression of an additional copy of the *hetR* gene elevated HetR protein expression and increased both the amount of fixed nitrogen in the recombinant strain and the rice plant fresh weight (Fig. 4.2). The ability of *Anabaena* and *Nostoc* strains to grow and be sustained photoautodiazotrophically (being both photosynthetic and nitrogen fixing) under tropical light, moisture, and temperature means these strains can be used as nitrogen biofertilizers in flooded rice fields. This clever and simple strategy works well under laboratory conditions; however, it remains to be seen whether regulatory agencies in various countries will permit this recombinant cyanobacterial strain to be used in the open field environment.

In addition to providing fixed nitrogen, some cyanobacteria may also benefit crop plants by providing various growth-promoting substances such as gibberellins, auxins, vitamins, free amino acids, and various carbohydrates and sugars. Moreover, many cyanobacteria are able to solubilize inorganic phosphate and thereby make it available for crop growth.

Given their ability to fix both carbon and nitrogen, there has recently been considerable interest in utilizing cyanobacteria as microbial cell factories as an alternative means of bioenergy production. Also, it has been assumed that ancient cyanobacteria played a key role in the formation of the world's petroleum deposits. Some of the perceived advantages of using cyanobacteria to synthesize various biofuels include: (1) they grow relatively rapidly; (2) under favorable weather conditions, they can be grown during the entire year; (3) they can be cultivated in brackish water and on nonarable land; (4) they can sometimes be grown on wastewater; (5) they do not require fertilizers, pesticides, or herbicides; and (6) after the biofuel has been extracted, the residual biomass may sometimes be useful as a cattle feed. Unfortunately, to be cost effective, the large-scale growth of cyanobacteria is often performed in open shallow tanks that may be subject to contamination. While there are not as yet any commercial products that have been produced from either wild-type or by genetically engineered cyanobacteria, several products have been produced on a small scale by cyanobacteria under laboratory conditions including ethanol, isobutyraldehyde, isobutanol, 1-butanol, isoprene, hydrogen, fatty acids, and fatty alcohols.

Interestingly, until recently it was thought that the biological nitrogen fixation that occurred in marine systems occurred primarily in subtropical waters, e.g., by the cyanobacterium *Trichodesmium* and the diatom (single-celled algae with cell walls of transparent silica) symbiont *Richelia*. This is because, when the nitrogen-fixing cyanobacteria isolated from warmer climates were transferred to cooler temperatures, they were unable to fix nitrogen. However, it has now been observed that a unicellular symbiotic cyanobacterium can fix nitrogen in the cold waters of the Western Arctic. The nitrogen fixing organisms that function efficiently at colder temperatures are largely found in surface samples in ice-melt waters that have a low level of salinity. This observation has important implications in terms of understanding the nature of global environmental changes.

4.1.2 Rhizobia

Rhizobia are bacteria that form nodules on the roots of legumes and subsequently fix nitrogen within the nodule and provide it to the plant. The Dutch microbiologist and botanist, Martinus Beijerinck was (arguably) the first person to isolate and cultivate a bacterium from the nodules of legumes in 1888. He named the bacterium *Bacillus radicicola*; however, this organism was later reclassified as belonging to the genus *Rhizobium*. Rhizobial inoculants are among the oldest commercial crop plant inputs. In fact, pure cultures of *Rhizobium* spp. were first produced in Germany in 1896. Subsequently, in Australia, the New South Wales Department of Agriculture provided *Rhizobium* spp. cultures to farmers for use as legume inoculants in 1914.

At the present time, ever-increasing amounts of *Rhizobia* inoculants are being utilized in conjunction with the growth of legume crops as an alternative to the use of chemically fixed nitrogen. In addition, the extensive biochemical and molecular biological studies of symbiotic diazotrophs, such as *Rhizobia*, have served as a conceptual starting point for understanding some of the mechanisms of plant growth promotion by other PGPB.

Rhizobia are Gram-negative, aerobic, flagellated, and (generally) rod shaped, and they form symbiotic relationships with legumes. Thousands of different strains of rhizobia have been isolated and characterized with many new strains being described each year. All of the rhizobial species that have been described so far belong to 11 genera of Alpha-proteobacteria (i.e., *Aminobacter, Azorhizobium, Bradyrhizobium, Devosia, Ensifer* (formerly *Sinorhizobium*), *Mesorhizobium, Methylobacterium, Microvirga, Ochrobactrum, Phylobacterium, Rhizobium,* and *Shinella*) or three genera of Beta-proteobacteria (i.e., *Burkholderia, Cupriavidus,* and *Herbaspirillum*). Proteobacteria is a major phylum of Gram-negative bacteria and Alpha- and Beta-proteobacteria are major classes of proteobacteria. Generally, each rhizobial species is specific for a limited number of plants and will not interact with plants other than its

Table 4.1 Plant specificities of some rhizobial species

Bacterial species	Host plant(s)
Azorhizobium caulinodans	West African legume (*Sesbania rostrata*)
Bradyrhizobium elkanii	Soybean (*Glycine max*), black-eyed pea (*Vigna unguiculata* subsp. *dekindtiana*), mung bean (*Vigna radiata*)
Bradyrhizobium japonicum	Soybean (*G. max*)
Mesorhizobium amorphae	Desert false indigo (*Amorpha fruticosa*)
Mesorhizobium ciceri	Chickpea (*Cicer arietinum*)
Mesorhizobium chacoense	White carob tree (*Prosopis alba*)
Mesorhizobium huakuii	Chinese milk vetch (*Astragalus sinicus*)
Mesorhizobium loti	Lotus (*Lotus japonicus*)
Mesorhizobium meditteraneum	Chickpea (*C. arietinum*)
Mesorhizobium tianshanense	7 legume species
Rhizobium sp. strain NGR234	>100 Tropical legume species
Rhizobium etli	Kidney bean (*Phaseolus vulgaris*), mung bean (*V. radiata*)
R. etli bv. mimosae	Mimosa (*Mimosa affinis*)
Rhizobium galegae	Goat's rue (*Galega officinalis*, *Galega orientalis*)
Rhizobium gallicum	Common bean (*P. vulgaris*)
Rhizobium huautlense	Danglepod (*Sesbania herbacea*)
Rhizobium leguminosarum bv. phaseoli	Kidney bean, mung bean
R. leguminosarum bv. trifolii	Clover (*Trifolium* spp.)
R. leguminosarum bv. viciae	Pea (*Pisum sativum*)
Rhizobium sullae	Sweetvetch (*Hedysarum coronarium*)
Rhizobium tropici	Mimosoid trees (*Leucaena* spp.) and some tropical legume trees (*Macroptilium* spp.)
Sinorhizobium fredii	Soybean (*G. max*)
Sinorhizobium meliloti	Alfalfa (*Medicago sativa*)
Sinorhizobium morelense	White popinac (*Leucaena leucocephala*)

natural hosts (Table 4.1). Scientists typically isolate new strains of *Rhizobia* by culturing the bacteria that are found within legume root nodules.

Rhizobia typically have large and complex genomes which often occur as several distinct pieces of, generally circular, DNA. The sizes of known rhizobial genomes range from 5.4 to 9.2×10^6 base pairs while the number of plasmids found in these strains varies between 0 and 7. The very large size of many rhizobial plasmids (often called megaplasmids with sizes more than 5×10^6 base pairs), together with the fact that some of these plasmids may carry essential genes, has blurred the distinction in these bacterial strains between plasmids and chromosomes. Interestingly, the strain *Bradyrhizobium japonicum* USDA 6T contains a larger genome and at the same time

Fig. 4.3 Schematic representation of red-colored bacteroids (not drawn to scale) located within a plant root nodule

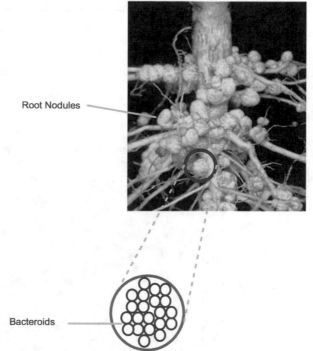

Root Nodules

Bacteroids

includes no plasmids. Many rhizobial genomes have two main components, the core and the accessory genome where the core genome includes essential genes and is conserved in related bacteria while the accessory genome is typically more variable and is often found on plasmids. Genes that are involved in the symbiotic relationship between the bacterium and its plant host are often found on plasmids or within so-called chromosomal islands within the core genome.

As part of their life cycle, rhizobial bacteria invade plant root cells and initiate a complex series of developmental changes that lead to the formation of a root nodule. Inside the root nodule, the bacteria proliferate and persist in a form called a bacteroid that has no cell wall (Fig. 4.3). The bacteria within the nodules fix atmospheric nitrogen by means of the enzyme nitrogenase. The structural and biochemical interactions between a symbiotic rhizobacterium and a host plant are quite complex and mutually beneficial (Fig. 4.4). Inside a nodule, nitrogenase is protected from the toxic effects of atmospheric oxygen in two ways. First, oxygen does not readily diffuse into the thick-walled nodule. Second, the oxygen content within a nodule is regulated (i.e., kept relatively low) by the protein leghemoglobin. The heme moiety of this oxygen-binding protein is synthesized by the bacterium. On the other hand, the globin portion of the leghemoglobin molecule is encoded within the plant genome. The leghemoglobin imparts a red color to the inside of the nodule (in much the same way that mammalian hemoglobin imparts a red color to red blood cells). As a consequence of this symbiotic relationship, the plant provides the

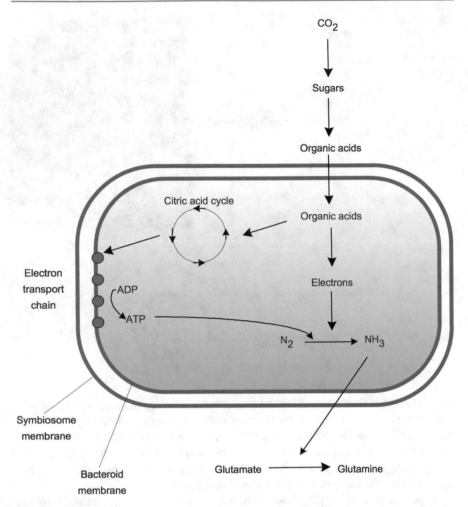

Fig. 4.4 Bacteroid metabolism. Dicarboxylic acids that were synthesized as a consequence of photosynthesis are taken up by the bacteroid. These acids are metabolized by the enzymes from the citric acid cycle. As a consequence, and with the involvement of the electron transport chain proteins (embedded into the bacteroid membrane), ATP, which provides the energy for nitrogen fixation, is generated. The ammonia that is synthesized is transferred to glutamic acid to form glutamine, which is then transported throughout the plant. Oxygen is slowly released from a complex with bacterial leghemoglobin and is used in several of the abovementioned enzymatic steps (not shown in this figure). Any excess sugars are converted to polyhydroxybutyrate, a carbon storage polymer, to be used at a later time

bacterium with photosynthetically fixed carbon (generally in the form of organic acids), which the bacterium requires for growth. In turn, nodule bacteria provide the plant with fixed nitrogen (ammonia that is used to convert the amino acid glutamic acid to the amino acid glutamine).

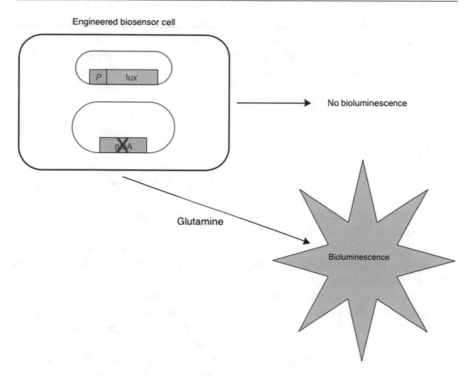

Fig. 4.5 Use of an engineered biosensor bacterial cell to detect the presence (and amount) of glutamine (gln) in leaf extracts. First, the *glnA* gene on the bacterial cell chromosomal DNA (not shown to scale) is mutagenized so that the cell can no longer synthesize glutamine. Then the mutant is transformed with a plasmid containing bacterial *lux* genes under the control of a promoter that responds to the presence of glutamine. To perform the assay, leaf disks of a specific size are removed from target plants, homogenized and then used to make plant tissue extracts. The plant tissue extract and the engineered biosensor cells are mixed together and the amount of biolumines-cence is measured in a luminometer giving an indirect measure of the amount of newly fixed nitrogen in that plant

When researchers isolate new strains of rhizobia, one of the things that they would like to assess is how effective the newly isolated strains are at fixing nitrogen. Traditionally, researchers would either measure (1) the ability of rhizobial strains to convert acetylene to ethylene over a given period of time or (2) the incorporation of ^{15}N into macromolecules. One simple and inexpensive means of monitoring nitro-gen fixation includes following the production of glutamine by the potential new inoculant (Fig. 4.5). This is because of the fact that following the conversion of atmospheric nitrogen into ammonia, the ammonia is rapidly assimilated into gluta-mine. Thus, measuring how much glutamine is present in a leaf extract of a rhizobia infected plant tells the researcher how active that strain of rhizobia has been in fixing nitrogen in that particular plant.

In the past few years, some scientists have suggested that the majority of rhizobia do not form nodules on legumes (or any other plants). This is probably because they lack the genetic determinants to induce nodulation. These rhizobia can exist in bulk soil, in the rhizosphere, inside root tissues as endophytes or they may enter into a nodule that has been formed by a nodulating strain of rhizobia. In fact, non-nodulating strains may sometimes reduce the effectiveness of nodulating strains of rhizobia.

4.1.3   Free-Living Bacteria

In addition to the well-known association between rhizobia and legumes, a large number of different free-living soil bacteria, especially various bacilli and proteobacteria, are capable of fixing atmospheric nitrogen and providing it to plants. Many species of diazotrophic bacteria can not only colonize plant surfaces (e.g., *Azospirillum* spp.), they can also multiply and spread within plant tissues as endophytes (e.g., *Azoarcus* spp., *Gluconacetobacter* spp. and *Herbaspirillum* spp.). In addition, some cyanobacteria such as *Nostoc* spp. are often found within plant tissues. While there is absolutely no doubt that the ability to fix atmospheric nitrogen contributes to the ability of some free-living bacteria to promote plant growth, the amount that different bacteria contribute to the promotion of plant growth is often quite difficult to quantify and may vary significantly according to the amount of available soil nitrogen. In one recent study, when soil nitrogen levels were low, inoculation of wheat, maize, and cucumber plants with a diazotrophic endophytic strain of *Paenibacillus beijingensis* increased plant shoot dry weight from 50 to 100% and root dry weight from 20 to 50%. At the same time, the total nitrogen in the inoculated plants was increased by 50 to 90%. However, when the level of fixed nitrogen in the soil was high, as expected, much smaller increases in both dry weights and plant nitrogen content were observed.

Nearly all of the characterized free-living diazotrophs contain other major mechanisms of plant growth promotion. For example, in addition to being diazotrophic, *Azospirillum* spp. are typically major producers of the phytohormone IAA. In addition, as discussed below, the process of nitrogen fixation is both highly energy intensive and extremely sensitive to the presence of oxygen. Thus, a free-living diazotroph likely only fixes nitrogen when no usable other source of nitrogen is available and when the bacterium is at least somewhat protected from inhibitory oxygen.

Several recent studies have found a number of different free-living bacteria to be present within legume nodules (presumably formed by nodulating rhizobia). These bacteria included members of the genera *Enterobacter, Chryseobacterium, Bacillus, Paenibacillus, Pseudomonas*, and *Sphingobacterium*. The presence of these bacteria was found to be influenced by the soil type, and to a lesser extent, by the plant genotype. While no specific role in nodulation or plant growth promotion has been ascribed to these endophytic bacteria, it has been suggested that they could enter the nodule through the junction between the nodule and the root. Nevertheless, some of these endophytic strains possess one or more plant growth-promoting traits.

4.1.4 Nitrogenase

The increased interest in the use of diazotrophs as biological fertilizers has overlapped the development of techniques for gene isolation and manipulation (approximately the past 50 years) thereby providing the impetus for studying the biochemical and molecular biological aspects of nitrogen fixation. Initially, these studies were undertaken in the belief that they would lead to the development of improved nitrogen-fixing organisms that would enhance crop yields to a greater extent than the wild-type strains. Some researchers even went so far as to suggest that bacterial genes for nitrogen fixation might be introduced directly into plants to enable them to fix their own nitrogen. While these ideas have not yet been realized, a consequence of this research is that a detailed understanding of the process of nitrogen fixation has emerged. And with this understanding, the possibility of improving the nitrogen-fixing activity of some diazotrophs by genetic manipulation is a little closer to becoming a reality.

All known nitrogenases are bacterial in origin and have two oxygen-sensitive components. Component I is a complex of 2 identical α-protein subunits (approximately 50,000 Da each), 2 identical β-protein subunits (approximately 60,000 Da each), 24 molecules of iron, 2 molecules of molybdenum, and an iron–molybdenum cofactor, sometimes called FeMoCo (Fig. 4.6). Component II has two α-protein subunits (approximately 32,000 Da each), which are not the same as the α-protein subunits of component I, and a number of associated iron molecules. The catalysis of nitrogen to ammonium ion requires the combination of components I and II, a complex of magnesium and ATP, and a source of reducing equivalents (Fig. 4.7). In addition to fixing nitrogen, the nitrogenase can reduce the gas acetylene to ethylene. The measurement by gas chromatography of the increase in ethylene production (or the decrease in the acetylene level) as a function of time provides a convenient (and commonly used) assay for nitrogenase activity. Component I catalyzes the actual reduction of N_2, and component II donates electrons to

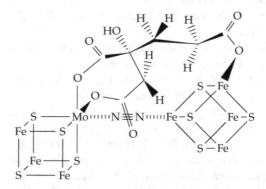

Fig. 4.6 Structure of the iron–molybdenum cofactor (FeMoCo) bound to a molecule of dinitrogen (N_2) prior to its conversion to ammonia. © 2010 American Society for Microbiology. Used with permission. No further reproduction or distribution is permitted without the prior written permission of American Society for Microbiology

Fig. 4.7 Chemical
conversion of N_2 (dinitrogen)
to NH_3 (ammonia) and
acetylene to ethylene. All of
the protons and electrons
involved in this process are
shown

Nitrogen Fixation

$$N_2 + 8H^+ + 8e^- + 16MgATP \rightarrow$$
$$2NH_3 + H_2\uparrow + 16MgADP + 16P_i$$

Acetylene Reduction

$$H—C≡C—H + 2H^+ \rightarrow H_2=C=C=H_2$$

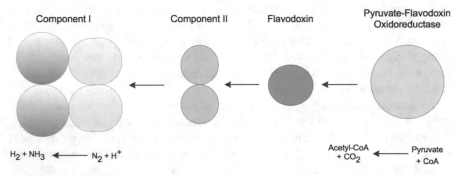

Fig. 4.8 Schematic representation of some of the key reactions in the process of nitrogen fixation. The five polypeptides shown are all part of the nif gene cluster. Nitrogenase consists of component I and component II. Flavodoxin and the pyruvate-flavodoxin oxidoreductase are largely involved in the transfer of electrons to the nitrogenase. The arrows between the proteins represent electron flow

component I (Fig. 4.8). Both components are extremely sensitive to oxygen and can be rapidly and irreversibly inactivated when the oxygen concentration is too high. In addition to components I and II, the activity of a complete, functional nitrogenase depends on 15–20 additional accessory proteins. The roles of most of the accessory proteins have been delineated and include the transfer of electrons to component II and the biosynthesis of the iron–molybdenum cofactor (FeMoCo) that is a part of component I.

Nitrogen fixation is a very complicated process requiring the concerted actions of a large number of different proteins. Therefore, it was not realistic to expect either that an intact single DNA fragment containing all the genetic information for nitrogen fixation could be readily cloned from a diazotrophic microorganism and transferred into a nondiazotrophic organism or that a recipient organism could easily maintain the physiological conditions necessary for nitrogenase activity. In addition, the nitrogenase and many of the other components of this system are highly sensitive to the presence of oxygen so that it is extremely difficult to purify and then biochemically characterize these proteins. In addition, the first time that genes involved in nitrogen fixation (*nif* genes) were isolated it was not possible to use previously isolated *nif* genes as DNA hybridization probes. Consequently, at that time, the most direct way to isolate *nif* genes was to identify and characterize those clones of a wild-type genomic DNA library of the original organism that restore

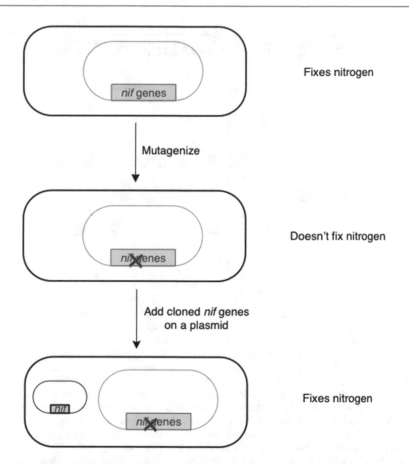

Fig. 4.9 Cloning *nif* genes by genetic complementation. The wild-type nitrogen-fixing bacterium contains a cluster of *nif* genes on either its chromosomal DNA or on a very large (mega) plasmid. Following chemical or uv mutagenesis, cells that can no longer fix nitrogen are selected. These cells typically have a mutation in one or more of the *nif* genes. A clone bank of the wild-type DNA is constructed on a small plasmid and then the clone bank is introduced into the mutant bacteria. Those bacteria that are now able to fix nitrogen have had their mutation complemented by the gene(s) on the plasmid. Thus, a functional gene on an introduced plasmid replaces the inactivated (mutagenized) gene on the chromosomal or mega-plasmid DNA

nitrogen fixation to various mutants that were unable to fix nitrogen (Fig. 4.9). This process is called gene cloning by genetic complementation.

To isolate genes, other than those that complement the abovementioned mutation, that are involved in the nitrogen fixation process, once they have been isolated, *nif* genes have been used as DNA hybridization probes, which have then been used to screen a chromosomal DNA clone bank that carries large (7- to 10-kb) inserts (Fig. 4.10). This scheme is based on the observation that in prokaryotic organisms the genes coding for proteins in a pathway are typically clustered on the chromosomal DNA and arranged in operons. Thus, DNA hybridization enables

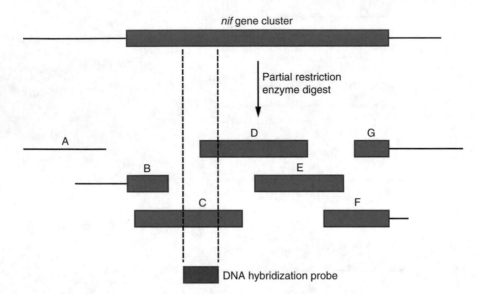

Fig. 4.10 Walking down the chromosome to isolate genes adjacent to an isolated *nif* gene. Partial restriction endonuclease digestion of the region of the chromosomal DNA that encodes *nif* genes. The *nif* gene that was isolated by genetic complementation is used as a DNA hybridization probe is shown in red. The DNA hybridization probe and the DNA fragments that result from the partial enzyme digest of the chromosomal DNA are ordered as they would be within the chromosomal DNA. The technique of colony hybridization of a clone bank consisting of the illustrated fragments, using the probe shown, would be expected to identify clones containing plasmids with fragments C and D. Once these fragments have been isolated and characterized, fragments C and D may be used, separately, as DNA hybridization probes to screen the same clone bank, in this case identifying clones containing plasmids with fragments B and E, respectively. In this manner, a large contiguous section of the host chromosome may be isolated as a set of overlapping clones. © 2010 American Society for Microbiology. Used with permission. No further reproduction or distribution is permitted without the prior written permission of American Society for Microbiology

investigators to identify clones containing additional *nif* genes that are adjacent to the *nif* gene sequence that was initially isolated by complementation. And, once adjacent genes are identified they may be characterized by DNA sequencing. In addition to being used to facilitate the isolation of the entire *nif* gene cluster, this technique of "walking down the chromosome" has been used to isolate nodulation (*nod*) and hydrogenase (*hup*) genes (both of which are discussed below).

The entire set of *nif* genes has been isolated and characterized from a number of different diazotrophic bacteria including both *Rhizobia* spp. and numerous rhizospheric and endophytic PGPB. These genes are typically arranged in a single cluster that occupies ~24 kb of the bacterial genome, contains seven separate operons and encodes 20 distinct proteins. All of the *nif* genes are transcribed and translated in a concerted fashion, under the regulatory control of the *nifA* and *nifL* genes, to produce a functional nitrogenase. The NifA protein is a positive regulatory factor. It turns on the transcription of all of the *nif* operons, except its own, by binding to a part of the promoter of each *nif* operon. The NifA protein is an enhancer-binding protein that binds to specific DNA sequences located upstream of the *nif* genes and interacts with the transcription initiation protein sigma-54 (σ^{54}) subunit of RNA polymerase before transcription from the *nif* promoter is initiated. The NifL protein is a negative regulatory factor; it is a flavoprotein that senses cellular redox status and binds to NifA to block its activity when conditions are not optimal for N_2 fixation. In the presence of either too much oxygen or high levels of fixed nitrogen, NifL acts as an antagonist of the NifA protein and, as a result, turns off the transcription of all other *nif* genes. This prevents unneeded and energy wasting nitrogen fixation. Most diazotrophic organisms have a similar array of genes encoding their nitrogen-fixing apparatus, and the DNA sequences of these genes do not vary much from one organism to another. However, not all nitrogen-fixing organisms have a NifL protein. In some organisms, the essential regions of NifL are part of NifA.

Increasing the amount of nitrogen that an organism can fix also increases the amount of energy, usually ATP from the metabolism of various carbon compounds that is needed to power its metabolism. Consequently, a bacterium that has been engineered to fix a higher than normal level of nitrogen may lose its effectiveness as a plant growth-promoting agent because of the diminished growth rate that results from depleting the cell's ATP stores.

In one series of recent experiments, a strain of *Mesorhizobium* sp. was genetically manipulated so that the *nifA* gene was cloned and then expressed under the transcriptional control of the *E. coli lac* promoter (which acts as a constitutive promoter in *Mesorhizobium* sp.), with the entire construct being placed on a broad-host-range plasmid (Fig. 4.11). This resulted in the NifA protein being significantly overexpressed within the transformed bacteria. Interestingly, this genetic manipulation did not significantly alter the number of nodules formed on host chickpea plants compared to the nontransformed strain. However, this manipulation did result in the weight of the average nodule of the plants inoculated with the NifA overproducer being approximately three times what it was with the wild-type strain. In addition, the shoot weight of the plants inoculated with the NifA overproducing bacterial strain was approximately double the shoot weight of plants inoculated with the wild-type bacterium. While at first glance these results are quite exciting and even perhaps spectacular, it remains to be seen whether this sort of genetic manipulation will result in a strain that is highly effective under field conditions.

Genetic modification of plants with the entire *nif* gene cluster would likely not be effective because the normal level of oxygen in the host cells would likely inactivate nitrogenase. In addition, it is difficult to conceive how the regulation of nitrogen fixation could be achieved in plants, since there are no plant promoters that respond to the NifA protein. Consequently, *nif* genes would not be turned on in such a transgenic plant unless it became possible to engineer plant promoters that were under the transcriptional control of bacterial NifA protein. Moreover, to enable their expression in plants, each of the *nif* genes would have to be under the control of separate promoters because plant cells cannot process multigene transcripts. Thus, at this stage, the introduction of a functional nitrogen fixation capability into plants is extremely unlikely.

NifA overproducing Mesorhizobium

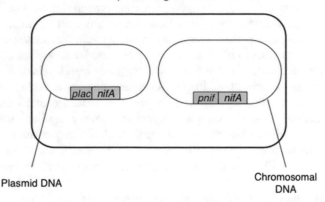

Plasmid DNA

Chromosomal
DNA

Fig. 4.11 An engineered strain of *Mesorhizobium* sp. that overproduces NifA. The bacterial chromosomal DNA contains a single copy of the *nifA* gene under the transcriptional control of its native promoter. The plasmid DNA contains a copy of the *Mesorhizobium* sp. *nifA* gene under the transcriptional control of the (stronger) *E. coli lac* promoter

While there may not be any straightforward and obvious way to increase nitrogen fixation by manipulating *nif* genes, several indirect strategies are worth considering. For example, since nitrogen fixation is a highly energy intensive process, any genetic manipulation that increases the cell's energy supply might also indirectly increase the amount of fixed nitrogen. Normally, in the presence of high levels of carbon (glucose) sources, the excess carbon is converted into glycogen, a carbon storage compound (Fig. 4.12). However, a deletion mutation in the gene that codes for the enzyme glycogen synthase prevents glycogen synthesis so that the glucose enters the tricarboxylic acid (TCA) cycle where the acetyl group of acetyl-CoA is degraded to form carbon dioxide and hydrogen. The hydrogen (or the corresponding electrons) is fed into the electron transport chain and the energy that is released is conserved by phosphorylation of three molecules of ADP to ATP. In this case, the mutated *Rhizobium* strain increases the nodule number and plant biomass (in the field as well as in the lab). Unfortunately, because the metabolism of this glycogen synthase minus mutant is geared toward "burning off" a significant portion of the carbon intake and synthesizing more ATP, the modified *Rhizobium* strain is therefore less persistent in the soil than the wild-type bacterium. Thus, the use of this mutant is not a practical solution for increasing the amount of fixed nitrogen.

The concentration of oxygen within a nodule is an important factor in determining the amount of nitrogen that is fixed by a rhizobial strain. On the one hand, oxygen is inhibitory to nitrogenase and is also a negative regulator of *nif* gene expression. On the other hand, oxygen is required for bacteroid respiration. This conundrum can be resolved by the introduction of the protein leghemoglobin, which binds free oxygen tightly so that both the transcription of *nif* genes and the functioning of nitrogenase can proceed unimpaired. In fact, the addition of exogenous leghemoglobin to isolated bacteroids results in a dramatic increase in nitrogenase

Fig. 4.12 A schematic representation of a strategy to indirectly increase nitrogen fixation by increasing the amount of ATP available for nitrogen fixation. © 2010 American Society for Microbiology. Used with permission. No further reproduction or distribution is permitted without the prior written permission of American Society for Microbiology

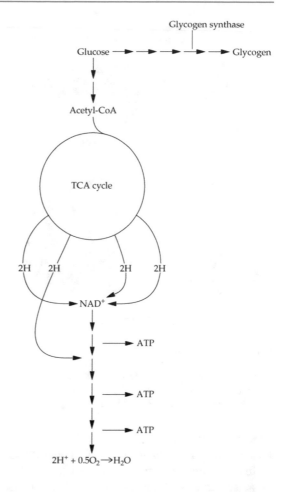

activity. Thus, it is possible to engineer more efficient strains of *Rhizobium* by overproducing leghemoglobin. Alternatively, since the globin portion of the leghemoglobin molecule is produced by the plant, it may be more efficient to transform rhizobial strains with genes encoding a bacterial equivalent of leghemoglobin.

In one experiment, following the transformation of a strain of *Rhizobium etli* with a broad-host-range plasmid carrying the *Vitreoscilla* sp. (a Gram-negative microaerophilic bacterium) hemoglobin gene at low levels of dissolved oxygen in the growth medium, the rhizobial cells had a two- to threefold higher respiratory rate than the nontransformed strain (Fig. 4.13). In subsequent greenhouse experiments, when bean plants were inoculated with either nontransformed or bacterial hemoglobin-containing *R. etli*, the plants inoculated with the hemoglobin-containing strain had approximately 68% more nitrogenase activity. This difference in nitrogenase activity led to a 25–30% increase in leaf nitrogen content and a 16% increase in the nitrogen content of the seeds that are produced. It remains to be seen whether this

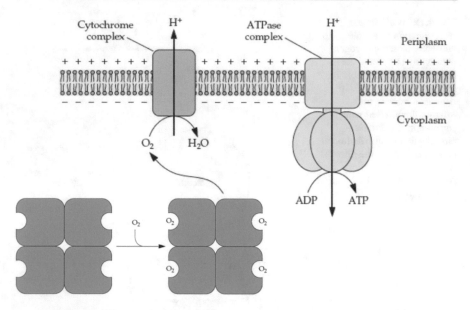

Vitreoscilla sp. hemoglobin

Fig. 4.13 Engineering *Rhizobium etli* to express a *Vitreoscilla* sp. hemoglobin gene to bind low levels of dissolved oxygen. Following transformation of *R. etli* with a broad-host-range plasmid carrying the *Vitreoscilla* sp. hemoglobin gene, then the expressed *Vitreoscilla* sp. hemoglobin binds oxygen, the oxygen is utilized to facilitate pumping H^+ from the cytoplasm to the periplasm, and the H^+ is subsequently taken up thereby generating ATP

genetically modified strain is as effective in the field as in the greenhouse and whether the regulatory authorities will approve its use in the field.

In Latin America, a majority of the bean plants are nodulated by *R. etli*. One strain of *R. etli* that is in common use in Latin America encodes three copies of the nitrogenase component II, which is also called nitrogenase reductase, encoded by the *nifH* gene(s), with each of these genes under the transcriptional control of a separate promoter. To increase the amount of nitrogenase in *R. etli*, the strongest of the three *nifH* promoters (i.e., *PnifHc*) was coupled to the *nifHcDK* operon, which encodes the nitrogenase structural genes (where *nifHc* is one of the three *nifH* genes in this bacterium). The *nifHc* promoter is typically induced during nodule development. The *PnifHc–nifHcDK* construct was cloned into a broad-host-range plasmid and introduced into the wild-type strain of *R. etli*. The net result of this genetic manipulation was a significant increase in nitrogenase activity, plant dry weight, seed yield, and the nitrogen content of the seeds (Table 4.2). Moreover, this genetic manipulation worked as well or better when the *PnifHc–nifHcDK* construct was introduced into the large Sym plasmid from *R. etli* that contains all of the genetic information for nodulation and nitrogen fixation.

Biological nitrogen fixation requires a large amount of energy in the form of ATP (Fig. 4.7). Thus, any mutation or genetic manipulation that increases the flux of

Table 4.2 Symbiotic behavior of genetically modified strains of *R. etli* in concert with common beans (*Phaseolus vulgaris*)

Bacterial strain	Nitrogenase activity, mmol ethylene/h/g of nodule	Plant dry weight, g/plant	Seed yield, g/plant	Seed N content, mg of N/g of seed
Wild-type	64.5	0.54	1.43	33.9
Wild-type + extra *nifHDK*	72.7	0.66	1.56	41.4
Wild-type + *nifH$_c$*	77.3	0.75	1.73	31.2
Wild-type + P*nifH$_c$*– *nifH$_c$DK*	108.2	0.81	2.50	43.6

Adapted from Peralta et al., *Appl. Environ. Microbiol.* **70**:3272–3281, 2004

carbon sources consumed by a bacterium through the citric acid cycle should be beneficial for nitrogen fixation (Fig. 4.14). This is because metabolism of glucose via the citric acid cycle results in the production of ATP. Consistent with this principle, it was observed that expression of the P*nifH$_c$*–*nifH$_c$DK* construct in a poly-β-hydroxybutyrate-negative strain of *R. etli* enhanced plant growth to an even greater extent than when this construct was expressed in a wild-type poly-β-hydroxybutyrate-positive strain. Poly-β-hydroxybutyrate is a polymer produced as a form of energy storage that bacteria use when more common energy sources are not available. Finally, since no foreign genes were introduced into *R. etli,* the scientists who constructed these strains hope that the regulatory bodies in their country will view the manipulated strains as benign and approve them for widespread environmental use.

As a direct result of worldwide industrialization and the burning of fossil fuels, the amount of carbon dioxide in the atmosphere has been increasing for the past 100 years or so and contributing to global climate change. Recently, researchers reported that by directed mutagenesis they were able to alter the specificity of the nitrogenase enzyme so that it was able to reduce carbon dioxide to methane (Fig. 4.15a). This change in the specificity of nitrogenase occurred as a consequence of changing two amino acid residues in the alpha subunit of component I; the Val at position 75 was changed to Ala and the His at position 201 was changed to Gln. These two amino acid substitutions altered the protein environment around the FeMo cofactor. Theses amino acid changes were made on the basis of previous experiments that showed that the protein environment immediately surrounding the FeMo cofactor control both the size of compounds that can be nitrogenase substrates and the reactivity of the FeMo cofactor toward those compounds. In addition to converting carbon dioxide to methane, the modified nitrogenase was able to catalyze the synthesis of propylene through the reductive coupling of carbon dioxide and acetylene (Fig. 4.15b). These chemical reactions occur at faster rates than any other enzymatic conversions of carbon dioxide and require only a single enzyme whereas other systems utilize a consortium of enzymes. However, the modified nitrogenase,

Fig. 4.14 The conversion of glucose to acetyl-CoA that in turn may either be converted into poly-β-hydroxybutyrate or used to generate ATP

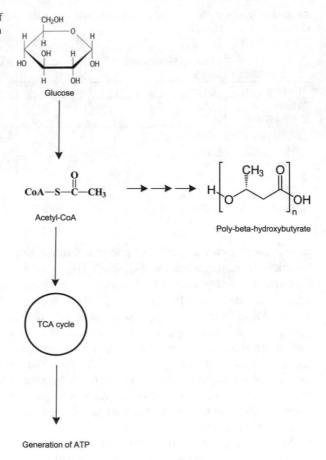

Fig. 4.15 (a) The conversion of carbon dioxide to methane by a genetically modified nitrogenase. (b) The conversion of carbon dioxide and acetylene to propylene by a modified nitrogenase

A

$$CO_2 + 8H^+ + 8e^- \longrightarrow CH_4 + 2H_2O$$

B

$$CO_2 + HC{\equiv}CH + 8H^+ + 8e^- \longrightarrow H_2C{=}CH\text{-}CH_3 + 2H_2O$$

like the native enzyme, uses considerable energy in the form of ATP. To get around this problem, researchers employed the photosynthetic bacterium *Rhodopseudomonas palustris* as an expression vehicle for the modified nitrogenase. This bacterium can fix nitrogen, and a mutation in the *R. palustris nifA* gene (encoding a transcriptional activator of all other nitrogenase genes) that had previously been identified that was able bypass the regulatory networks that repress nitrogenase and produce nitrogenase constitutively. In addition, the photosynthetic

R. palustris is able to generate considerable amounts of the required ATP to energetically drive the reaction. In this genetic background, *R. palustris* efficiently reduces carbon dioxide to methane. The two major advantages of this system are (1) that the remodeled nitrogenase can be produced constitutively (which allows the bacterium to grow with ammonium as a nitrogen source, a situation that would normally shut down nitrogenase synthesis) and (2) that the bacterium chosen can meet the high ATP requirement for nitrogenase-catalyzed methane production. This work describes early steps in what is hoped will eventually become an efficient and clean method for the conversion of an unwanted greenhouse gas (carbon dioxide) into an industrially useful compound.

4.1.5 Nodulation

A major goal of agricultural biotechnology research is the development, by genetic manipulation, of rhizobium strains that can increase plant productivity more effectively than naturally occurring strains. In this regard, many commercial inoculant strains that have been developed, by traditional mutation and selection procedures, to be superior nitrogen fixers are not very effective at establishing nodules on host plant roots when placed in competition with rhizobium strains that are already present in the soil. Conversely, although many of the strains that are indigenous to the soil are highly successful in establishing nodules in competitive situations, they are often not especially efficient at nitrogen fixation. It would therefore be advantageous to be able to increase the nodulation capability of commercial inoculant strains that were previously selected for their ability at nitrogen fixation. The question then arises, what is the genetic basis of this effective "competitiveness" at nodulation.

When scientists first attempted to isolate nodulation (*nod*) genes, the absence of any specific information about the biochemical or genetic basis of nodulation meant that a strategy had to be devised for the identification of the genes. Therefore, the technique of genetic complementation was used (see Sect. 4.1.4). Nodulation-defective (Nod⁻) mutants of *Sinorhizobium meliloti* (a strain that nodulates alfalfa, *Medicago*; sweet clover, *Melilotus*; and fenugreek, *Trigonella*) were transformed with a clone bank of wild-type *S. meliloti* chromosomal DNA constructed with the use of a broad-host-range plasmid, and those colonies that had acquired the ability to nodulate alfalfa roots were isolated. Once a single nodulation gene was identified, it was then used as a DNA hybridization probe to "walk down the chromosome" and identify adjacent regions of the *S. meliloti* chromosomal DNA (see Sect. 4.1.4).

The isolation and characterization of the complete repertoire of nodulation genes from *S. meliloti* (and subsequently from many other rhizobial species) revealed that, like nitrogen fixation, nodulation and its regulation is a complex process that requires the functioning of a large number of genes. Some nodulation genes are highly conserved (i.e., common) among nodulating bacteria, and others are species specific. The *nod* genes are grouped into three separate classes: common genes, host-specific genes, and the regulatory *nodD* gene. Thus, for example, the *nodABC* genes are

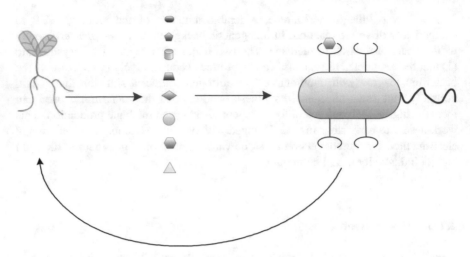

Fig. 4.16 The binding of small molecule chemo-attractants secreted by plants by bacterial receptors—eight different chemo-attracts exuded by alfalfa plants that can bind to Sinorhizobium meliloti receptors have been identified—and the subsequent movement of the bacteria toward the plant and the sites of chemo-attractant secretion

common to all *Rhizobium* species, are structurally interchangeable, and in most species they are found on a single operon.

The bacterium *S. meliloti* has been shown to utilize chemotaxis to optimize its movement toward host plant-secreted nutrients that act as chemo-attractants. This enables this bacterium (and others) to locate infection sites along emerging roots and, as a result, to compete more effectively for nodulation. Germinating seeds exude a wide range of organic compounds including some flavonoids, and alfalfa seed exudates elicit a chemotactic response from *S. meliloti* (Fig. 4.16). The early recruitment of the microbial symbiont to the growing root, e.g., during seed germination, is one strategy for maximizing the rhizobia–legume interaction. Previous research had suggested that flavonoids act as host-specific chemo-attractants; however, they elicit only a weak chemotactic response and have only a short diffusion range in aqueous soil due to their hydrophobic nature. Thus, amino acids, organic acids, and sugars are more likely to function as recruiting agents thereby facilitating the initial binding of the bacterium to the plant and the movement of the bacteria toward the plant.

Subsequently, a number of events occur during nodulation (Fig. 4.17a). First, the *nodD* gene protein product, which is constitutively expressed, recognizes and binds to a flavonoid molecule, which is excreted by the roots of the potential host plants. Flavonoids are a class of plant secondary metabolites that perform a number of different functions for the plant, such as pigmentation and defense against fungi or insects. The binding of flavonoids to the NodD protein is the major determinant of rhizobial host specificity, because each rhizobial species recognizes and binds to only a limited number of flavonoid structures and each plant species produces its

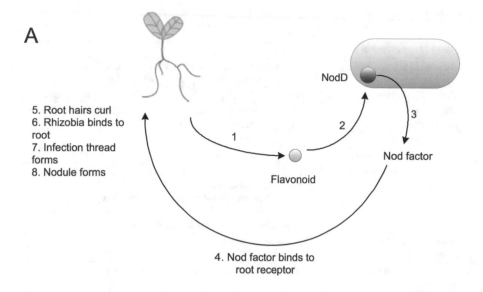

A

5. Root hairs curl
6. Rhizobia binds to root
7. Infection thread forms
8. Nodule forms

NodD

1

2

3

Flavonoid

Nod factor

4. Nod factor binds to root receptor

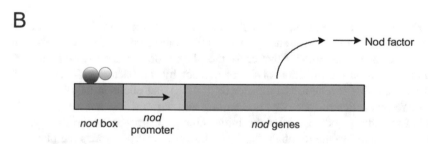

B

Nod factor

nod box

nod promoter

nod genes

Fig. 4.17 Overview of the nodulation process. (**a**) The plant secretes/releases several specific flavonoid molecules (1), the flavonoid is taken up by the rhizobial strain and then is bound to the NodD protein (2), the flavonoid–NodD complex activates the synthesis of the Nod factor (3), the flavonoid–NodD complex binds to a root receptor (4), the root hair tips swell and curl (5), the *Rhizobium* strain attaches to the root hair tips (6), an infection thread forms (7) and once inside the root cells, the nodule forms (8). (**b**) The flavonoid–NodD complex binds to the *nod* box that is a part of the *nod* promoter and activates the synthesis of the Nod factor and transcription of other *nod* genes

own specific set of flavonoid molecules (Table 4.3). Some rhizobial strains, such as *R. leguminosarum* biovar (bv.) trifolii, have a very narrow host range, responding to only a few kinds of flavonoids, while others, such as *Rhizobium* sp. strain NGR234, have a very broad host range and respond to a much larger number of different flavonoids.

The binding of a flavonoid molecule activates the NodD protein and enables the flavonoid–NodD complex to attach to a nodulation promoter element called a *nod*

Table 4.3 Some legumes and the *nodD* gene inducers that they produce

Legume	Compound
Lupin (*Lupinus albus*)	Erythronic acid, tetronic acid
Alfalfa (*Medicago sativa*)	Stachydrine, trigonelline, luteolin, chrysoeriol, 4,4′-dihydroxy-2′-methoxychalcone, liquiritigenin, 7,4′-dihydroxyflavone
Clover (*Trifolium repens*)	7,4′-Dihydroxyflavone, geraldone, 4′-hydroxy-7-methoxyflavone
Common bean (*Phaseolus vulgaris*)	Delphinidin, kaempferol, malvidin, myricetin, petunidin, quercetin, eriodictyol, genistein, naringenin
Pea (*Pisum sativa*)	Apigenin, eriodictyol
Soybean (*Glycine max*)	Daidzein, genistein, coumestrol
Vetch (*Vicia sativa* subsp. *nigra*)	3,5,7,3′-Tetrahydroxy-4′-methoxyflavanone, 7,3′-dihydroxy-4′-methoxyflavanone, naringenin, 4,4′-dihydroxy-2′-methoxychalcone, liquiritigenin, 7,4′-dihydroxy-3′-methoxyflavanone, 5,7,4′-trihydroxy-3′-methoxyflavanone, 5,7,3′-trihydroxy-4′-methoxyflavanone naringenin
Barrel Medic (*Medicago truncatula*)	7,40-Dihydroxyflavone
Black locust (*Robinia pseudoacacia*)	Apigenin, naringenin, chrysoeriol, isoliquiritigenin
Cowpea (*Glycine max*)	Daidzein, genistein, coumestrol

These inducers are commonly released by either roots or germinating seeds

box (Fig. 4.17b). This promoter element is located upstream from all the nodulation genes, except the *nodD* gene, and it activates the transcription of these genes. The *nodABC* genes encode proteins that cause the plant root hair tips to swell and curl, the initial steps in the infection of the plant root by the bacterium. The bacteria synthesize an oligosaccharide Nod factor (Fig. 4.18) that elicits in the plant a host-specific response that includes root hair curling and deformation and is essential for rhizobia to induce nodules. After the initial change in the root morphology, the bacterium attaches to the root hair. Next, the bacterial cell penetrates the plant cell through an infection thread. As the bacterial cell divides, it moves through the infection thread. Finally, a number of additional *nod* gene products are synthesized. These proteins, together with some plant-encoded proteins, contribute to the formation of the nodule.

It has become clear that the process of nodulation is quite complicated. Thus, considerable additional effort will be required before it is possible to further enhance the competitiveness of rhizobial strains by genetic engineering. To date, despite the fact that *nod* genes have been isolated and characterized from numerous rhizobial strains, no simple genetic means has been devised for using *nod* genes to enable inoculated strains of *Rhizobium* to outcompete indigenous strains. Nevertheless, it is possible to alter host specificity by the transfer of a *nodD* gene from a broad-specificity rhizobial strain to one with narrow specificity.

Low levels of ethylene are often produced by a plant following the plant's infection by a soil microorganism such as a strain of rhizobia. Thus, nodule formation engenders a small rise in the plant ethylene level that is generally localized to a portion of the root and can inhibit, and therefore limit, subsequent rhizobial infection

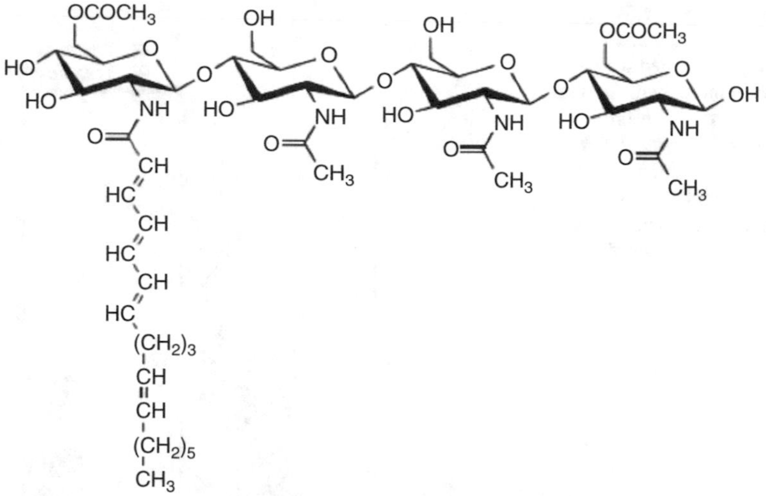

Fig. 4.18 Chemical structure of a Nod factor from *Rhizobium leguminosarum* bv. viciae. © 2010 American Society for Microbiology. Used with permission. No further reproduction or distribution is permitted without the prior written permission of American Society for Microbiology

and nodulation. One way in which some strains of rhizobia naturally increase the number of nodules that they can form on the roots of a host legume is to limit the rise in ethylene that occurs following the initial infection. Different rhizobia species can decrease ethylene levels either by synthesizing a small molecule called rhizobitoxin that chemically inhibits ACC synthase, one of the ethylene biosynthetic enzymes, or by producing the enzyme ACC deaminase and cleaving, and hence removing, some of the ACC before it can be converted to ethylene. The result of lowering the local level of ethylene in plant roots is that both the number of nodules and the biomass of the plant are increased by 25–40%. Assays of newly isolated rhizobial strains indicate that in the field approximately 1–10% of them naturally possess ACC deaminase. It should therefore be possible to increase the nodulation efficiency of rhizobia strains that lack ACC deaminase by genetically engineering these strains with isolated rhizobia ACC deaminase genes (and their regulatory regions). In fact, insertion of a single copy of an ACC deaminase gene from *R. leguminosarum* bv. *viciae* into the chromosomal DNA of a strain of *S. meliloti* that lacked this enzyme significantly increased both the nodule numbers and biomass of host alfalfa plants (Fig. 4.19). Although genetically engineered strains of rhizobia are generally not acceptable for use in the field in most jurisdictions throughout the world at this time, as a result of this work, several commercial inoculant producers have recently begun screening newly isolated rhizobia strains for active ACC deaminase.

To get around the problem of using genetically engineered rhizobia strains, some scientists have co-inoculated plants with both rhizobia and PGPB strains that are active producers of ACC deaminase. In fact, both in the lab and in the field, using both rhizospheric and endophytic PGPB, co-inoculation was as effective as

Fig. 4.19 The increased
ability of *Sinorhizobium
meliloti* transformed with an
ACC deaminase gene (and its
regulatory region) from
Rhizobium leguminosarum
bv. viciae to nodulate alfalfa
(nodules per plant) and
promote plant growth (shoot
dry weight in mg per plant)

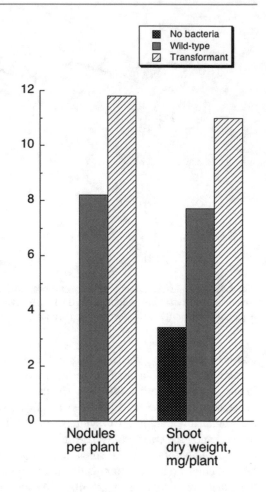

genetically engineering rhizobial strains to produce ACC deaminase. Thus, co-inoculation significantly increased both the shoot and root dry weight of treated plants, their number of nodules and the average weight per nodule, their resistance to certain common fungal phytopathogens, and the amount of nitrogenase activity per nodule. This co-inoculation approach has very recently been adopted commercially and appears to be quite successful.

In addition to insuring that rhizobial inoculants contain ACC deaminase and thereby insuring an increased number of nodules and a higher plant biomass, dual inoculation of pea plants with strains of *R. leguminosarum* and a PGPB strain, *Pseudomonas putida,* synergistically improves plant growth, nodulation, and seed yield (Table 4.4). In this case, despite the fact that both strains contained ACC deaminase, the presence of the two strains always produced the largest and healthiest plants and the greatest yield of peas. This may reflect the ability of the PGPB strain to promote plant growth by a variety of (undefined in this study) mechanisms. More-over, this inoculation was more effective than the addition of optimal levels of

Table 4.4 Growth of pea plants in the field following different treatments

Treatment	Dry weight		N content		Seed yield, g/plant
	Roots, g/plant	Shoots, g/plant	Roots, mg/g	Shoots, mg/g	
Control	0.35	1.2	30.0	40.0	5.4
P. putida	0.55	1.8	32.0	46.3	7.8
R. leguminosarum	0.65	1.8	32.3	48.0	8.2
P. putida + *R. leguminosarum*	0.80	2.3	39.5	59.3	10.2
N + P fertilizer	0.85	2.1	38.3	52.7	9.2

nitrogen and phosphorus fertilizer, and this effect was significant both in the greenhouse and in the field. Thus, not only can rhizobial inoculants be improved by selecting for the presence of ACC deaminase, additional growth improvements are possible when a dual inoculation strategy is employed.

In one recent study, scientists sought to determine whether the overexpression of NifA or NodD, regulators of the processes of nitrogen fixation and nodulation, respectively, could be used to improve the behavior of mesorhizobia in facilitating the growth of chickpea plants. To test this hypothesis several different *Mesorhizobium* sp. strains from different regions of Portugal were transformed with additional copies of *nifA* or *nodD* genes, and then the growth of chickpea plants that were grown in the presence of these strains (wild-type and transformants) was assessed. While there is some scatter in the data, the results strongly indicate that with strain V-15, *nifA* overexpression significantly improved plant growth and bacterial effectiveness. In addition, the overexpression of *nodD* led to an improvement in plant growth with strains ST-2 and PM-6 (Table 4.5). Although this result is still preliminary and the efficacy of this approach needs to be proven in the field, this is an important step in improving the symbiotic performance of mesorhizobia. And, while genetically engineered bacteria may not be readily approved for field use, it would be possible to manipulate bacterial strains to overproduce NifA or NodD by using CRISPR to modify the regulation of one or the other or both of those genes.

Very recently, scientists in China observed that some strains of rhizobia are able to form nodules on either the roots or stems of *Aeschynomene indica* (a shrub in the legume family that grows widely in warm climates) and that this occurs using a Nod-factor independent pathway. Of 19 strains examined, all of the strains formed root nodules but only 9 strains formed shoot nodules. At this point, the detailed mechanism of Nod-factor independent nodulation is not known.

Other very recent reports have addressed the evolutionary origins of legume root nodules. These reports suggest that some essential components of lateral root organogenesis have been used in legumes to coordinate nodule formation. Moreover, it was suggested that this understanding might eventually lead to approaches in which some nonlegumes could be engineered to interact with rhizobia thereby allowing for the development of nitrogen fixation in nonlegume crops.

Table 4.5 Growth of chickpea plants treated with wild-type and transformed strains (with either *nifA* or *nodD*) of *Mesorhizobium* sp. isolated from different locations in Portugal

Bacterial strain	Shoot dry weight, g	Root dry weight, g	Nodule number per plant	Weight per nodule, mg
V15	0.385	0.274	117	0.576
V15 + *nifA*	0.716	0.362	76	1.684
V15	1.404	0.472	71	1.650
V15 + *nodD*	1.529	0.604	138	2.042
PM16	1.511	0.608	33	6.459
PM16 + *nifA*	1.675	0.730	72	2.912
PM16	1.008	0.317	28	4.737
PM16 + *nodD*	1.337	0.461	39	4.235
ST2	1.336	0.850	41	3.591
ST2 + *nifA*	1.641	0.912	57	4.253
ST2	1.227	0.539	51	2.114
ST2 + *nodD*	1.556	0.633	62	2.544

All plants were grown under controlled greenhouse conditions in pots with vermiculite as a substrate and harvested after 8 weeks of growth

4.1.6   Hydrogenase

An undesirable side reaction of the fixation of nitrogen by nitrogenase is the reduction of H^+ to hydrogen gas (Fig. 4.20). In this reaction, energy in the form of ATP is wasted on the production of hydrogen (hydrogen gas), which is eventually lost to the atmosphere. This unavoidable side reaction significantly lowers the overall efficiency of the nitrogen-fixing process. However, if H_2 could be recycled to H^+, the extent of energy loss could be diminished, and the nitrogen-fixing process would become more efficient. This side reaction cannot be prevented directly because it is a consequence of the chemistry of the active site of the nitrogenase; hence, it is not possible to alter nitrogenase as a means of blocking hydrogen production.

On the other hand, some strains of *Bradyrhizobium japonicum* (as well as other *Rhizobia*) have an enzyme called hydrogenase that can take up H_2 from the atmosphere and convert it into H^+ and in the process produce ATP that can be used to fix more nitrogen (Fig. 4.20). In addition, plants inoculated with strains that produce hydrogenase (they are said to be hydrogen uptake positive or Hup⁺) have more biomass and nitrogen content than plants that are treated with strains that do not produce hydrogenase (i.e., they are Hup⁻), independent of the level of nitrogenase activity in the Hup⁻ strains (Table 4.6). Thus, the presence of a hydrogen uptake system in a symbiotic diazotroph, such as *B. japonicum,* improves the ability of a bacterium to stimulate plant growth, presumably by binding and then recycling the hydrogen gas that is formed inside the nodule by the action of nitrogenase

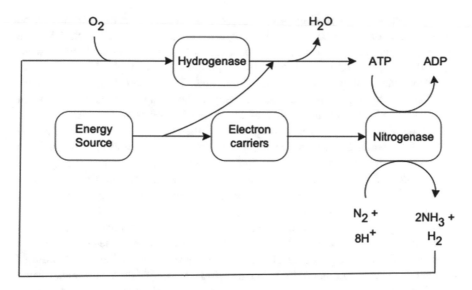

Fig. 4.20 Recycling of the hydrogen gas that is produced as a byproduct of nitrogen fixation. When it is present, hydrogenase captures the hydrogen gas and converts it into H^+ and ATP

Table 4.6 Relative enzyme activities and growth-stimulating performance of a parental Hup^+ *B. japonicum* strain (SR) and three Hup^- mutants of the same strain (SR1, SR2, and SR3)

B. japonicum strain	Relative nitrogenase activity	Relative hydrogenase activity	Relative plant dry weight	Relative nitrogen content
SR	1.00	1.00	1.00	1.00
SR1	1.27	0.01	0.81	0.93
SR2	1.13	0.01	0.74	0.91
SR3	1.23	0.01	0.65	0.85

Adapted from Albrecht et al., *Science* 203:1255–1257, 1979. Nitrogenase activity was assessed by monitoring the amount of acetylene that was reduced to ethylene by gas chromatography as a function of time. Hydrogenase activity was measured by means of a hydrogen electrode. Plant dry weight included the entire plant (roots and shoots). The nitrogen content was calculated as the fraction of the plant dry weight that was nitrogen. All values have been normalized relative to the parental strain

(Fig. 4.20). Although it is clearly beneficial to the plant to obtain its nitrogen from a symbiotic diazotroph that has a hydrogen uptake system, this trait is not commonly found in naturally occurring rhizobial strains (only 0.1–10% of naturally occurring rhizobial strains contain a hydrogen uptake system) so that Hup^- rhizobial strains are prime candidates for genetic transformation to a Hup^+ phenotype.

However, the conversion of a Hup^- strain of *Rhizobium* into a Hup^+ strain may not be readily achieved by the introduction of just a single hydrogenase gene. Rather, the introduced gene(s) must encode all of the enzyme's subunits and must be able to interact with the appropriate electron transport molecule within the host organism.

The most common strategy for isolating hydrogenase genes has been genetic complementation followed by walking down the chromosome (see Sect. 4.1.4). In this regard, *hup* genes from *B. japonicum* were isolated from a clone bank of wild-type DNA constructed in the broad-host-range cosmid vector pLAFR1 by genetic complementation of *B. japonicum* Hup⁻ mutants. The presence of a hydrogenase that takes up hydrogen from the atmosphere in the complemented Hup⁻ mutant strains was indicated by the ability of the active hydrogenase to reduce the dye methylene blue in a hydrogen atmosphere and thereby change its color. Detailed studies of the isolated *B. japonicum hup* genes showed that they were organized into several transcriptional units covering approximately 20 kb of the bacterial genome and included 18 separate genes. Subsequent studies, on the *hup* genes from *Rhizobium leguminosarum*, have indicated that these genes are similar in both DNA sequence and gene organization to the *hup* genes from *B. japonicum*.

Following the isolation of the complete complement of *R. leguminosarum hup* genes, and despite the complexity of this system, it is possible to use cosmid vectors to transfer all of these genes, originally from a naturally occurring Hup⁺ strain of *R. leguminosarum*, to a naturally occurring Hup⁻ strain (Table 4.7). Plants treated with *R. leguminosarum* that had been transformed to become Hup⁺ grew larger and contained more nitrogen than the plants inoculated with the Hup⁻ parental strain. Thus, the ability of a diazotroph to stimulate plant growth may be improved, albeit directly, by genetic manipulation.

As indicated above, in *R. leguminosarum*, 18 genes are associated with hydrogenase activity. There are 11 *hup* genes (including the operon *hup SLCDEFGHIJK*) responsible for the structural components of the hydrogenase, the processing of the enzyme, and electron transport (Fig. 4.21). There are also seven *hyp* (hydrogenase pleiotropic) genes (*hypABFCDEX*), which are involved in processing the nickel that is part of the active center of the enzyme. The *hup* promoter is under the transcriptional control of the NifA protein (which is also required to activate the synthesis of *nif* genes), so that *hup* genes are expressed only within bacteroids and at the same time as *nif* genes. On the other hand, the *hyp* genes are transcriptionally regulated by an FnrN-dependent promoter (Fnr = fumarate-nitrate reduction regulator), which is switched on when the cellular levels of oxygen are low (i.e., under microaerobic conditions). Thus, the *hyp* genes are typically expressed in bacteroids.

Modifying the chromosomal DNA of *R. leguminosarum* and exchanging the *hup* promoter for an FnrN-dependent promoter (Fig. 4.21) enabled researchers to create a derivative of the wild-type bacterium with an increased level of hydrogenase

Table 4.7 Plant growth and nitrogen assimilation after the introduction of the complete panel of *R. leguminosarum hup* genes into a Hup⁻ strain of *R. leguminosarum*

Bacterial phenotype	Relative plant dry weight	Relative nitrogen amount	Relative leaf area	Relative nitrogen concentration
Hup⁻	1.00	1.00	1.00	1.00
Hup⁺	1.35	1.52	1.53	1.15

Adapted from Brewin and Johnston, U.S. patent 4,567,146, January 1986. The data have been normalized relative to the Hup⁻ parental strain

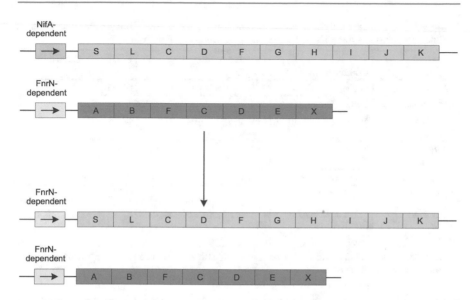

Fig. 4.21 Replacement of the *R. leguminosarum* NifA-dependent *hup* promoter with the *R. leguminosarum* FnrN-dependent *hyp* promoter so that the *hup(SLCDEFGHIJK)* genes are controlled by the *hyp* promoter. The *R. leguminosarum hyp* genes [i.e., *hyp(ABFCDEX)*] are already under the transcriptional control of the FnrN-dependent *hyp* promoter. The arrows indicate the direction of transcription

activity. The engineered *R. leguminosarum* strain displays a twofold increase in hydrogenase activity compared to the wild-type and a sixfold decrease in the amount of hydrogen gas that is produced as a by-product of nitrogen fixation. In the transformed strain, the amount of hydrogen evolved from nitrogen-fixing nodules was extremely low, indicating that virtually all of the hydrogen produced by nitrogenase was recycled. This is expected to make this strain of *R. leguminosarum* much more effective at promoting plant growth and increasing plant nitrogen content. When the effectiveness of the engineered strain of *R. leguminosarum* bv. *viciae* was tested with a variety of different legume hosts, all of the legumes tested, with the notable exception of lentils, displayed an increased level of hydrogen-uptake hydrogenase activity in bacteroids and as a consequence a decreased level of hydrogen evolution (Table 4.8). At this point, it is not clear why different legume hosts respond so differently to the presence of the introduced *R. leguminosarum*.

It has been calculated that in the field, every hectare of a legume crop in association with a nitrogen fixing Hup⁻ rhizobial strain might produce around 5000 L of hydrogen gas per day during the peak growing season. This is the equivalent of a loss of approximately 5% of the crop's net photosynthetic gain. However, in reality only a very small fraction of the anticipated hydrogen gas escapes from the soil surface. This is likely a consequence of the presence of a large number of hydrogen gas oxidizing bacteria being present within a few centimeters of the root nodules. Moreover, the hydrogen gas oxidizing bacteria

Table 4.8 Hydrogenase activity in bacteroids and hydrogen evolution from nodules with different legume host plants in symbiosis with *Rhizobium leguminosarum* bv. *viciae*

Legume	Common name	Relative hydrogenase activity	Relative hydrogen evolution
Pisum sativum	Pea	140	0.095
Lens culinaris	Lentil	1	1
Lathyrus sativus	White pea	80	0.02
Lathyrus odoratus	Sweet pea	45	0.615
Vicia sativa	Common vetch	87	0.095
Vicia villosa	Hairy vetch	382	0.045
Vicia ervilia	Bitter vetch	92	0.002
Vicia monanthos	Single-flowered vetch	473	0.024

Both hydrogenase activity and hydrogen evolution values have been normalized to the value reported for *Lens culinaris*

have been found to promote plant root growth although the mechanism of this growth promotion is currently not known. Nevertheless, one might speculate that the hydrogen gas oxidizing bacteria might provide some H^+ and ATP to the bacteria within the nodule (Fig. 4.22). This could decrease the energy stress within the nodule that occurs as a consequence of depleting ATP levels because of the use of large amounts of ATP in nitrogen fixation. Decreased energy stress might decrease the stress ethylene that is formed thereby slowing the rate of nodule senescence.

In addition to the more well-documented role of ethylene in promoting nodule senescence, it has been suggested that some gibberellins may have a negative effect on nodule senescence. In pea nodules senescence begins at around 4 weeks after inoculation and is most active at about 6 weeks after inoculation. At the same time (4–6 weeks after inoculation), the transcription of gibberellin biosynthesis genes (and the actual amount of gibberellin) was observed to decrease in the inoculated pea plants. It has been suggested that gibberellin might enhance α-amylase production and thereby facilitate the hydrolysis of starch located in the nodule which in turn supports the energy requirements of the nitrogen fixing rhizobia (Fig. 4.23). Thus, in this model, 6-week-old nodules with lower levels of gibberellin would have less energy for nitrogen fixation and nodule maintenance.

4.2   Sequestering Iron

Living organisms, animals, plants, and microorganisms, all require iron as a component of proteins involved in important processes such as respiration, photosynthesis, and nitrogen fixation. Despite the abundance of iron on the earth's surface, soil organisms such as plants and microbes cannot readily assimilate enough iron to support their growth because the iron in soil is largely present as insoluble, ferric

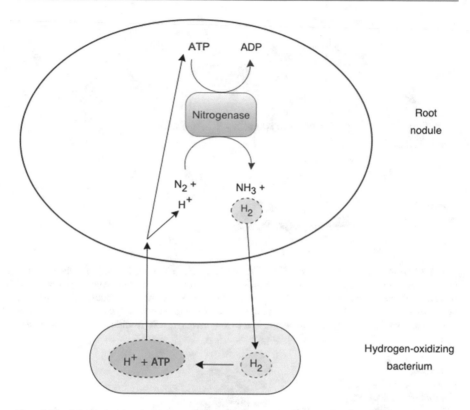

Fig. 4.22 Schematic representation of how hydrogen-oxidizing bacteria might promote plant growth. The bacterium takes up the hydrogen gas that is produced with a nodule as a by-product of nitrogen fixation. The hydrogen gas is converted into H^+ and ATP (as well as other compounds which are not shown). The H^+ and ATP are taken up by the nodule and used to fuel additional nitrogen fixation. Moreover, energy within the nodule is not depleted so that ethylene-catalyzed nodule senescence is delayed

(Fe^{+3}) hydroxides, which is only sparingly soluble and cannot be readily transported into cells. To solve this problem, bacteria, fungi and some plants secrete low molecular mass (~400–1000 Da), specialized iron-binding molecules called siderophores into the soil to scavenge iron. Siderophores bind to Fe^{+3} with an exceptionally high affinity (i.e., $Kd = 10^{-20}$ to 10^{-50}). Once bound, the now soluble iron–siderophore complex is taken up by specific receptors on the exposed surfaces of either microorganisms or plants (Fig. 4.24); following either reduction to the ferrous state (Fe^{+2}) or lowering the external pH, the iron is released from the siderophore.

The promotion of plant growth by siderophore-producing bacteria occurs by the bacteria either directly supplying iron for plant utilization and/or by removing iron from the environment of nearby phytopathogens thereby reducing the competitiveness of those phytopathogens.

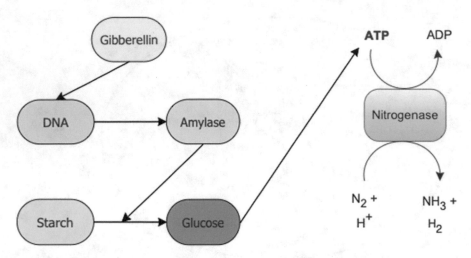

Fig. 4.23 A model of how gibberellin decreases nodule senescence. High levels of gibberellin turn on the transcription of genes encoding α-amylase a starch degrading enzyme. The α-amylase digestion of the starch results in a large amount of glucose whose metabolism facilitates the synthesis of ATP. The ATP drives both nitrogen fixation and other (energy-requiring) processes in nodules. Low levels of gibberellin would lead to low levels of ATP resulting in energy stress and subsequent ethylene-catalyzed nodule senescence

Fig. 4.24 Schematic representation of a bacterial siderophore molecule (star shape) sequestering iron from the environment. The siderophore–iron complex may be taken up either by the siderophore-synthesizing bacteria or by plants. In this depiction, the red bacteria are bound to the plant root surface

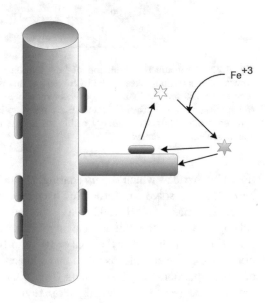

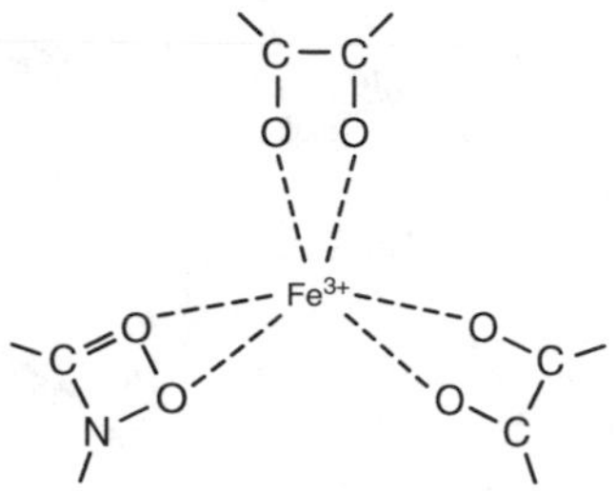

Fig. 4.25 A six coordinate iron–siderophore complex. Three bidentate functional groups on a siderophore molecule bind to the ferric iron. © 2010 American Society for Microbiology. Used with permission. No further reproduction or distribution is permitted without the prior written permission of American Society for Microbiology

| Hydroxamate Group | Catecholate Group | Carboxylate Group | Ethylenediamine Group |

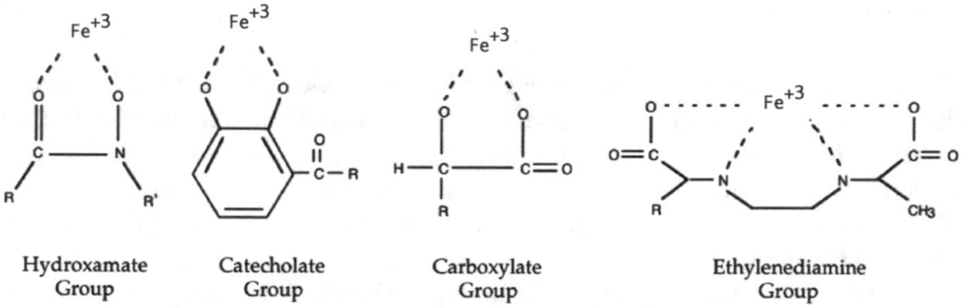

Fig. 4.26 Some of the major iron-binding groups that may be present in a microbial siderophore. © 2010 American Society for Microbiology. Used with permission. No further reproduction or distribution is permitted without the prior written permission of American Society for Microbiology

4.2.1 Siderophore Structure

Siderophores are low molecular weight molecules, usually less than 1 kDa, with three functional, or iron-binding, groups connected via a flexible backbone (Fig. 4.25). Each functional group has two oxygen atoms, or less commonly, nitrogen, that bind to iron. The functional groups are bidentate, and trivalent ferric iron can accommodate three of these groups to form a six-coordinate complex. The functional groups on microbial siderophores are typically hydroxamates or catecholates; however, other functional groups including carboxylate moieties such as citrate, and ethylenediamine are also commonly used (Fig. 4.26). Different combinations of these functional groups may be present on a single siderophore molecule. In general, hydroxamate- and carboxylate-type siderophores are more common in fungi, while catecholates, which bind iron more tightly than hydroxamates or carboxyl, are common in bacterial siderophores. Plant siderophores, which bind iron much less tightly than microbial siderophores, are often linear hydroxy- and amino-substituted iminocarboxylic acids, such as

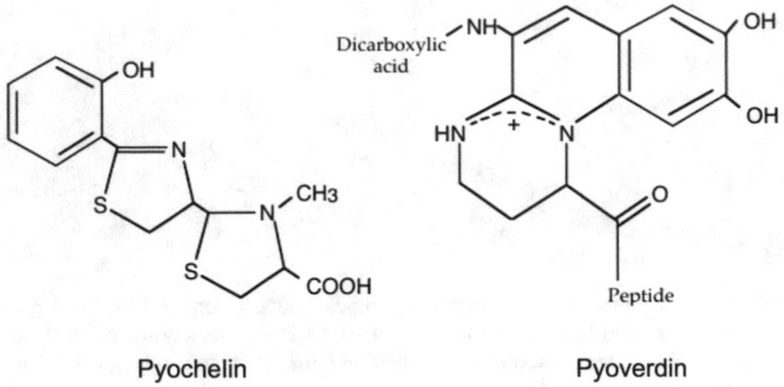

Fig. 4.27 The chemical structures of two well characterized classes of *Pseudomonas* spp. siderophores

mugineic acid and avenic acid. Although many molecules with negatively charged donor groups are able to bind to iron, they do so with a much lower affinity than bacterial siderophores. Other trivalent metal ions, such as aluminum, can also bind bacterial siderophores, albeit with a lower affinity than iron. Most environmental metals however are divalent and less electronegative and therefore do not bind tightly to bacterial siderophores.

Enormous structural diversity exists among the hundreds of known siderophores, even among the multiple siderophores synthesized by a single organism. Thus, for example various *Pseudomonas* spp. typically produce two major types of siderophores, pyochelin and pyoverdin (Fig. 4.27). Pyochelins are phenolate siderophores derived from salicylic acid and cysteine; two pyochelin molecules bind one molecule of ferric iron with a relatively low efficiency. On the other hand, pyoverdins have a much greater affinity for iron; they are water-soluble pigments that fluoresce yellow-green under ultraviolet light giving the fluorescent pseudomonads their characteristic appearance.

4.2.2   Siderophore Biosynthesis Genes

Given the structural diversity that exists between siderophores from different bacteria and the possibility of several genes being involved in the synthesis of a single siderophore, the isolation of siderophore biosynthesis genes is not necessarily simple or straightforward. One reliable approach to the isolation of siderophore biosynthesis genes utilizes the strategy of cloning by complementation (discussed earlier). Thus, it is necessary to first isolate a series of mutants in siderophore biosynthesis. In one instance, the PGPB *P. putida* WCS358 was first mutagenized and 28 siderophore-deficient mutants were selected by (1) lack of fluorescence under ultraviolet light and (2) the inability to grow in the presence of bipyridyl, a molecule that sequesters most

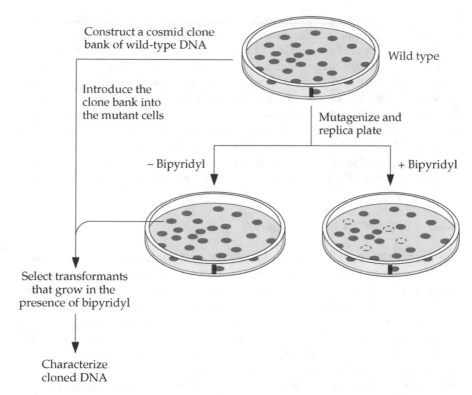

Fig. 4.28 Scheme for the isolation of genes involved in siderophore biosynthesis. Mutants unable to grow on bipyridyl (because they do not produce siderophores and therefore cannot take up the trace amounts of iron in the growth medium in completion with the bipyridyl) were complemented using a clone bank of wild-type DNA (from a siderophore-producing bacterium) in the broad-host-range cosmid pLAFR1. Transformants that can grow in the presence of bipyridyl do so because they can complement the mutation in one of the siderophore biosynthesis genes. © 2010 American Society for Microbiology. Used with permission. No further reproduction or distribution is permitted without the prior written permission of American Society for Microbiology

of the iron in the growth medium. When most of the iron in the vicinity of the target bacterium is unavailable, only a cell that produces siderophores can grow.

A clone bank of *P. putida* WCS358 DNA was then constructed in the broad-host-range cosmid vector pLAFR1 and was introduced by conjugation into each of the 28 siderophore-deficient mutants (Fig. 4.28). All of the resultant transformants were tested by mutant complementation for restoration of fluorescence and/or the ability to grow in the presence of bipyridyl (a compound that binds to traces of iron in the growth medium and makes it unavailable for bacterial growth). Thirteen unique complementing cosmid clones, with an average insert size of 26 kb, were identified. Following detailed analyses, these clones were found to represent at least five separate gene clusters. When one of these gene clusters was studied in detail, it was found to have a length of 33.5 kb and to contain five transcriptional units with at least seven separate genes. Additional siderophore biosynthesis genes were isolated using the technique of walking down the chromosome described earlier (see Sect. 4.1.4). Thus, like nitrogen fixation and nodulation, siderophore biosynthesis is a complex process involving a number of different genes. Therefore, genetically engineering bacteria to produce modified siderophores is not a simple matter. However, it may be possible to extend the range of iron–siderophore complexes that one bacterial strain can utilize so that a genetically altered plant growth-promoting biocontrol bacterial strain could take up and use siderophores synthesized by other soil microorganisms, thereby giving it a competitive advantage in the soil environment. In fact, this was achieved when the genes for iron–siderophore receptors from one PGPB were isolated and then introduced into other PGPB strains.

4.2.3 Regulation of Iron Uptake

The bound iron must also be taken up and released into bacterial or plant cells to become available for their metabolism. Because siderophores are typically hydrophilic, they therefore rely on membrane-bound proteins for transportation across cell membranes. Once inside the cell, iron is released from the siderophore–iron complex to become available for use in metabolic processes. For this to happen, either the siderophore is enzymatically cleaved or the ferric iron is reduced to the ferrous state. To this end, researchers have identified (1) some esterases that can cleave various siderophores (and release iron as a consequence of reduced affinity) and (2) some reductases that can reduce bound ferric iron (which is then released from the siderophore–iron complex).

Transcription of most of the genes whose products are involved in siderophore synthesis and iron transport is activated under conditions where the cellular level of iron is low (i.e., iron limitation). In pseudomonads, the cellular concentration of iron is sensed by "Fur," an iron-binding protein. When cellular levels of iron are sufficient, this cytoplasmic protein binds ferrous iron and represses the transcription of iron-regulated genes. A Fur-binding consensus sequence, called the Fur-box, has been found within the promoter regions of a number of *Pseudomonas* (and other PGPB) genes (consistent with the modulation of these genes by iron). In addition, siderophore biosynthesis and ferric-siderophore receptor genes are regulated and activated when iron is limited.

4.2.4 Siderophores in the Rhizosphere

In pseudomonads, as many as 15 enzymes are involved in the synthesis of some pyoverdin siderophores, and a number of additional proteins are required for the transport of ferric–siderophore complexes and in the regulation of siderophore and receptor expression. The simplest explanation for why microbes invest such a large amount of their resources for the synthesis and utilization of siderophores is that these resources are absolutely needed because iron is essential for bacterial survival.

In addition, the ability to obtain iron efficiently in an iron-limited environment and at the expense of other microbes provides a means for a bacterium to effectively compete for limited resources in the rhizosphere.

While the synthesis of siderophores enables microbes to scavenge low levels of environmental iron, several factors contribute to the success of those siderophores in binding the iron. The amount of siderophore produced is an important factor. Because the binding relationship between siderophore and iron is stoichiometric, the more siderophore molecules available, the more iron that can be bound. In addition, the greater the affinity the siderophore has for iron and the faster the rate of association of the iron and siderophore, the more success the bacterium will have in obtaining iron from the environment. However, in order to bind iron, siderophores must lose protons. Therefore, to form a stable complex with iron is, in part, a function of its tendency to lose protons that in turn is influenced by the rhizosphere pH. The association of a siderophore and a molecule of iron is thus reduced in acidic soils; however, ferric iron is reduced to a more soluble ferrous state in acidic conditions and therefore siderophore-mediated iron chelation is not as critical for iron acquisition under these conditions. The more protons that a siderophore has to lose, the more pH dependent it is.

In the rhizosphere, the capacity to produce large amounts of high affinity siderophores may not to be as important for bacterial competition and root colonization as the ability to utilize a variety of different ferric-siderophores. Thus, the expression of relatively nonspecific receptors that can bind a range of ferric-siderophores, or the production of several siderophore receptors each with different binding specificities gives a bacterium a competitive edge on the plant surface.

The best bacterial competitors therefore are those that can utilize a broad range of siderophores and at the same time produce siderophores that few other microbes can use. Some pseudomonads produce outer membrane receptors that recognize a variety of ferric-siderophores and also secrete siderophores that most other bacteria are unable to bind and thus are particularly effective biocontrol agents. By the use of genetic engineering to extend the spectrum of siderophores that a PGPB can recognize, it may be possible to increase the capacity of these bacteria to facilitate plant growth.

Siderophores may also help to alleviate the stress imposed on plants by high soil levels of heavy metals. In one report, a strain of *Kluyvera ascorbata*, a siderophore-producing PGPB, was able to protect canola plants from heavy metal (nickel, lead, and zinc) toxicity. Moreover, a naturally occurring siderophore-overproducing mutant of this bacterium provided an even greater level of protection, with three different plants (canola, Indian mustard, and tomato) cultivated in nickel-contaminated soil. In this case, the siderophores of *K. ascorbata* are directly supplying iron to the contaminated plants thereby relieving them of iron deficiency, a common symptom of heavy metal toxicity.

4.3   Solubilizing Phosphorus

Plant growth requires a significant amount of phosphorus, a key component in both cell membranes and nucleic acids and approximately 0.2% of a plant's dry weight. Fortunately, most soils contain quite a lot of phosphorus (~400–1200 mg/kg of soil). On the other hand, most of the phosphorus in soil is insoluble and is therefore not readily available to support plant growth. Thus, the amount of soluble phosphorus in many soils is only ~1 mg/kg of soil which is suboptimal for plant growth. Thus, farmers commonly add soluble inorganic phosphorus as a chemical fertilizer to increase the yield of their crops. However, an unfortunate consequence of applying soluble inorganic phosphorus as a chemical fertilizer is that a considerable fraction of the amount that is applied by farmers as a fertilizer is immobilized by minerals in the soil soon after it is applied. In practice, this means that much of the chemical phosphorus fertilizer that is added to soil soon becomes unavailable to plants and is therefore wasted and ultimately contributes to eutrophication (i.e., the run-off of nutrients from fields into bodies of water where these nutrients are likely to promote unwanted algal growth).

Fortunately, an alternative to the use of soluble inorganic phosphorus is possible. That is, a large number of soil microorganisms, including bacteria, fungi, and algae, are able to mobilize phosphorus in the soil, that is otherwise unavailable, and provide it to plants. Soil bacteria that have been reported to solubilize phosphorus include but are not limited to organisms from the following genera: *Pseudomonas, Bacillus, Agrobacterium, Arthrobacter, Azotobacter, Burkholderia, Enterobacter, Erwinia, Duganella, Pseudoduganella, Variovorax, Kushneria, Paenibacillus, Pantoea, Ralstonia, Rhizobium, Rhodococcus, Serratia, Bradyrhizobium Salmonella, Sinomonas,* and *Thiobacillus.* Despite the effectiveness of various soil bacteria in providing phosphorus to plants, sometimes the best option for improving plant phosphorus nutrition is to provide both a low level of a phosphorus fertilizer together with soil bacteria that have been shown to be effective phosphorus solubilizers.

The insoluble phosphorus found in the soil includes both inorganic and organic forms. The organic phosphorus in soil (typically ~30–50% of the total) is mostly in the form of inositol hexaphosphate (phytate) that is the principal storage form of phosphorus in plants (Fig. 4.29). Phytate is not digestible by humans or nonruminant

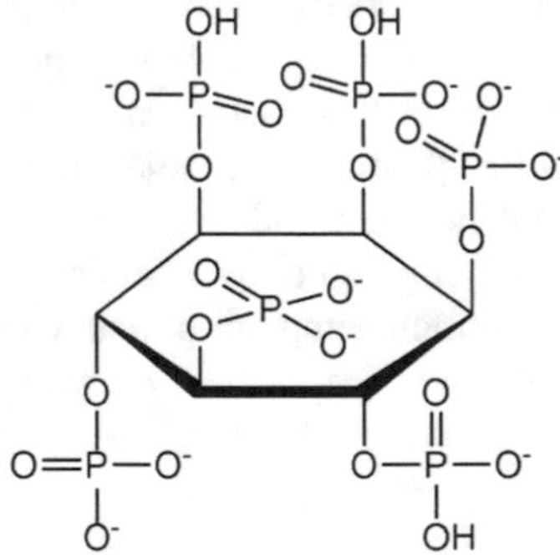

Fig. 4.29   The chemical structure of phytate (phytic acid)

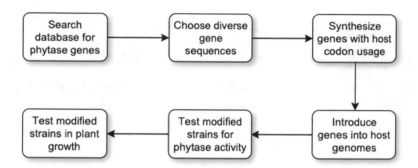

Fig. 4.30 Overview of a scheme that was used to introduce phytase activity into root binding PGPB

animals. Moreover, phytate is not bioavailable to plants as plant roots generally produce only very low levels of phytases, the enzymes that break down phytate. Nevertheless, a number of different soil microorganisms (both bacteria and fungi) contain enzymes that allow them to readily degrade phytate. In addition to phytate, some of the other forms of organic phosphate found in the soil include phosphomonoesters, phosphodiesters, phosphotriesters, and phospholipids that can be broken down by various phosphatases. Given that the pH of most soils ranges from acidic to neutral, it is likely that acid phosphatases are the enzymes that play the major role in the release of phosphorus from these compounds.

Recently, a group of researchers used a combinatorial synthetic biology-based approach to generate a number of PGPB that could efficiently hydrolyze phytate (Fig. 4.30). To start, these researchers chose as potential hosts for foreign phytase genes, one strain of *Pseudomonas simiae*, one of *P. putida* and one of *Ralstonia* sp. that were known to be stably associated with plant roots and had all been previously engineered to contain a "launching pad" DNA that facilitates the integration of foreign DNA into their genome. In addition, the three strains had previously been shown to produce and release organic acids that were responsible for solubilizing inorganic forms of phosphorus. Next, the database of sequenced bacterial genomes was searched for the presence of histidine acid phytases and cysteine phosphatases with the scientists selecting 82 genes that contained the maximum amount of sequence diversity. These DNA sequences were redesigned to reflect the codon usage and regulatory sequences used by the bacterial hosts (without altering the amino acid sequences of these enzymes). All of the redesigned genes were then transferred into the genomes of the three bacterial hosts yielding nearly 200 engineered strains that were then tested for the ability to solubilize phytate in liquid medium. Of the engineered strains, 41 strains release high levels of phosphate. When these 41 strains were tested to see if they could improve plant growth, 12 strains significantly improved the growth of *Arabidopsis* plants when the plants were grown with phytate as the sole source of phosphorus. It now remains to be seen how these 12 engineered strains behave in field trials. Assuming that at least some of them behave as expected, there is some hope that various regulatory agencies will

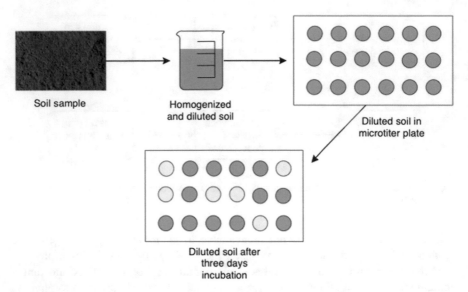

Soil sample

Homogenized
and diluted soil

Diluted soil in
microtiter plate

Diluted soil after
three days
incubation

Fig. 4.31 A rapid and simple procedure for selecting new inorganic phosphate solubilizing bacteria. Each soil sample is diluted in sterile water and then homogenized. The homogenized soil suspension is diluted and dispensed into wells of a microtiter plate. Following several days of incubation, the dye in the wells that contain bacteria that produce and secrete organic acids (and therefore are able to solubilize inorganic phosphate) turn yellow. Bacteria from the most diluted wells that turned yellow are cultured and subsequently characterized

permit the use of these engineered PGPB strains in the environment since their phytase genes are stably integrated into their genomes and cannot easily be lost or transferred to other bacteria in the soil.

The inorganic forms of phosphorus include minerals such as apatite. Apatite is really a group of phosphate minerals that includes hydroxyapatite, fluorapatite, and chloroapatite. Hydroxyapatite is also the major component of tooth enamel and bone mineral. Apatite is often ground into a powder and used as a fertilizer as a source of phosphorus. Plants and bacteria typically solubilize inorganic phosphorus through the production and secretion of small organic acids such as gluconic, citric, lactic, 2-ketogluconic, oxalic, tartaric, malonic, fumaric, glyoxalic, α-ketobutyric, malic, isobutyric, isovaleric, itaconic, propionic, succinic, gluconic, aspartic, maleic, glutamic, glycolic, and acetic acid. Not surprisingly, the amount and type of organic acid varies with the strain of bacterium. These organic acids act as chelating agents that solubilize minerals that contain phosphorus, and in the process of solubilizing inorganic phosphorus, they sometimes solubilize some other nutrients at the same time.

A recently described procedure makes it relatively easy to isolate new (and possibly unique) inorganic phosphate solubilizing bacteria (Fig. 4.31). In this procedure, small soil samples (typically 1 g) are added to 100mL of sterilized water and homogenized. The soil suspension is serially diluted (10- to 10^5-fold) and 1 μL aliquots of each dilution are added to individual wells of a 96-well microtiter plate

Fig. 4.32 Petri plate showing a zone of clearance around an inorganic phosphorus-solubilizing bacterium

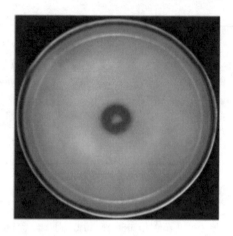

with each well also containing growth medium and bromocresol purple (a pH indicator). Following incubation of the microtiter plates at 30 °C for a minimum of 3 days, wells that change color from purple to yellow are selected since the yellow color is an indication that the bacterial growth that had occurred in those wells was accompanied by the production and secretion of one or more organic acids. Those wells that both turn yellow and where the soil suspension is diluted to the greatest extent are selected for further study since they are less likely to contain several different bacterial strains.

One simple way of testing whether a PGPB strain can solubilize inorganic phosphorus is to plate the bacterium on solid medium that contains inorganic phosphorus and, following growth of the bacterium, look for a halo or zone of clearance around the bacterial colony (Fig. 4.32). Such a zone of clearance indicates that the bacterium is synthesizing and secreting organic acids that can solubilize the inorganic phosphorus that is present in the growth medium. This simple plate assay may also serve to facilitate the isolation of genes involved in organic acid production.

It is important to note that sometimes the same bacterial strain can naturally solubilize both organic and inorganic forms of phosphorus. This also means that it is possible to simply engineer more efficient strains of phosphorus-solubilizing bacteria. For example, in one experiment, a phytase gene (actually a cDNA) was isolated from the fungus *Aspergillus fumigatus,* placed under the transcriptional control of a bacterial promoter and, using a transposon, inserted into the chromosomal DNA of a strain of *Bacillus mucilaginosus* that can normally dissolve inorganic phosphorus. When this genetically transformed bacterium was tested in both greenhouse and field experiments, it was observed to be more efficient at providing phosphorus to tobacco plants than the nontransformed strain of this bacterium (Table 4.9). While, it is currently unlikely that any genetically engineered strain will be approved for widespread environmental dissemination, this is an important step in trying to overcome the inconsistent and disappointing results observed so far in the employment of naturally occurring phosphorus solubilizing bacteria. Finally, it is important to bear

Table 4.9 Growth of tobacco plants for 90 days in sterilized soil pots in the greenhouse with different treatments

Characteristic	Treatment		
	No added bacteria	Wild-type *B. mucilaginosus*	Transformed *B. mucilaginosus*
Plant height (cm)	17.4	21.4	24.7
Plant dry weight (mg)	1297	1685	1870
Leaf P content (mg/g)	710	732	800

Transformed *Bacillus mucilaginosus* included a phytase gene from the fungus *Aspergillus fumigatus*

in mind that in nature, phosphorus solubilization is not preformed exclusively by naturally occurring phosphorus-solubilizing bacteria but rather includes a variety of soil fungi (such as mycorrhizae) as well.

Finally, it has been observed that there is a significant amount of cross talk between iron and phosphorus deficiency in some plants. That is, iron deficiency can induce the expression of phosphorus acquisition genes and phosphatase activity. Similarly, phosphorus deficiency can induce the expression of iron acquisition genes and ferric reductase activity. All indications are that this signaling is mediating by ACC/ethylene.

Questions
1. What are cyanobacteria and how might they be useful as a biological fertilizer?
2. What are heterocysts?
3. How can some strains of cyanobacteria be genetically improved so they fix more nitrogen?
4. Why is the process of nitrogen fixation energy intensive?
5. How is the nitrogen fixation apparatus with a bacteroid protected against the inhibitory effects of oxygen?
6. How is bacteroid nitrogen fixation coordinated with plant photosynthesis?
7. How can nitrogenase activity be assayed biochemically?
8. How can the level of glutamine in a cell be used to estimate the extent of nitrogen fixation?
9. Why is it unlikely that plants can be genetically engineered to fix nitrogen?
10. What is cloning by genetic complementation and how does it work?
11. What is walking down the chromosome and how does it work?
12. How do NifA and NifL regulate nitrogenase expression?
13. How can NifA be overproduced? What is the effect on plants treated with a rhizobial strain in which NifA is overproduced?
14. How might glycogen synthase affect nitrogen fixation?
15. How would the presence of an active *Vitreoscilla* sp. hemoglobin gene affect nitrogen fixation?
16. How can the *nifH* promoter be manipulated to increase nitrogen fixation?

17. How might the inability of a *Rhizobium* strain to produce the poly-β-hydroxybutyrate affect nitrogen fixation?
18. How can nitrogenase be used to provide a sink for atmospheric carbon dioxide?
19. How are *nod* genes turned on?
20. How can a Rhizobium strain be selectively attracted to a host legume?
21. What are some of the major events that occur during nodulation?
22. How does modifying ethylene levels affect nodulation efficiency?
23. How might the presence of hydrogen-oxidizing bacteria affect rhizobial nitrogen fixation?
24. How might PGPB that produce gibberellin affect rhizobial nitrogen fixation?
25. How does the presence of the enzyme hydrogenase affect nitrogen fixation?
26. How can the level of hydrogenase in a bacterium be increased?
27. What are bacterial siderophores and how do they affect plant growth?
28. How would you isolate siderophore biosynthesis genes from a newly isolated PGPB?
29. How do PGPB facilitate plant uptake of inorganic phosphate from the soil?
30. How would you easily isolate new inorganic phosphate solubilizing bacteria from soil samples?
31. How do PGPB facilitate plant uptake of organic phosphate from the soil?

Bibliography

Adams MWW, Mortenson LE, Chen JS (1981) Hydrogenase. Biochim Biophys Acta 594:105–176

Ahmad E, Khan MS, Zaidi A (2013) ACC deaminase producing *Pseudomonas putida* strain PSE3 and *Rhizobium leguminosarum* strain RP2 in synergism improves growth, nodulation and yield of pea grown in alluvial soils. Symbiosis 61:93–104

Albrecht SL, Maier RJ, Hanus FJ, Russell SA, Emerich DW, Evans HJ (1979) Hydrogenase in *Rhizobium japonicum* increases nitrogen fixation by nodulated soybeans. Science 203:1255–1257

Alori ET, Glick BR, Babalola OO (2017) Microbial phosphorus solubilization and its potential for use in sustainable agriculture. Front Microbiol 8:971

Brito B, Palacios JM, Imperial J, Ruiz-Argüeso T (2002) Engineering the *Rhizobium leguminosarum* bv. viciae hydrogenase system for expression in free-living microaerobic cells and increased hydrogenase activity. Appl Environ Microbiol 68:2461–2467

Brito B, Toffanin A, Prieto R-I, Imperial J, Ruiz-Argüeso T, Palacios JM (2008) Host-dependent expression of *Rhizobium leguminosarum* bv. viciae hydrogenase is controlled at transcriptional and post-transcriptional levels in legume nodules. Mol Plant Microbe Interact 21:597–604

Cantrell MA, Haugland RA, Evans HJ (1983) Construction of a *Rhizobium japonicum* gene bank and use in isolation of a hydrogen uptake gene. Proc Natl Acad Sci USA 80:181–185

Chaurasia AK, Apte SK (2011) Improved eco-friendly recombinant *Anabaena* sp. strain PCC7120 with enhanced nitrogen biofertilizer potential. Appl Environ Microbiol 77:395–399

Cheng Q (2008) Perspectives in biological nitrogen fixation. J Integr Plant Biol 50:784–796

da Silva JR, Menéndez E, Eliziário F, Mateos PF, Alexandre A, Oliveira S (2019) Heterologous expression of *nefA* or *nod* genes improves chickpea-*Mesorhizobium* symbiotic performance. Plant Soil 436:607–621

Duan J, Müller KM, Charles TC, Vesely S, Glick BR (2009) 1-Aminocyclopropane-1-carboxylate (ACC) deaminase genes in *Rhizobia* from southern Saskatchewan. Microb Ecol 57:423–436

Egamberdieva D, Wirth S, Jabborova D, Räsänen LA, Liao H (2017) Coordination between *Bradyrhizobium* and *Pseudomonas* relieves salt stress in soybean through altering root system architecture. J Plant Interact 12:100–107

Evans HJ, Harker AR, Papen H, Russell SA, Hanus FJ, Zuber M (1987) Physiology, bio-chemistry, and genetics of the uptake hydrogenase in rhizobia. Annu Rev Microbiol 41:335–361

Fernandez-Valiente E, Quesada A (2004) A shallow water ecosystem: rice fields. The relevance of cyanobacteria in the ecosystem. Limnetica 23:95–108

Fixen KR, Zheng Y, Harris DF, Shaw S, Yang Z-Y, Dean DR, Seefeldt LC, Harwood CS (2016) Light-driven carbon dioxide reduction to methane by nitrogenase in a photosynthetic bacterium. Proc Natl Acad Sci USA 113:10163–10167

Gano-Cohen KA, Stokes PJ, Blanton MA, Wendlandt CE, Hollowell AC, Regus JU, Kim D, Patel S, Pahua VJ, Sachs JL (2016) Nonnodulating *Bradyrhizobium* spp. modulate the benefits of legume-rhizobium mutualism. Appl Environ Microbiol 82:5259–5268

Gresshoff PM, Roth LE, Stacey G, Newton WE (eds) (1990) Nitrogen fixation: achievements and objectives. Chapman & Hall, New York

Harding K, Turk-Kubo KA, Sipler RE, Mills MM, Bronk DA, Zehr JP (2018) Symbiotic unicellular cyanobacteria fix nitrogen in the Arctic ocean. Proc Natl Acad Sci USA 115:13371–13375

Hennecke H (1990) Nitrogen fixation genes involved in the *Bradyrhizobium japonicum*-soybean symbiosis. FEBS Lett 268:422–426

Khan MS, Ahmad E, Zaidi A, Oves M (2013) Functional aspect of phosphate-solubilizing bacteria: importance in crop production. In: Maheshwari DK, Saraf M, Aeron A (eds) Bacteria in agrobiology: crop productivity. Springer, Berlin, pp 237–263

Laranjo M, Alexandre A, Oliveira S (2014) Legume growth-promoting rhizobia: an overview on the *Mesorhizobium* genus. Microbiol Res 169:2–17

Leite J, Fischer D, Rouws LFM, Fernandes-Júnior PI, Hofmann A, Kublik S, Schloter M, Xavier GR, Radl V (2017) Cowpea nodules harbor non-rhizobial bacterial communities that are shaped by soil type rather than plant genotype. Front Plant Sci 7:2064

Lemanceau P, Bauer P, Kraemer S, Briat J-F (2009) Iron dynamics in the rhizosphere as a case study for analyzing interactions between soils, plants and microbes. Plant Soil 321:513–535

Li Y, Li Y, Zhang H, Wang M, Chen S (2019) Diazotrophic *Paenibacillus beijingensis* BJ-18 provides nitrogen for plant and promotes plant growth, nitrogen uptake and metabolism. Front Microbiol 10:1119

Long SR, Buikema WJ, Ausubel FM (1982) Cloning of *Rhizobium meliloti* nodulation genes by direct complementation of Nod⁻ mutants. Nature 298:485–488

Lucena C, Porras R, Garcia MJ, Alcántara E, Pérez-Vicente R, Zamarreño AM, Bacaicoa E, Garcia-Mina JM, Smithe AP, Romera FJ (2019) Ethylene and phloem signals are involved in the regulation of responses to Fe and P deficiencies in roots of strategy I plants. Front Plant Sci 10:1237

Ma W, Guinel FC, Glick BR (2003a) The *Rhizobium leguminosarum* bv. *viciae* ACC deaminase protein promotes the nodulation of pea plants. Appl Environ Microbiol 69:4396–4402

Ma W, Sebestianova S, Sebestian J, Burd GI, Guinel F, Glick BR (2003b) Prevalence of 1-aminocyclopropaqne-1-carboxylate in deaminase in *Rhizobia* spp. Anton. Van Leeuwenhoek 83:285–291

Ma W, Charles TC, Glick BR (2004) Expression of an exogenous 1-aminocyclopropane-1-carbox-ylate deaminase gene in *Sinorhizobium meliloti* increases its ability to nodulate alfalfa. Appl Environ Microbiol 70:5891–5897

Maier RJ, Triplett EW (1996) Toward more productive, efficient, and competitive nitrogen-fixing symbiotic bacteria. Crit Rev Plant Sci 15:191–234

Maimaiti J, Zhang Y, Yang J, Cen Y-P, Layzell DB, Peoples M, Dong Z (2007) Isolation and characterization of hydrogen-oxidizing bacteria induced following exposure to hydrogen gas and their impact on plant growth. Environ Microbiol 9:435–444

Marroquí S, Zorreguieta A, Santamaría C, Temprano F, Soberón M, Megías M, Downie JA (2001) Enhanced symbiotic performance by *Rhizobium tropici* glycogen synthase mutants. J Bacteriol 183:854–864

Marugg JD, van Spanje M, Hoekstra WPM, Schippers B, Weisbeek PJ (1985) Isolation and analysis of genes involved in siderophore biosynthesis in plant-growth-stimulating *Pseudomonas putida* WCS358. J Bacteriol 164:563–570

Marugg JD, Nielander HB, Horrevoets AJG, van Megen I, van Genderen I, Weisbeek PJ (1988) Genetic organization and transcriptional analysis of a major gene cluster involved in siderophore biosynthesis in *Pseudomonas putida* WCS358. J Bacteriol 170:1812–1819

Mylona P, Pawlowski K, Bisseling T (1995) Symbiotic nitrogen fixation. Plant Cell 7:869–885

Nap J-P, Bisseling T (1990) Developmental biology of a plant-prokaryote symbiosis: the legume root nodule. Science 250:948–954

Nascimento FX, Brígido C, Glick BR, Oliveira AL (2012) *Mesorhizobium ciceri* LMS-1 expressing an exogenous 1-aminocyclopropane-1-carboxlyate (ACC) deaminase increases its nodulation abilities and chickpea plant resistance to soil constraints. Lett Appl Microbiol 55:15–21

Nascimento FX, Brígido C, Glick BR, Oliveira AL (2016) The role of rhizobial ACC deaminase in the nodulation process of leguminous plants. Int J Agron 2016:1369472

Nascimento FX, Tavares MJ, Glick BR, Rossi MJ (2018) Improvement of *Cupriavidus taiwanensis* nodulation and plant-growth promoting abilities by the expression of an exogenous ACC deaminase gene. Curr Microbiol 75:961–965

Nascimento FX, Tavares MJ, Franck J, Ali S, Glick BR, Rossi MJ (2019) ACC deaminase plays a major role in *Pseudomonas fluorescens* YsS6 ability to promote the nodulation of Alpha- and Betaproteobacteria rhizobial strains. Arch Microbiol. https://doi.org/10.1007/s00203-019-01649-5

Ortiz-Marquez JC, Do Nasciment M, Zehr JP, Curatti L (2013) Genetic engineering of multispecies microbial cell factories as an alternative for bioenergy production. Trends Biotechnol 31:521–529

Peralta H, Mora Y, Salazar E, Encarnación S, Palacios R, Mora J (2004) Engineering the *nifH* promoter region and abolishing poly-β-hydroxybutyrate accumulation in *Rhizobium etli* enhance nitrogen fixation in symbiosis with *Phaseolus vulgaris*. Appl Environ Microbiol 70:3272–3281

Peters JW, Fisher K, Dean DR (1995) Nitrogenase structure and function: a biochemical-genetic perspective. Annu Rev Microbiol 49:335–366

Ramírez M, Valderrama B, Arrendondo-Peter R, Soberón M, Mora J, Hernández G (1999) *Rhizobium etli* genetically engineered for the heterologous expression of *Vitreoscilla* sp. hemoglobin: effects on free-living and symbiosis. Mol Plant-Microbe Interact 12:1008–1015

Sarsekeyeva F, Zayadan BK, Usserbaeva A, Bedbenov VS, Sinetova MA, Los DA (2015) Cyanofuels: biofuels from cyanobacteria. Reality and perspectives. Photosynth Res 125:329–340

Scavino AF, Pedraza RO (2013) The role of siderophores in plant growth-promoting bacteria. In: Maheshwari DK, Saraf M, Aeron A (eds) Bacteria in agrobiology: crop productivity. Springer, Berlin, pp 265–285

Schiessl K, Lilley JLS, Lee T, Tamvakis I, Kohlen W, Bailey PC, Thomas A, Luptak J, Ramakrishnan K, Carpenter MD, Mysore KS, Wen J, Ahnert S, Grieneisen VA, Oldroyd GED (2019) *NODULE INCEPTION* recruits the lateral root development program for symbiotic nodule organogenesis in *Medicago truncatula*. Curr Biol 29:3657–3668

Seefeldt FC, Hoffman BM, Dean DR (2009) Mechanism of Mo-dependent nitrogenase. Annu Rev Biochem 78:701–722

Serova TA, Tsyganova AV, Tikhonovich IA, Tsyganov VE (2019) Gibberellins inhibit nodule senescence and stimulate nodule meristem bifurcation in pea (*Pisum sativum* L.). Front Plant Sci 10:285

Shulse CN, Chovatia M, Agosto C, Wang G, Hamilton M, Deutsch S, Yoshikuni Y, Blow MJ (2019) Engineered root bacteria release plant-available phosphate from phytate. Appl Environ Microbiol 85:e01210-19

Soyano T, Shimoda Y, Kawaguchi M, Hayashi M (2019) A shared gene drives lateral root development and root nodule symbiosis pathways in *Lotus*. Science 366:1021–1023

Spaink HP, Wijffelman CA, Pees E, Okker RJH, Lugtenberg BJJ (1987) *Rhizobium* nodulation gene *nodD* as a determinant of host specificity. Nature 328:337–340

Sprent JI (1986) Benefits of *Rhizobium* to agriculture. Trends Biotechnol 4:124–129

Stacey G (1995) *Bradyrhizobium japonicum* nodulation genetics. FEMS Microbiol Lett 127:1–9

Tavares MJ, Nascimento FX, Glick BR, Rossi MJ (2018) The expression of an exogenous ACC deaminase by the endophyte *Serratia grimesii* BXF1 promotes the early nodulation and growth of common bean. Lett Appl Micorbiol 66:252–259

Thilakarathna MS, Moroz N, Raizada MN (2017) A biosensor-based leaf punch assay for glutamine correlates to symbiotic nitrogen fixation measurements in legumes to permit rapid screening of rhizobia inoculants under controlled conditions. Front Plant Sci 8:1714

van Rhijn P, Vanderleyden J (1995) The *Rhizobium*-plant symbiosis. Microbiol Rev 59:124–142

Wang D, Yang S, Tang F, Zhu H (2012) Symbiosis specificity in the legume-rhizobial mutualism. Cell Microbiol 14:334–342

Webb BA, Hildreth S, Helm RF, Scharf BE (2014) *Sinorhizobium meliloti* chemreceptor McpU mediates chemotaxis toward plant exudates through direct proline sensing. Appl Environ Micorbiol 80:3404–3415

Yang Z-Y, Moure VR, Dean DR, Seefeldt LC (2012) Carbon dioxide to methane and coupling with acetylene to form propylene catalyzed by remodeled nitrogenase. Proc Natl Acad Sci USA 109:19644–19648

Zafar-ul-Hye M, Ahmad M, Shahzad SM (2013) Synergistic effect of rhizobia and plant growth promoting rhizobacteria on the growth and nodulation of lentil seedlings under axenic conditions. Soil Environ 32:79–86

Zhang Z, Li Y, Pan X, Shao S, Liu W, Wang E-T, Xie Z (2019) *Aeschynomene indica*-nodulating rhizobia lacking Nod factor synthesis genes: diversity and evolution in Shandong Peninsula, China. Appl Environ Microbiol 85:e00782-19

Zheng B-X, Ibrahim M, Zhang D-P, Bi Q-F, Li H-Z, Zhou G-W, Ding K, Peñuelas J, Zhu Y-G, Yanf X-R (2018) Identification and characterization of inorganic-phosphate-solubilizing bacteria from agricultural fields with a rapid isolation method. AMB Exp 8:47

Modulating Phytohormone Levels

5

Abstract

In this chapter, the role of bacterial phytohormones in plant growth and development is addressed and the many mechanisms that plant growth-promoting bacteria use to modulate phytohormone levels are discussed in some detail. Many of the biochemical pathways used by plant growth-promoting bacteria to synthesize and regulate the synthesis of auxin (primarily indoleacetic acid, IAA) are examined. Included in this overview is consideration of the fact that many bacterial strains contain multiple IAA biosynthesis pathways. Key to understanding what bacteria do to facilitate plant growth and development is the functioning of the enzyme ACC deaminase (and its interaction with IAA signal transduction pathways). This enzyme lowers plant ethylene levels and thereby facilitates growth, especially in the presence of various abiotic and biotic stresses. Finally, the chapter includes some discussion of the synthesis and functioning of bacterial cytokinin, gibberellin and volatile organic compounds.

Most, if not all, of the physiological activities of a plant are regulated by one or more plant hormones (phytohormones) including auxin, cytokinin, gibberellin, abscisic acid, ethylene, salicylic acid, jasmonic acid, and brassinosteroids. In addition to plants, many soil bacteria are also capable of synthesizing and/or modulating the level of some of these hormones including auxin, cytokinin, gibberellin, and ethylene (Fig. 5.1). It should be emphasized here that phytohormones are generally not nutrients, but rather are chemicals that are typically present in plants in very low concentrations and promote and/or influence the growth, development, and differentiation of plant cells and tissues. Phytohormones help plants to adjust their growth rates and patterns in response to a wide range of environmental and developmental conditions. In order to modulate different aspects of a plant's environmental and developmental responses, phytohormones turn on (or off) the expression of various plant genes that ultimately help to optimize a plant's growth and development.

© Springer Nature Switzerland AG 2020

B. R. Glick, *Beneficial Plant-Bacterial Interactions*,

https://doi.org/10.1007/978-3-030-44368-9_5

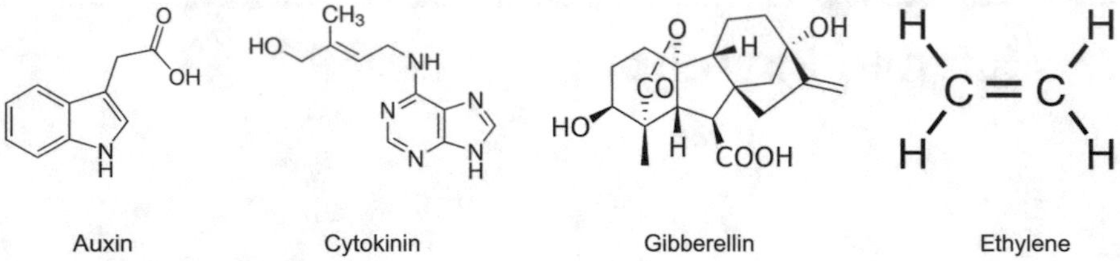

Auxin   Cytokinin   Gibberellin   Ethylene

Fig. 5.1 Structures of phytohormones whose concentration in plants is directly affected by some PGPB

Complicating this picture, many soil bacteria inactivate and consume various phytohormones notwithstanding the fact that the biosynthetic pathways have been studied to a much greater extent than the biodegradative pathways.

It is necessary to bear in mind that despite having a wide range of bioassays as well as precise analytical techniques for measuring various phytohormone levels, when studying systems that contain both plants and PGPB, it is often extremely difficult to separate the contribution to the total amount of phytohormone made by each of the partners. In this regard, one biological approach to this issue is the construction and testing of either plant or bacterial mutants in phytohormone synthesis or degradation.

Much of what we know about the mode of action of specific phytohormones comes from varying amounts of purified hormones to plants at different stages of their growth. However, it is difficult to know precisely how much of an exogenously added hormone has been taken up by the plant. Moreover, when studying the effect of phytohormones produced by PGPB, it is difficult to separate the effects of adding a particular phytohormone when many PGPB simultaneously produce or modulate the levels of several phytohormones. One of the best accepted ways of pinpointing the role of a particular PGPB-produced phytohormone is to develop underproducing and/or overproducing mutants of that PGPB and compare the effects of the mutants to the effects of the wild-type PGPB strain.

5.1 Auxin

The mechanism that is most often used to explain the positive effects of plant growth-promoting bacteria is the production of phytohormones, and the phytohormone that has received most of the attention is auxin. Several naturally occurring auxins are known although indole-3-acetic acid (IAA) is by far the most studied and the most common (Fig. 5.2). In fact, the terms auxin and IAA are often used interchangeably in the scientific literature. Some precursors of IAA, for example, indole-3-acetonitrile, may also have auxin activity. Other endogenous auxins include indole-3-butyric acid, which is synthesized from IAA. Several synthetic auxins with enhanced efficacy or specialized application are commercially available. For example, 1-naphthaleneacetic acid (NAA) and 2,4-dichlorophenoxyacetic acid (2,4-D)

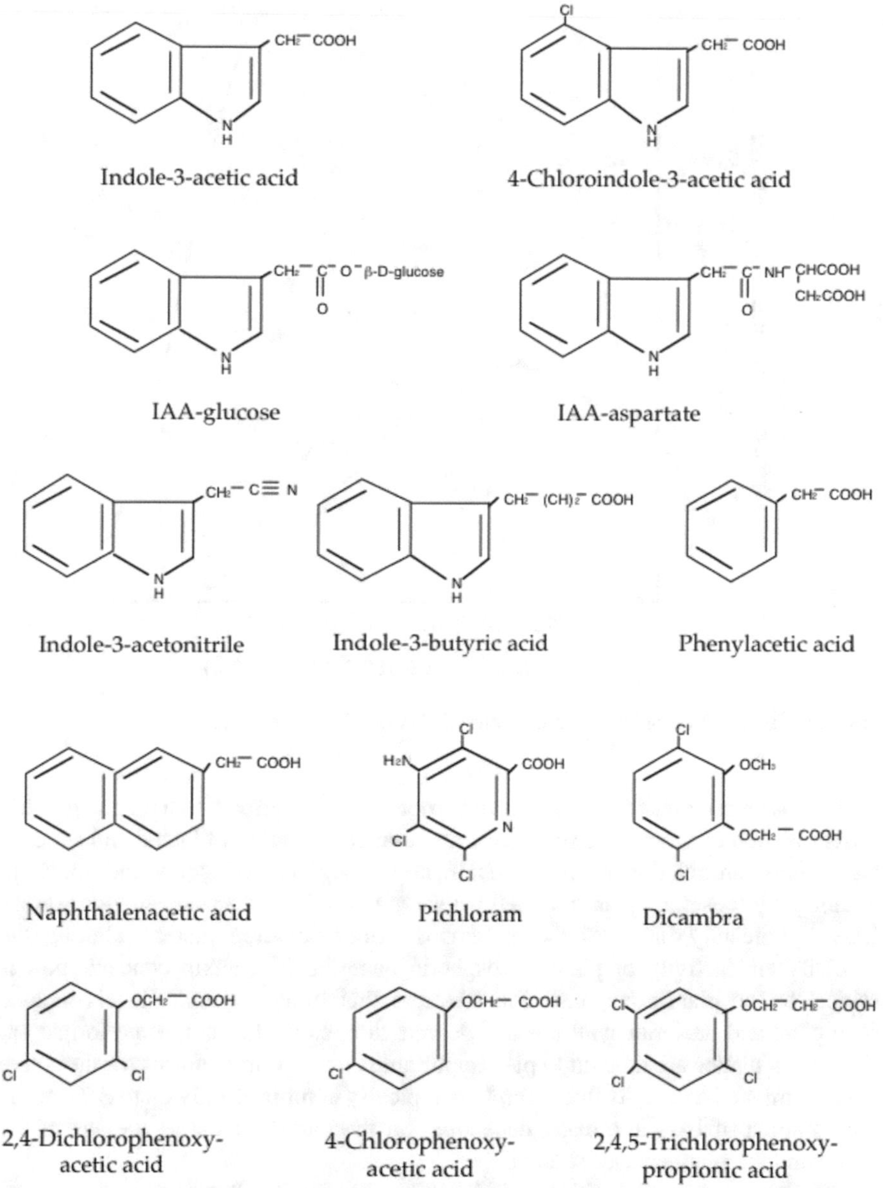

Fig. 5.2 Structures of some natural and synthetic auxins

are often used in tissue culture to stimulate plant organogenesis from callus cells, and in horticulture to promote the rooting of cuttings. Application of the auxins 4-chlorophenoxyacetic acid and 2,4,5-trichlorophenoxypropionic acid in orchards can stimulate fruit set and prevent fruit abscission, respectively.

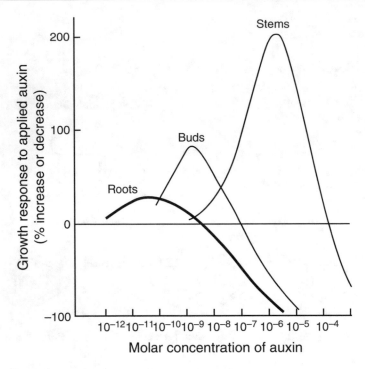

Fig. 5.3 The effect of various concentrations of IAA on plant tissue growth

Auxins are involved in the response of root and shoot growth to light and gravity, differentiation of vascular tissue, apical dominance, initiation of lateral and adventitious roots, stimulation of cell division and elongation of stems and roots. In addition, by loosening plant root cell walls (as part of its effect on cell elongation), IAA significantly increases the amount of root exudation thereby altering the diversity and activity of plant rhizospheric bacteria. The auxin concentration is critical to the plant's response. Plant factors that influence auxin levels include both plant and bacterial synthesis and degradation, as well as conjugate formation. Auxin conjugates are thought to play important roles as storage forms for the active plant hormone IAA. In its free form, IAA typically comprises only up to 25% of the total amount of IAA in a plant, depending on the particular tissue, the age of the tissue and the plant species studied.

Different plants are sensitive to different levels of auxin. In addition, the optimal level of auxin to promote plant growth is ~5 orders of magnitude lower for roots than it is for shoots (Fig. 5.3). The existing level of plant-synthesized IAA is important in determining whether bacterial IAA will stimulate or inhibit growth. Thus, plant IAA levels may be either suboptimal or optimal (i.e., too little or just right). Input of additional IAA that is synthesized by bacteria may change the level of IAA within a plant to either optimal or supraoptimal (just right or too much). Thus, the observation that inoculation of plant cuttings with a PGPB strain that produced IAA stimulated root development in black currant cuttings but inhibited root development in sour cherry cuttings has been interpreted as suggesting that IAA from the bacterium elevated IAA in the black currant cuttings to an optimal level for root growth while sour cherry cutting IAA levels became inhibitory.

A very large percentage of rhizosphere bacteria (often ~85%) have been reported to synthesize IAA. This is probably an indication that IAA biosynthesis is an essential (or at least very important) component of the functioning and metabolism of these bacteria, possibly for both plant growth promotion as well as functions other than plant growth promotion. To ensure that they produce sufficient levels of IAA, many soil bacteria have a high affinity for tryptophan (typically the biochemical precursor of IAA) thereby allowing them to scavenge tryptophan from the environment even when only very low levels are available.

5.1.1 Biosynthetic Pathways

Although IAA was first discovered at the end of the nineteenth century, it is only in recent years, with the complete sequencing of a large number of bacterial genomes, that the numerous pathways of IAA biosynthesis have been elaborated. Interestingly, in many of the bacterial strains that have been examined in detail, there are often multiple pathways within a single organism seemingly encoding the biosynthesis of IAA. Moreover, these pathways sometimes intersect with one another, thereby posing difficulties when trying to characterize them. Of course, such a seemingly redundant multi-route system compensates for the loss of any particular pathway, as IAA production can readily be fluxed through an alternate biosynthetic route.

IAA biosynthetic pathways may be characterized either genetically or biochemically. Genetic characterization of a given pathway involves the development of either IAA-deficient mutants or IAA overproducing strains with the under- or overproduction of a particular biosynthetic enzyme leading to a decrease or increase in IAA production. On the other hand, biochemical approaches aim to isolate and characterize the enzymes that catalyze IAA biosynthesis. This approach includes both substrate feeding assays as well as structure–function studies of the relevant enzymes.

5.1.1.1 The IAM Pathway

IAA biosynthesis via the indole-3-acetamide (IAM) pathway occurs in both pathogenic bacteria and PGPB but is best characterized in pathogens. Notwithstanding the presence of this pathway in a number of pathogenic bacteria, many PGPB also produce IAA via this pathway. In this pathway, IAA is produced in a two-step reaction from a tryptophan precursor (Fig. 5.4). The first enzyme is a tryptophan 2-monooxygenase that converts tryptophan to the indole-3-acetamide intermediate. The second reaction is catalyzed by an IAM-specific hydrolase/amidase that hydrolyzes indole-3-acetamide to the final IAA product. The main genes of the IAM pathway are typically organized in an operon and have been found to be

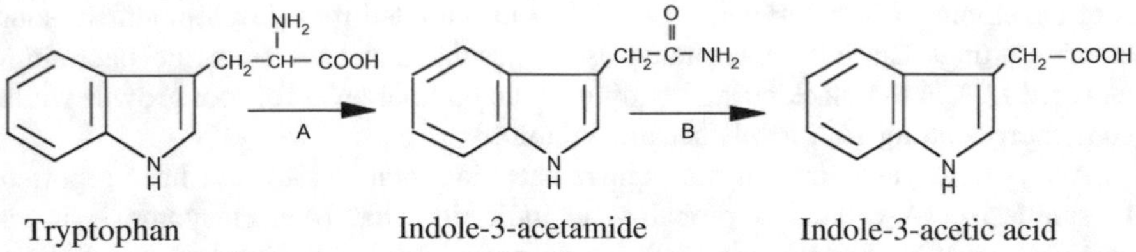

Fig. 5.4 The indoleacetamide (IAM) pathway. Two enzymes are involved in this pathway: (A) tryptophan monooxygenase which converts tryptophan to indole-3-acetamide and indoleacetamide hydrolase which converts indole-3-acetamide to indole-3-acetic acid (IAA)

conserved between the phytopathogens *P. syringae* pv. syringae, *P. syringae* pv. savastanoi, and *A. tumefaciens*.

In *A. tumefaciens*, which infects host plants and causes crown gall formation, the genes are not functional inside the bacteria, but are transferred into a host plant cell where they are inserted into the plant chromosome resulting in very high levels of IAA production, which in turn leads to uncontrollable tumor growth. In contrast to *A. tumefaciens*, in *P. syringae* pv. savastanoi these two genes are only functional inside the bacterium and are not directly transferred into plant cells. In this case, the pathogen-induced plant tumors that are caused by this bacterial pathogen depend on the continuous production of IAA by the infecting bacteria, with IAA overproducing mutants displaying increased virulence and larger galls (tumors) on the host plant. Conversely, mutants of this bacterium that have lost these genes are less virulent than the wild-type strain.

A phylogenetic analysis of the genes of the IAM pathway suggests that homologues of the *iaaM* gene exist in several bacterial genera and these sequences fall into two groups. Group I sequences are known to participate in IAA synthesis, such as those encoding tryptophan 2-monooxygenases in the plant pathogens *A. tumefaciens, P. syringae* pv. savastanoi, *P. agglomerans,* and *D. dadantii.* Homologous *iaaM*-like sequences from *Burkholderia* spp., *Agrobacterium* spp. and a few other *P. syringae* strains also fall into this cluster. The Group II sequences are not well characterized and, for the most part, their involvement in the synthesis of IAA remains to be determined experimentally.

5.1.1.2 The IPA Pathway

This pathway is common among various strains of PGPB, but has also been found to exist in some phytopathogens. Initially, L-tryptophan is deaminated to indole-3-pyruvic acid (IPA) by an aminotransferase. Subsequently, a decarboxylase converts indole-3-pyruvic acid into indole-3-acetylaldehyde (IAAld), which is then oxidized to IAA by aldehyde dehydrogenase, mutase or oxidase enzymes (Fig. 5.5). There is an alternative pathway in which tryptophan is directly converted into indole-3-acetylaldehyde by a tryptophan side chain monooxygenase enzyme (Fig. 5.5). This is known as the tryptophan side chain oxidase (TSO) pathway. The genes for the key enzymes of the IPA pathway, tryptophan aminotransferase and IPA decarboxylase, have been found encoded in numerous bacterial genomes. However, there

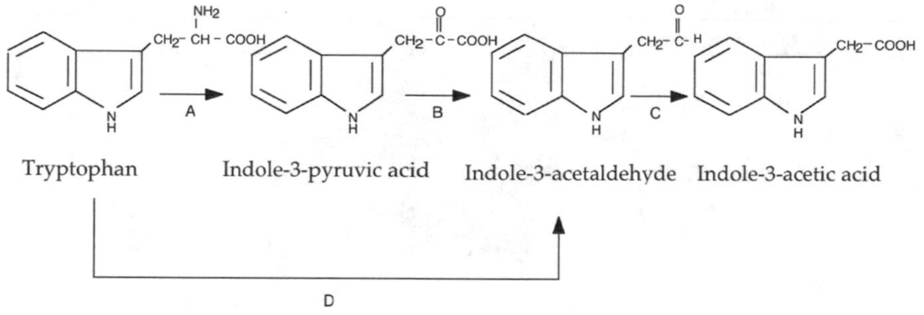

Fig. 5.5 Indole-3-pyruvic acid (IPA) pathway (A–C) and tryptophan side chain pathway (D) for IAA biosynthesis in bacteria. The IPA pathway is found both in PGPB and some pathogens. A = tryptophan aminotransferase; B = indole-3-pyruvic acid decarboxylase; C = indole-3-acetaldehyde oxidase; D = tryptophan side chain oxidase

Table 5.1 Synthesis of IAA by a wild-type and an *ipdC* insertion mutant of a PGPB strain in the presence of increasing concentrations of tryptophan

Tryptophan concentration (µg/ml)	IAA produced (µg/ml/OD$_{600}$)	
	Wild-type	Mutant
0	0.5	0.5
50	15.5	0.5
100	22.5	0.8
200	26.2	1.0
500	32.7	2.0

are no literature reports on the genes in bacteria that are responsible for the aldehyde oxidation step (i.e., step C in this pathway).

The second enzyme of the IPA pathway is an indole-3-pyruvic acid decarboxylase (IPDC), which is encoded by the *ipdC* gene. This gene has been cloned from several different PGPB including, *Enterobacter, Azospirillum* and *Erwinia*. In one experiment, to prove that bacterial IAA is directly responsible for promoting plant growth, a strain of *Enterobacter cloacae* was genetically engineered to create a stable mutant deficient in IAA production (Table 5.1). This mutant was created by inserting a foreign gene encoding resistance to the antibiotic kanamycin into the reading frame of the *ipdC* gene (Fig. 5.6). Following construction of this mutant strain, IAA production was quantified by HPLC in both wild-type and mutant strains of the PGPB (Table 5.2), with the mutant strain producing only very low levels of IAA at all added tryptophan concentrations. This result confirmed that the tested PGPB strain produced IAA predominantly via the IPA pathway. Moreover, rendering the *ipdC* gene inactive significantly decreased the ability of the bacterium to stimulate root growth in canola seedlings. Since this initial demonstration of the effectiveness of bacterial IAA in promoting plant growth, several other groups of researchers have reported similar results with different PGPB strains. Interestingly, in several instances low IAA levels were still detected in the culture supernatant of

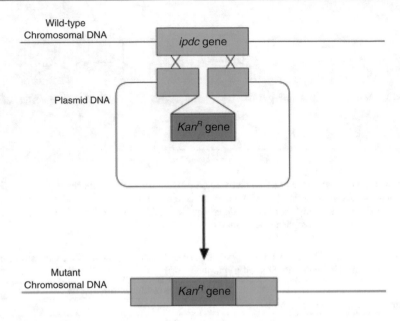

Fig. 5.6 Construction of a knockout mutant of a bacterial indole-3-pyruvic acid decarboxylase (*ipdC*) gene so that the bacterium no longer synthesizes IAA by this pathway. Kan^R = a kanamycin resistance gene

Table 5.2 Root and shoot lengths of canola plants following the treatment of canola seeds with either wild-type or an *ipdC* insertion mutant of a PGPB strain following 5 days of growth in gnotobiotic growth pouches in a controlled environment growth chamber

Bacterial treatment	Root length (mm)	Shoot length (mm)
None	54.5 ± 2.8	35.2 ± 1.0
Wild-type	74.2 ± 4.4	32.4 ± 2.1
IAA minus mutant	61.9 ± 1.9	32.8 ± 0.9

The numbers represent the mean of approximately 60 seedlings per treatment ±1 S.E.M

the *ipdC* mutant. A possible explanation for this observation is that these low levels of IAA are from alternative IAA pathways coexisting in these strains.

Azospirillum brasilense is a Gram-negative, nitrogen-fixing bacterium. It is a rhizosphere inhabitant that associates with the roots of grasses and cereals. It promotes plant growth by several different mechanisms including increasing the number of root hairs and lateral roots as a consequence of its synthesis of IAA. This bacterium mainly produces IAA via the IPA pathway so that an engineered mutation in the *ipdC* gene resulted in a mutant producing less than 10% of the IAA produced by the wild-type strain. In another experiment, researchers were able to increase IAA production in *A. brasilense* Sp245 by introducing recombinant plasmid-based copies of the *ipdC* gene under the control of the constitutive promoter P^{nptII} or the root exudate-responsive promoter P^{sbpA}. Both modified strains maintained high cell densities on plant roots 1 month after sowing. More importantly, winter wheat

Fig. 5.7 The Indole-3-
acetonitrile/indole-3-
acetaldoxime pathway.
A = oxidoreductase;
B = acetyldoxime
dehydratase; C = nitrilase.
Trp = tryptophan;
IAOx = indole-3-
acetyldoxime; IAN = indole-
3-acetonitrile; IAA = indole-
3-acetic acid

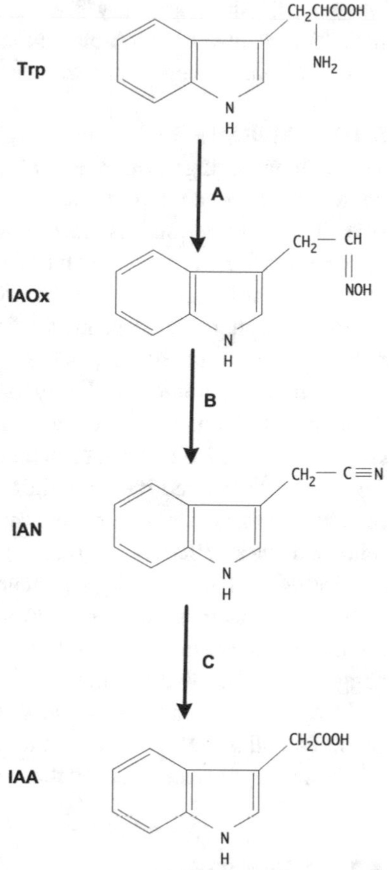

inoculated with either of the IAA-overproducing strains displayed enhanced shoot biomass and slightly decreased root biomass.

5.1.1.3 The IAN Pathway

IAA may also be synthesized via the indole-3-acetonitrile (IAN)/indole-3-acetaldoxime (IAOx) pathway (Fig. 5.7). The first step of this pathway is the conversion of tryptophan into indole-3-acetaldoxime. The microbial enzyme responsible for this conversion is likely an oxidoreductase. In the second step, the indole-3-acetaldoxime intermediate is converted into indole-3-acetonitrile by an indoleacetaldoxime dehydratase. The indole-3-acetonitrile intermediate is consequently converted to IAA by a nitrilase enzyme in a single step, or by a nitrile hydratase and an amidase in a two-step process.

Nitrilases are part of a large superfamily of enzymes with over 180 known members classified into 13 branches. Branch I enzymes are known to have true nitrilase activity, converting nitriles into their corresponding carboxylic acids and ammonia. Although this branch of nitrilase enzymes is implicated in IAA biosynthesis, there are very few literature reports of bacterial nitrilases producing IAA. This pathway of IAA biosynthesis is not well defined as the enzymes involved have not been studied to any great extent.

5.1.1.4 Multiple Pathways

There is mounting evidence that a number of soil bacteria, both PGPB and phytopathogens, contain more than one pathway encoding the biosynthesis of IAA. This observation has led to considerable speculation as to why this might be the case. On the one hand, soil bacteria that associate with the roots of plants use IAA to loosen plant cell walls and thereby increase the extent of root exudation. Thus, when rhizosphere bacteria synthesize IAA, they can increase the level of plant nutrients to which they have access.

When the genome of the well-studied PGPB *Pseudomonas* sp. UW4 was sequenced and analyzed, seven potential IAA biosynthetic genes were identified (Fig. 5.8). Based on DNA sequence homology with the IAA biosynthesis genes from other sequenced bacteria, these genes were assigned to the IAM and the IAN pathways with the two pathways intersecting and overlapping. In fact, a search of other sequenced *Pseudomonas* spp. genomes for a UW4-like IAA pathway revealed the presence of six orthologous genes in *P. fluorescens* SBW25 and *P. putida* F1, suggesting that these bacterial strains contain similar IAA biosynthesis pathways compared to strain UW4. Whether or not these inferred pathways operate as suggested; and how this complex biosynthetic pathway might be regulated remains to be elaborated. One way to do this would be to generate mutations in the genes involved in this pathway and test their activity, and/or to biochemically characterize all of the enzymes encoded by these genes.

5.1.2   Regulation

Given the fact that IAA is produced both by plants and by the microorganisms with which the plants associate, it is sometimes difficult to sort out how these two sources of IAA interact and affect one another. Here, the regulatory mechanisms of bacterial IAA biosynthesis are emphasized. Interestingly, regardless of whether IAA is delivered to plants via rhizospheric or endophytic bacterium, the effect on plants is essentially the same.

5.1.2.1 Gene Localization

The amount of IAA produced by a bacterium is subject to several means of regulation. For example, whether the IAA biosynthesis genes are located on the bacterial chromosome or a plasmid can affect the amount of IAA produced by that organism. For example, genes present on multi-copy plasmids will be present in more copies than when the same genes are present as a single copy in the chromosome so that IAA biosynthesis genes present on plasmids are likely to produce more IAA. For example, the phytopathogen *P. savastanoi* pv. savastanoi has plasmid-borne IAA biosynthetic genes and produces more IAA than the phytopathogen *P. syringae* pv.

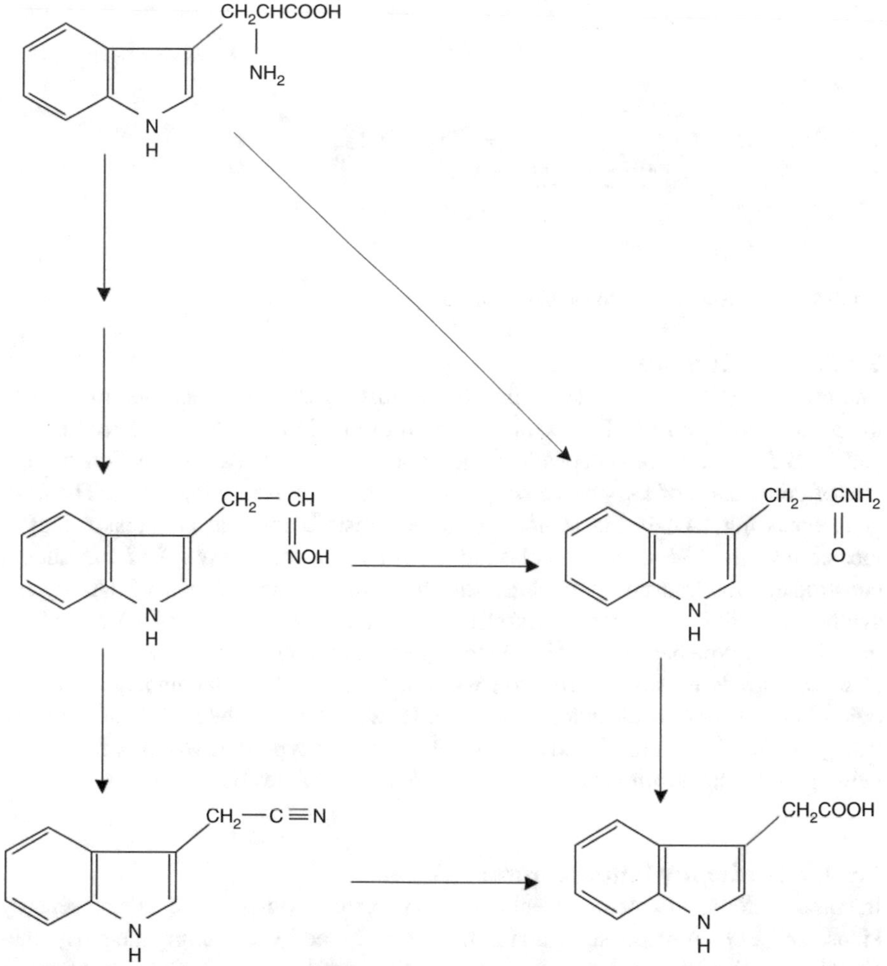

Fig. 5.8 Two potential overlapping pathways of IAA biosynthesis inferred from an analysis of the sequence of the complete genome of *Pseudomonas* sp. UW4. The two potential IAA biosynthesis pathways are the IAM and IAN pathways. Genes encoding enzymes catalyzing each of the steps shown here were identified in the genome of this bacterium. TRP = tryptophan; IAOx = indole-3-acetyldoxime; IAN = indole-3-acetonitrile; IAM = indole-3-acetamide; IAA = indole-3-acetic acid

syringae, whose homologous genes are on the chromosome. In addition, different IAA biosynthesis pathways may be more readily expressed in certain microenvironments such as on plant surfaces or inside plant tissues. Thus, with the phytopathogen, *Erwinia herbicola* pv. gypsophilae, the genes for the IAM pathway, located on the pPATH plasmid are more highly transcribed inside the plant stems as opposed to on the leaf surfaces.

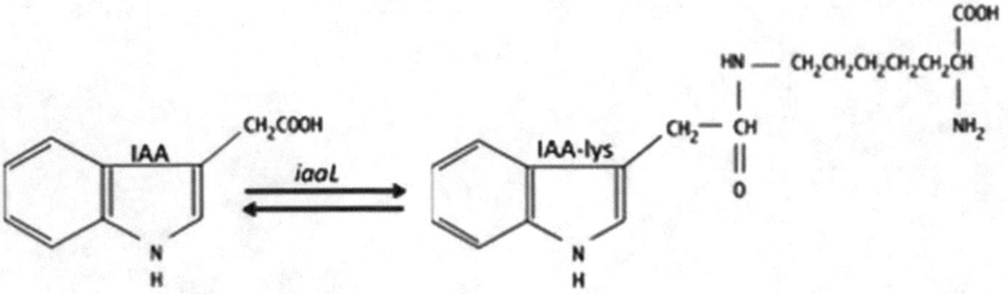

Fig. 5.9 Conversion of IAA to the IAA-lysine conjugate

5.1.2.2 IAA Conjugation

The presence of the *iaaL* gene within a bacterium's genome is presumed to regulate the amount of free IAA. For example, the amount of free IAA synthesized by the pathogen *P. savastanoi* is dependent upon both the rate of IAA production and the rate of conversion of IAA to the conjugated form, IAA-lysine (Fig. 5.9). The *iaaL* gene encoding IAA-lysine synthetase is responsible for this conversion and is located near the IAA operon on plasmid pIAA1 in this bacterium. An *iaaL* mutant, constructed via transposon mutagenesis, had undetectable levels of IAA-lysine synthetase activity and consequently accumulated five-fold more IAA in culture than the wild-type bacterial strain. On the other hand, olive gall isolates of wild-type *P. savastanoi* do not convert IAA to IAA-lysine, naturally producing approximately twice as much IAA as oleander isolates do. However, when the IAA-lysine synthetase gene (*iaaL*) was introduced into an isolate of wild-type *P. savastanoi* from olive galls, IAA-lysine accumulated in culture, reducing IAA levels by 30%.

5.1.2.3 Stationary Phase Control

In some bacteria, a two-component regulatory system composed of a GacS sensory kinase and a GacA response regulator may be involved in controlling the biosynthesis of IAA. This system has previously been shown to regulate a variety of genes involved in the production of bacterial secondary metabolites. For example, when *Enterobacter cloacae* CAL2 or *Pseudomonas* sp. UW4 were genetically transformed with a plasmid harboring either *rpoS* or *gacS* genes from a strain of *P. fluorescens*, there was a ten- and twofold increase in IAA levels, respectively. Prior to this work, the GacS system was found to upregulate the biosynthesis of RpoS, the stationary phase sigma factor. In nature, soil bacteria mostly exist in the stationary phase, and IAA synthesis has previously been associated with the stationary phase. Thus, RpoS and GacS work together in the regulation of IAA synthesis and the observed increase in IAA production is most likely a direct consequence of RpoS accumulation.

The PGPB, *Azospirillum brasilense* is typically found in soils with low oxygen concentrations so that it is not surprising that the IPA pathway, which is commonly used by *A. brasilense* strains for IAA biosynthesis, is enhanced under anaerobic conditions. Under aerobic conditions, IAA production and *ipdC* gene expression are greatly reduced compared to microaerobic conditions, while the bacterial biomass remains relatively unaffected. The available data suggests that carbon limitation and a reduction in bacterial growth rate promote *ipdC* gene expression, and eventually IAA synthesis, in *A. brasilense*, conditions that are characteristic of the stationary phase of growth. In fact, the inability of *Azospirillum* strains to produce IAA in the presence of abundant amounts of carbon may explain the failure of this bacterium to promote plant growth in rich or highly fertilized soils.

In one recent study, researchers demonstrated that when a PGPB strain (*Serratia* sp.) was grown in the laboratory under low nitrogen conditions (i.e., nitrogen starvation) the level of IAA was increased. Interestingly, the nitrogen stressed *Serratia* sp. promoted the growth of *Populus euphratica* (Euphrates poplar) seedlings to a significantly greater extent than when the bacterium was grown in rich medium. The reason for this effect is not known; however, the authors of this work speculated that IAA might be acting as a stress-response signaling molecule in this instance. Alternatively, the nitrogen starvation stress may have triggered a general stress response in the bacterium. This would have the effect of increasing the production of RpoS, along with a concomitant decrease in RpoD (the logarithmic phase sigma factor), which would alter the transcription of a large number of bacterial genes thereby altering the biosynthesis of a large number of bacterial molecules (including increasing the level of IAA).

5.1.2.4 Regulatory Sequences

The promoter region of the *ipdC* gene of the PGPB strain *Enterobacter cloacae* UW5 contains a sequence (TGTAAA-N_6-TTTACA) that is similar to the TyrR recognition sequence of *E. coli*. The *tyrR* gene encodes a regulatory protein that is both an activator and a repressor of genes involved in aromatic amino acid transport and metabolism. In fact, an *E. cloacae* UW5 *tyrR* insertion mutant, similar to an *ipdC* insertion mutant, did not produce any IAA in the presence of tryptophan in the stationary phase. Using real-time qRT-PCR it was possible to show that *ipdC* transcript levels in the *tyrR* mutant were very low in the stationary phase even in the presence of high levels of added tryptophan. These results indicate that, in this PGPB strain both the TyrR recognition sequence and an intact *ipdC* gene are required for IAA production.

The DNA sequence of the 5′ upstream region of the *ipdC* gene of *A. brasilense* strains SM, Sp45 and Sp7 has been found to be similar to the auxin responsive elements that have been found in some plants. This promoter region extends to position −81 relative to the *ipdC* transcription initiation site. Within this upstream region, located between positions −2 and −23 is the plant-like TGTCNC element, which is essential for *ipdC* expression. In addition, between positions −58 and −38 there are two 8-bp inverted repeats separated by a 4-bp spacer that is also required for full activation of the *ipdC* promoter and for its IAA inducibility.

In *A. brasilense*, there is also an open reading frame 99 bp downstream of the *ipdC* gene that is co-transcribed in an operon with *ipdC* and appears to reduce the level of IAA production. This open reading frame is known as *iaaC*. Moreover, a putative mRNA stem-loop structure that may function as a transcription terminator

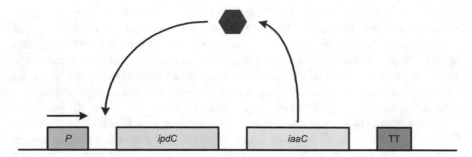

Fig. 5.10 Possible regulation of the expression of the *Azospirillum brasilense ipdC* gene by a protein encoded by the *iaaC* gene. The red hexagon represents the iaaC gene product that binds to the DNA upstream of the *ipdC* gene and prevents it from being transcribed. The arrow above the promoter region indicates the direction of transcription. Abbreviations: *P* promoter, *TT* transcription termination site

was found in the region downstream of *iaaC*. With an *iaaC* insertion mutant, there was no difference in *ipdC* expression compared to the wild-type strain; however, the IAA levels in the *iaaC* insertion mutant were dramatically increased (up to sixfold) when measured at the stationary growth phase. These findings suggest that the protein encoded by *iaaC* acts as an inhibitor/regulator of IAA production (Fig. 5.10).

5.1.3   Effects on PGPB Survival

In addition to directly affecting plant gene expression, bacterial IAA has been found to have some important modulating effects on the metabolism of some PGPB. In one set of experiments, *Sinorhizobium meliloti* was either treated with a high level of IAA or else it was genetically modified, by introducing an exogenous IAM pathway (from a PGPB strain), to overexpress IAA. In *S. meliloti*, IAA biosynthesis typically occurs through the IPA pathway. In both cases, the activity of some key tricarboxylic acid (TCA) cycle enzymes and the level of acetyl-CoA were all increased compared to the wild-type untreated strain (Fig. 5.11). Moreover, the levels of enzymes involved in the biosynthesis of the carbon storage compound polyhydroxybutyrate (PHB), which is generally metabolized when other common energy sources are not available, were also increased. The results of these experiments were confirmed by a transcriptomic analysis using a microarray of *S. meliloti* genes. In addition, following a month-long survival experiment, the presence of an increased amount of IAA resulted in more *S. meliloti* cells surviving. Based on these results, it was speculated that the increased level of PHB might provide carbon and energy sources to *S. meliloti* that facilitated its conversion from a free-living bacterium into differentiated bacteroids. In addition, it is thought that the increase in the long-term survival of the bacterium might be a consequence of the presence of a higher level of PHB in the bacterium. Besides increasing bacterial long-term survival, increased IAA has been found to upregulate genes coding for the high-affinity P

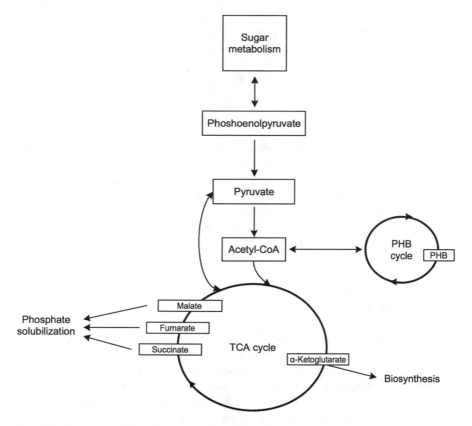

Fig. 5.11 Overview of *Sinorhizobium meliloti* metabolism showing the relationship between the TCA cycle phosphate solubilization and polyhydroxybutyrate synthesis and degradation

transport system, induce acid phosphatase activity, and the increase secretion into the growth medium of malate, succinate, and fumarate. Since organic acids such as malate, succinate, and fumarate are well known for their ability to solubilize inorganic phosphate, the IAA-stimulated increased flux through the TCA cycle appears to be directly responsible for this effect (Fig. 5.11). More recently, researchers modified the *rolA* promoter from *Agrobacterium rhizogenes* and used the modified promoter to drive the transcription of *iaaM* and *tms2* genes (encoding IAA biosynthesis). They found that increasing the IAA concentration in *S. meliloti* resulted in a significant increase in the shoot fresh weight of alfalfa plants nodulated by the modified *S. meliloti*, compared to plants nodulated by the wild-type strain. This work could be important from a practical perspective as it suggests that the approach of mutagenizing some endogenous rhizobial promoters without introducing foreign DNA may be an effective means of increasing the ability of some rhizobial strains to promote plant growth. That is, it should be possible to specifically mutagenize promoters of rhizobial (or other PGPB) IAA biosynthesis genes (by either direct mutagenesis or through the use of CRISPR) and thereby

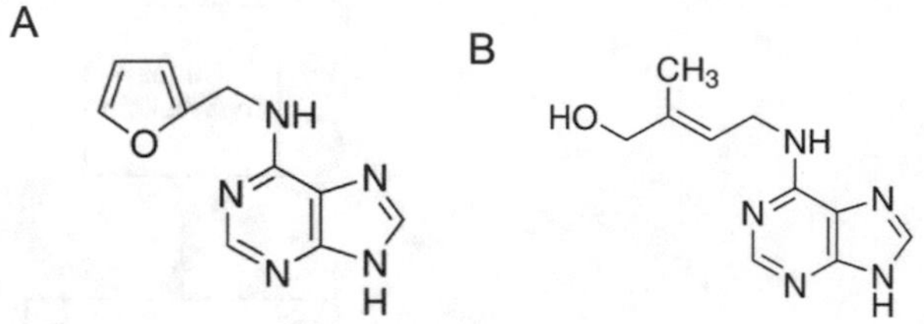

Fig. 5.12 Chemical
structures of (**a**) kinetin and
(**b**) zeatin

increase bacterial production of IAA. If these constructed IAA overproducing mutants increased their promotion of plant growth compared to the wild-type strains, according to rules in some countries, they could be deliberately released to the environment and would not necessarily be considered to be genetically engineered strains (see Chap. 11). Finally, whether the effects that have been observed here reflect a phenomenon that is common to many different PGPB or is specific to *S. meliloti* remains to be determined.

5.2   Cytokinin

Cytokinins are a class of compounds that promote plant cell division along with some other similar functions and have a structure that resembles adenine. Kinetin (Fig. 5.12a) was the first cytokinin to be discovered (in 1955) and was so named because of the compound's ability to promote cytokinesis or cell division. Today, kinetin is usually considered to be a "synthetic" cytokinin because it is not synthesized in plants per se; it was originally found in yeast extract. The most common form of cytokinin in plants is zeatin (Fig. 5.12b) which was originally isolated from corn (*Zea mays*) in 1963. Other bacterially-synthesized cytokinins include zeatin riboside and isopentenyladenine. Plant-produced cytokinins are synthesized primarily within roots although the hormone is found distributed throughout the plant.

Cytokinins affect a range of different plant cells. They have been shown to modulate cell division, seed germination, apical dominance, root elongation, xylem and chloroplast differentiation, transition to reproductive growth phase, flower and fruit development, leaf senescence, nutritional signaling, and plant–pathogen interactions. The wide variety of effects observed suggests that cytokinins might have different mechanisms in different plant tissues. Cytokinins can stimulate cell division/growth and, consequently, they simultaneously inhibit plant aging/senescence. Under natural conditions, a higher ratio of cytokinin to auxin favors shoot, over root, formation.

Cytokinins have been detected in the cell-free medium of some PGPB strains including: *Azotobacter* spp., *Rhizobium* spp., *Pantoea agglomerans*, *Rhodospirillum rubrum*, *Pseudomonas* spp., *Bacillus* spp., *Vibrio* spp., *Aeromonas* spp., *Achromobacter* spp., *Chromobacterium* spp., *Acinetobacter* spp., *Flavobacterium* spp., and *Paenibacillus polymyxa*. In addition, cytokinin biosynthesis genes have also been detected in a number of different phytopathogens including: *Agrobacterium tumefaciens* (outside of the T-DNA which is actually plant-derived DNA), *Pseudomonas savastanoi*, *Erwinia herbicola* pv. gypsophilae, and *Streptomyces turgidiscabies*. Thus, as with IAA, many, but not all, PGPB strains (and some strains of phytopathogens) synthesize cytokinins; however, the data that is currently available suggests that it is less common for a PGPB strain to synthesize cytokinin than IAA (keeping in mind that cytokinin biosynthesis in soil bacteria is much less studied than IAA biosynthesis). In addition, based on limited data, it is currently thought that PGPB probably produce lower levels of cytokinin compared to phytopathogens so that the effect of the cytokinin-producing PGPB on plant growth is generally stimulatory while the effect of the pathogens is "over-stimulatory" (i.e., tumor-inducing).

It has been shown, in transgenic plants that overexpress cytokinin, especially in response to certain stresses like drought or the presence of high levels of salt, high levels of cytokinin can significantly delay the plant senescence response that often occurs as a direct consequence of the stress. Thus, under the right conditions, salt- or drought-stressed transgenic plants that overexpress cytokinin will remain green and healthy while non-transformed plants are completely wilted under the same conditions. In addition, there are reports that adding purified cytokinins to some plants can increase the level of resistance against the phytopathogen *Pseudomonas syringae* pv. *tabaci*.

A number of PGPB have been shown to both possess and express cytokinin genes, and in some instances it has been shown that the addition of PGPB that secrete cytokinins to growing plants can significantly alter the plant's composition of several other phytohormones. Unfortunately, notwithstanding these observations, researchers have not as yet directly demonstrated that bacterial cytokinin may be as useful to plants as was demonstrated with the overexpression of foreign cytokinin genes in transgenic plants. To do this will require the isolation and testing of mutants of PGPB in which the wild-type strain but not the (nonproducing) cytokinin mutant strain promotes specific aspects of plant growth and development. An important first step in this regard is the development of a simple and reproducible means of assaying bacterial strains for the ability to synthesize cytokinins. Of course, cytokinin concentrations may readily be assayed by HPLC. However, despite the availability of this equipment in many labs, it is not always available in developing countries and is not an efficient means of assaying very large numbers of biological samples. As an alternative, some workers have developed a rather simple biological assay that might be useful in this endeavor (Fig. 5.13). Notwithstanding its simplicity, this nonquantitative assay is able to detect cytokinin concentrations as low as 10^{-7} M. In addition, cytokinins may be measured by immunoanalysis where the competition of cytokinins present in a sample with a known quantity of labeled or immobilized cytokinin for binding to an anti-cytokinin antibody. The degree of competition is measured using either radiolabeled cytokinin (RIA) or enzyme-linked immunosorbent assays (ELISA) using anti-cytokinin monoclonal antibodies. These immunoassays have detection limits in the femtomole range. Thus, in the future it should be possible for researchers to use this assay to both screen large numbers of

Leaf   Bacteria

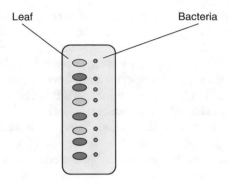

Fig. 5.13 Schematic representation of a plate assay for bacterial cytokinin production. In this assay, cucumber cotyledons are excised from etiolated plants and placed on an agar plate ~2 mm from a PGPB culture that may synthesize cytokinin. Following a 20 h incubation in the dark, the plates are exposed to light for 3 h; as a result, the cotyledons that are exposed to cytokinin secreted by the PGPB cells turn a dark green color

Fig. 5.14 Chemical structure of gibberellic acid (GA3), the active and most common form of the phytohormone gibberellin

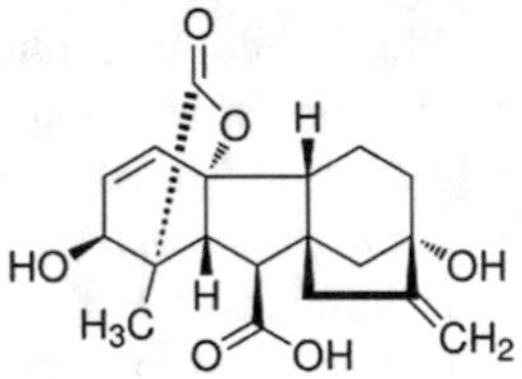

bacteria for the presence of cytokinin minus mutants and, subsequently, to complement those mutants with recombinant clones that complement those mutations and encode the biosynthesis of active cytokinin. As well, with either this assay or by using HPLC, it should also be possible to select naturally occurring bacterial strains that overproduce cytokinins. If the benefits of high levels of exogenous cytokinin are to be realized by treated plants this can occur (albeit inefficiently) if the cytokinin is added as a foreign chemical or (more efficiently) if the cytokinin is added as part of a naturally occurring PGPB. Both at the present time and in the foreseeable future, the widespread use of genetically engineered PGPB strains (that e.g. overproduce cytokinin) is not acceptable to the regulatory authorities in most jurisdictions.

5.3   Gibberellin

Gibberellins are a large family of closely related *ent*-kaurene-derived diterpenoid phytohormones (Fig. 5.14). Of the 126 known gibberellins, only 4 have been shown to have biological activity (i.e., GA1, GA3, GA4, and GA7). Gibberellic acid (GA3; the most common gibberellin) is synthesized by some PGPB (Table 5.3) and is produced commercially by fermentation of the fungus *Giberella fujikuroi*. Despite the fact that there are large number of different gibberellins, the biological activity and role of gibberellin molecules other than GA3 remains unresolved. Since its

Table 5.3 Some soil bacteria that are able to synthesize gibberellin

Acetobacter diazotropicus	Bacillus licheniformis
Acinoetobacter calcoaceticus	Bacillus megaterium
Agrobacterium radiobacter	Bacillus macrolides
Agrobacterium tumefaciens	Bacillus pumilis
Arthrobacter atrocyaneous	Bacillus subtilis
Arthrobacter aurescens	Bradyrhizobium japonicum
Arthrobacter citrens	Burkholderia sp.
Arthrobacter globiforis	Pseudomonas aeruginosa
Arthrobacter tumescens	Pseudomonas fluorescens
Arthrobacter sp.	Pseudomonas fragi
Azospirillum brasilense	Pseudomonas perolens
Azospirillum lipoferum	Pseudomonas striata
Azotobacter chroococcum	Pseudomonas sp.
Bacillus aryabhattai	Rhizobium leguminosarum
Bacillus aquimaris	Rhizobium meliloti
Bacillus brevis	Rhizobium phaseoli
Bacillus cereus	Rhizobium trifoli

discovery, scientists have ascertained that gibberellins are involved in increasing plant stem growth, germination dormancy, flowering, and leaf and fruit senescence. More recently, it has been shown that gibberellin primarily affects shoot growth; however, root growth is also regulated by gibberellin, but in a lower concentration range than is required for shoot growth (Fig. 5.15). Thus, it is generally recognized that gibberellin plays an indispensable role in both shoot and root development in many plants. The concentration effect observed with gibberellin is very similar to what has been observed for IAA. Finally, despite the fact that a number of different gibberellin-producing PGPB have been identified, PGPB are not generally selected, or even tested, for gibberellin synthesis, and no specifically gibberellin-producing PGPB strains have been commercialized, although some commercialized strains may produce gibberellins. Most of what is currently believed to be the role of bacterially produced gibberellins (and cytokinins as well) is based on the response of plants to the exogenous addition of purified gibberellins (or cytokinins) to growing plants. At present, while the gibberellin biosynthetic pathways in plants and fungi are known, much less is known regarding how bacteria produce gibberelins. Plants generally utilize two diterpene synthases to form *ent*-kaurene, while fungi use only a single bifunctional diterpene synthase. However, relatively recently, it was found that some strains of *Bradyrhizobium japonicum* encode two sequentially acting diterpene cyclases, *ent*-copalyl diphosphate synthase and *ent*-kaurene synthase. Together, these two enzymes transform geranylgeranyl diphosphate to *ent*-kaurene, the olefin precursor to gibberellin. The operon encoding these two genes is selectively expressed during nodulation. Moreover, the operon encoding these genes has been found in several additional species of Rhizobia suggesting that gibberellin biosynthesis may be common in many strains of Rhizobia (and ironically in some pathogen strains as well).

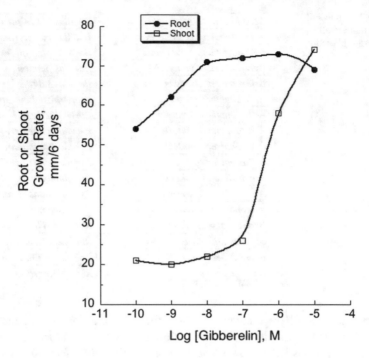

Fig. 5.15 Gibberellin stimulates the growth of both roots and shoots but is active at different concentrations in the two types of tissue

In one series of experiments, scientists isolated a strain of the bacterium *Enterococcus faecium* and tested its growth-promoting ability in a gibberellin-deficient rice dwarf mutant. This bacterium, originally isolated from a soil sample from the rhizosphere of oriental melon (*Cucumis melo* L.), was shown to synthesize both IAA and several gibberellins (i.e., GA1, GA3, GA7, GA8, GA9, GA12, GA19, GA20, GA24, and GA53) with GA3, GA7, and GA12 being produced to the greatest amount. In initial experiments, it was observed that the isolated bacterial strain significantly promoted increases in oriental melon shoot length, root length, plant fresh weight, and chlorophyll content. Subsequently, the isolated strain of *Enterococcus faecium* was used to treat a gibberellin-deficient rice dwarf mutant cultivar (that displayed a decrease in shoot length, root length, plant fresh weight, and chlorophyll content compared to a normal rice cultivar). Based on the observation that the isolated bacterium promoted increases in all of the abovementioned parameters, it was surmised that this bacterium was able to provide gibberellin to the gibberellin-deficient rice cultivar. Interestingly, the addition of the isolated *Enterococcus faecium* strain to a normal rice cultivar promoted even greater increases in shoot length, root length, plant fresh weight, and chlorophyll than were observed with the gibberellin-deficient rice cultivar. The researchers concluded that the isolated *Enterococcus faecium* strain could promote plant growth by producing both gibberellin and IAA. Finally, these researchers noted that the application of cell-free extracts of the bacterium also promoted the growth of rice plants, a result that is perhaps not surprising given the fact that both gibberellin and IAA are secreted by the bacterium.

5.4   Ethylene

Ethylene is one of the simplest known molecules with biological activity. In the Hebrew Bible, the prophet Amos is described as being a "herdsman and a nipper of figs." This description indicates that as early as the ninth century B.C.E. (when the Hebrew Bible is thought to have been written) it was recognized that nipping (or scoring or piercing) figs, that were traditionally picked when they were still unripe, produced ethylene gas thereby hastening the ripening process and making the figs sweeter. Despite this very early observation of the consequences of ethylene action, it was not until the late nineteenth century (C.E.) that ethylene was first recognized (by Dimitri Neljubov) as an active plant hormone.

Ethylene displays a wide range of biological activities and has been estimated to be active at concentrations as low as 0.05 µL/L; however, in ripening fruit ethylene concentrations may reach levels as high as 200 µL/L. Ethylene, which is produced in almost all higher plants, mediates a wide range of plant responses and developmental steps. Ethylene is involved in seed germination, tissue differentiation, formation of root and shoot primordia, root branching and elongation, lateral bud development, flowering initiation, anthocyanin synthesis, flower opening and senescence, fruit ripening and degreening, production of volatile organic compounds responsible for aroma formation in fruits, storage product hydrolysis, leaf senescence, leaf and fruit abscission, Rhizobia nodule formation, mycorrhizae-plant interaction, and (importantly) the response of plants to various biotic and abiotic stress. In some instances, ethylene is stimulatory while in others it is inhibitory, an effect that is often the consequence of the amount of ethylene that is produced.

5.4.1   Ethylene Biosynthesis and Mode of Action

In plants, the biosynthesis of ethylene begins with the compound S-adenosylmethionine (SAM), a compound that is also a precursor of several other pathways and is, therefore, relatively abundant within plant tissues (Fig. 5.16). For the synthesis of ethylene, first the enzyme 1-aminocyclopropane-1-carboxylic acid (ACC) synthase converts SAM to ACC and 5'- methylthioadenosine (MTA). The MTA is recycled to L-methionine (the immediate precursor of SAM) enabling the level of L-methionine in a cell to remain relatively unchanged even during fairly high rates of ethylene production. Researchers have suggested that this is the committal step in the ethylene biosynthesis pathway, since the extremely labile ACC synthase enzyme may be rate limiting and also rises proportionally with the increase in ethylene levels within the tissues of some plants. The gene that encodes this enzyme is part of a multigene family, and considerable evidence indicates that

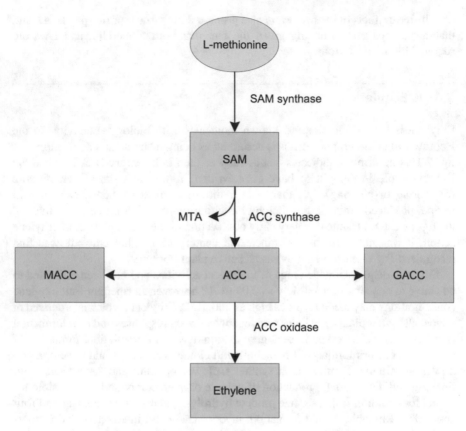

Fig. 5.16 Overview of the biosynthesis of ethylene. Abbreviations: *SAM* S-adenosyl methionine, *MTA* 5′-methylthioadenosine, *ACC* 1-aminocylopropane-1-carboxylate, *MACC* 1-(malonyl)-ACC, *GACC* 1-(glutamyl)-ACC

the transcription of different forms of this enzyme are induced under different environmental or physiological conditions.

The next step is the conversion of ACC to ethylene by the plant enzyme ACC oxidase, which is typically present in most plant tissues at very low levels. Similar to ACC synthase, ACC oxidase genes are also part of a multigene family and different isoforms of ACC oxidase appear to be active under different environmental or physiological conditions. Thus, other than in ripening fruit, both ACC synthase and ACC oxidase are inducible enzymes.

The conjugation of ACC with malonate, glutathione, or jasmonate is thought to occur via the action of the enzymes ACC N-malonyl transferase, γ-glutamyl transpeptidase and jasmonic acid resistance 1, with the production of 1-(malonyl)-ACC (MACC), 1-(glutamyl)-ACC (GACC), and jasmonyl-ACC (JACC), respectively, Moreover, it has been speculated that the linkage between the precursors of ethylene and the active form of the plant hormone jasmonic acid (to form J-ACC) might provide a mechanism to control the levels of both of these hormones. Kinetic studies have indicated that the K_m of ACC N-malonyl transferase for ACC is much higher than the K_m of ACC oxidase for ACC so that if these two enzymes are present in similar amounts the ACC will bind preferentially to ACC oxidase and be converted into ethylene. [It should be kept in mind that a high K_m means less tight binding of an enzyme to a substrate than a low K_m.] The conjugation reactions create a sink for ACC only when it is present in high levels (i.e., there is more ACC than is needed for ethylene synthesis), or when ACC oxidase is either saturated or absent.

Ethylene perception occurs through a chain of events that are conserved among higher plants. The ethylene receptor proteins bind ethylene gas with the help of a copper cofactor that is bound in the transmembrane domain of the receptor. In the absence of ethylene, the receptor activates a downstream kinase called CTR1 that, through several additional kinase-mediated phosphorylation steps, inhibits ethylene responsive gene expression. Thus, the kinase cascade serves as a negative regulator of ethylene responsive gene expression. When ethylene is bound to the receptor, the receptor complex undergoes a conformational change causing the receptor to be "inhibited" as is the CTR1 kinase. This causes the inactivation of the negative signaling events and, thereby permits the ethylene response.

Some recent evidence suggests that 1-aminocyclopropane-1-carboxylate (ACC), in addition to its well-known role as a precursor to ethylene, may have biological activity on its own. This evidence suggests that ACC may play a role of signaling in plants, affecting a number of functions including plant root elongation. However, given the indirect nature of much of this data, the signaling role of ACC independent of ethylene is considered to remain a matter of some debate. Regardless of the precise role(s) played by ACC, the enzyme ACC deaminase is expected to have a similar effect on ethylene- and ACC-catalyzed processes.

5.4.2 ACC Deaminase

The enzyme ACC deaminase is capable of degrading the ethylene precursor, ACC, to ammonia and α-ketobutyrate (Fig. 5.17). Since its initial discovery in 1978, this enzyme has been reported to be widespread in soil bacteria and to also be present in some soil fungi. Moreover, the functioning of this enzyme has been implicated as one of the key mechanisms employed by PGPB to facilitate plant growth. Simply put, by breaking down some of the ACC in plants, this enzyme can lower plant

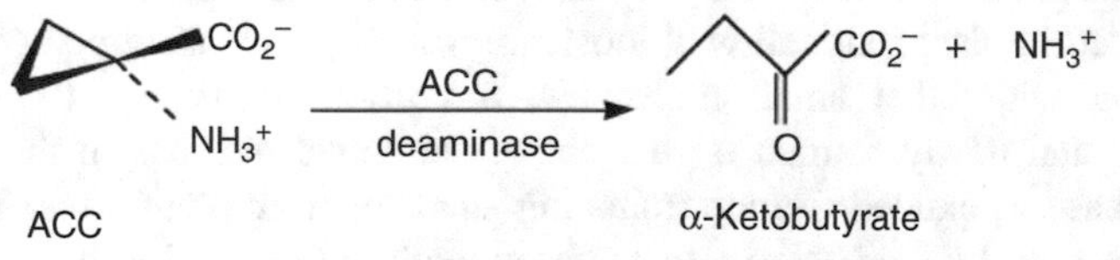

Fig. 5.17 Chemical cleavage of ACC to α-ketobutyrate and ammonia by ACC deaminase

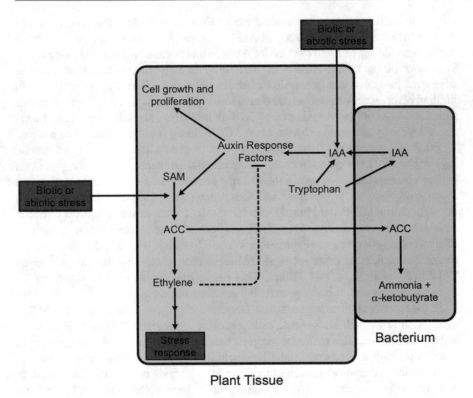

Fig. 5.18 Schematic representation of a model explaining how ACC deaminase-containing PGPB promote plant growth. © 2010 American Society for Microbiology. Used with permission. No further reproduction or distribution is permitted without the prior written permission of American Society for Microbiology

ethylene levels thereby decreasing the extent of ethylene inhibition of plant growth, especially following various biotic and abiotic stresses.

The model that was developed (and then modified) to explain the ability of ACC deaminase-containing PGPB to facilitate plant growth may be described as follows (Fig. 5.18). At the outset, PGPB must efficiently colonize either plant roots (either internally or externally) or seeds. If the bound bacteria synthesize and secrete IAA (a phenomenon that is very common in PGPB), that IAA is taken up by the plant, and together with the IAA synthesized by the plant itself, can either stimulate plant growth (cell proliferation and/or elongation) or induce the transcription of the gene encoding the plant enzyme ACC synthase, which converts SAM to ACC. The IAA is also involved in plant root cell wall loosening thereby increasing plant root exudation. Along with other small molecules, a portion of the plant's ACC, both preexisting and newly formed by the abovementioned increase in the amount of ACC synthase, is exuded from germinating seeds or from plant roots. The exuded ACC is taken up by the bacteria, and subsequently converted by the enzyme ACC deaminase to ammonia and α-ketobutyrate. As a result of this activity, the level of

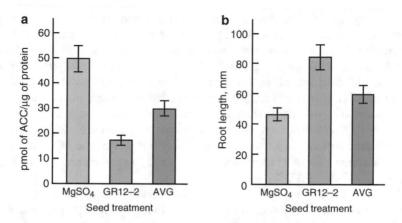

Fig. 5.19 The effect of treating canola seeds with either MgSO₄, the PGPB *P. putida* GR12–2 (suspended in MgSO₄) or the chemical ethylene inhibitor 1-aminovinylglycine (AVG) in MgSO₄ on (**a**) the ACC content of root cells and (**b**) the root length in mm. The seeds were treated for 1 h prior to growing them for four and a half days at 20 °C in sterile growth pouches in a growth chamber. The values shown are the means of the values for 50–60 seeds ± one standard error of the mean

ACC within plant tissues and hence the amount of ethylene produced by the plant is lowered and does not reach levels that are inhibitory for plant growth. The lowering of plant ethylene levels is particularly important when the plant is subjected to either biotic or abiotic stress, which can result in an increase in ACC synthesis in the plant. Moreover, lowering the level of ACC will decrease the extent of ethylene inhibition of the IAA signaling pathway thereby enabling bacterial IAA to more readily facilitate plant growth. The main short-term visible effect of seed inoculation with ACC deaminase-producing bacteria, under gnotobiotic conditions, and therefore a simple and reliable biological assay for this enzyme, is the enhancement of plant seedling root length. For example, when the amount of ACC present in canola roots and the root length were assessed following treatment of seeds with an ACC deaminase-containing PGPB or a chemical inhibitor of ethylene synthesis (AVG), both treatments lowered the amount of ACC in the roots and increased the root length of the seedlings, with the bacterial treatment being more effective than the chemical treatment (Fig. 5.19). In addition, when an ACC deaminase minus mutant of the same bacterial strain was substituted for the wild-type bacterium, the ACC content within the roots was not lowered and the root length was not increased. Thus, the observed effects are a direct consequence of the functioning of the PGPB enzyme ACC deaminase. Importantly, the growth promotion effects that are seen in short-term canola growth pouch assays (typically done for 4–5 days) typically persist throughout the life of the plant and are readily visible in soil (Fig. 1.16).

One problem with the abovementioned model is that the enzyme ACC oxidase has ~100 times greater affinity for ACC than does ACC deaminase. If the two enzymes were present at the same time in comparable amounts (assuming that the diffusion of ACC from plant cells is not rate limiting), nearly all of the ACC would be bound by the ACC oxidase and then converted into ethylene. Thus, for ACC deaminase to be effective at lowering plant ethylene, it must be present *prior* to the induction of ACC oxidase so that it remains in a 100- to a 1000-fold molar excess over ACC oxidase (notwithstanding the fact that ACC oxidase is present in plant cells and ACC deaminase is present in bacterial cells). The net result of the cleavage of exuded ACC by bacterial ACC deaminase is that the bacterium is de facto acting as a sink for ACC with the amount of ACC within the plant tissues decreasing. The amount of ACC that is present within plant tissues can be determined by (1) making a soluble extract of the plant tissue, (2) chemically derivatizing the ACC (a technique that is well known for other amino acids) and then (3) separating and quantifying the derivatized ACC by HPLC.

Recently, several groups of researchers have reported that the ethylene precursor ACC, independent of ethylene, also plays a role in regulating plant gene expression that affects several plant processes including root to shoot communication. Thus, the interaction of plants with ACC deaminase-producing bacteria would be expected to decrease the level of ACC regulation of specific plant functions. However, since ACC deaminase lowers both plant ACC and ethylene levels, until researchers can experimentally separate the effects of ACC and ethylene, treatment of plants with ACC deaminase-producing bacteria is likely to be attributed to the lowering of plant ethylene concentrations.

Interestingly, a group of researchers reported that bacteria that contain ACC deaminase stimulate the growth of mushrooms in a manner that is similar to its effect on plants. They observed that an ACC deaminase-containing pseudomonad could attach to fungal hyphae, reduce fungal ethylene concentrations and thereby stimulate hyphal growth and primordium initiation. At the same time, an ACC deaminase minus strain of the same bacterium did not affect fungal growth. The mushroom species that was utilized in these experiments is *Agaricus bisporus*, the most common edible mushroom with a commercial market value in the billions of dollars.

The enzyme ACC deaminase has been biochemically characterized in several studies. These studies have indicated that the enzyme (1) is found in both bacteria and fungi; (2) requires pyridoxal phosphate as a cofactor; (3) has a monomeric molecular mass of approximately 35–45 kD; (4) is a multimeric enzyme (with a native size estimated to be 100–115 kD) that has been suggested to be either a dimer or a trimer; (5) binds the substrate, ACC, only poorly with a Km typically in the range of ~1.5–6 mM; (6) has only very limited activity with substrates other than ACC; (7) is a sulfhydryl enzyme; (8) is cytoplasmically localized; and (9) although a low level of the enzyme is constitutively synthesized, is induced by low levels of ACC (~100 nM).

Various species of *Trichoderma*, a genus of plant beneficial fungi, have been shown to produce both ACC deaminase and IAA. Moreover, it has recently been suggested that the combination of ACC deaminase and IAA in a strain of *Trichoderma longbrachiatum* is responsible for the ability of this fungal strain to enhance the tolerance of wheat seedlings to NaCl stress.

Ironically, the presence of a functional ACC deaminase has been detected by scientists in several different pathogenic organisms including the fungus *Verticillium*

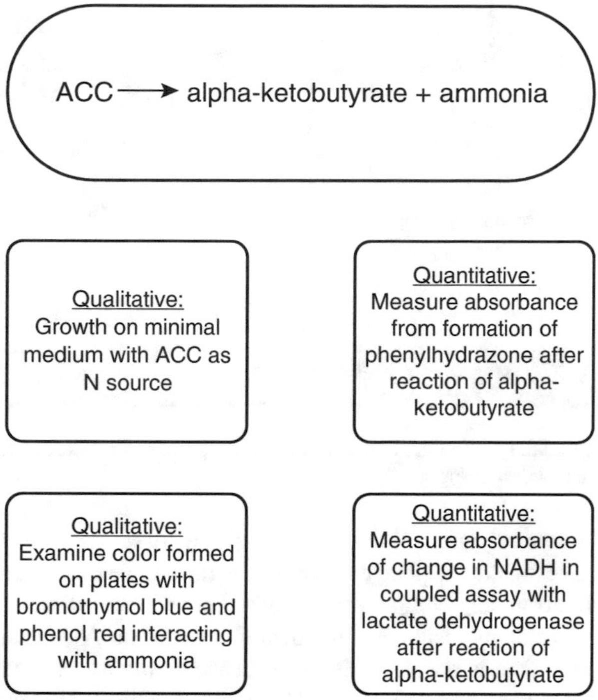

Fig. 5.20 Overview of some qualitative and quantitative assays of ACC deaminase activity. The conversion of ACC by ACC deaminase to α-ketobutyrate and ammonia is shown at the top of this figure

ACC ⟶ alpha-ketobutyrate + ammonia

Qualitative:
Growth on minimal medium with ACC as N source

Quantitative:
Measure absorbance from formation of phenylhydrazone after reaction of alpha-ketobutyrate

Qualitative:
Examine color formed on plates with bromothymol blue and phenol red interacting with ammonia

Quantitative:
Measure absorbance of change in NADH in coupled assay with lactate dehydrogenase after reaction of alpha-ketobutyrate

dahlia and the bacteria *Bacillus anthracis*, *Pseudomonas aeruginosa*, and *Serratia marcescens*. In these instances, the ability to lower plant ethylene levels appears to have provided an advantage to these pathogens in their interaction with plants.

5.4.2.1 Enzyme Assays

A variety of methods have been devised in an effort to measure microbial levels of ACC deaminase (Fig. 5.20). The simplest, qualitative, method includes assessing whether or not the microorganism being tested is able to grow on minimal medium using ACC as a sole source of nitrogen. In this case, growth is relatively slow, and care must be taken to ensure that highly purified agar that does not contain any traces of available nitrogen is employed. Another simple plate assay includes growing the test microorganisms on minimal medium with ACC as the sole source of nitrogen with each organism grown separately on (i) minimal medium with ACC plus the dye bromothymol blue and (ii) minimal medium with ACC plus the dye phenol red. In this case, the plates with colonies that yield more intense colors produce greater amounts of ACC deaminase. Thus, this method may be said to be semiquantitative. Alternatively, a reliable quantitative assay includes spectrophotometrically measuring the color change following the reaction between α-ketobutyrate and dinitrophenylhydrazine to form a dinitrophenylhydrazone. Another reliable quantitative assay includes measuring the color change in the conversion of NADH to NAD⁺ in the coupled assay of the enzyme lactate dehydrogenase reacting with

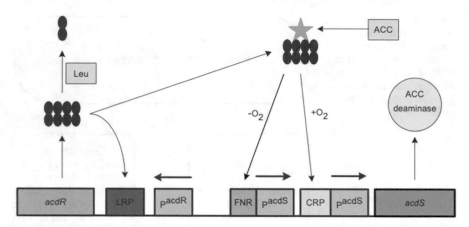

Fig. 5.21 A model of the transcriptional regulation of ACC deaminase in some free-living bacteria and *Rhizobia* spp.

α-ketobutyrate. Notwithstanding the ease of the quantitative assays, many investigators depend exclusively on the simple qualitative assays and they are unable to accurately compare the activity between one strain and another.

In many instances, the identification of an ACC deaminase gene in a particular organism is based on DNA sequence similarity between the putative ACC deaminase gene and known ACC deaminase genes. However, sequence homology by itself does not suffice to define a DNA open reading frame as an ACC deaminase encoding sequence. This is because several other pyridoxal phosphate-containing enzymes have amino acid sequences that are similar to ACC deaminase. For instance, in one study, based on enzyme activity, changing only two amino acids was sufficient to convert a D-cysteine desulfhydratase gene to one encoding ACC deaminase and an ACC deaminase gene to one encoding D-cysteine desulfhydratase.

5.4.2.2 Transcriptional Regulation

ACC deaminase activity is transcriptionally regulated in a similar manner in a number of different free-living soil bacteria including most Rhizobia. In the model shown, the *acdR* (ACC deaminase regulatory) gene encodes the LRP (leucine-rich regulatory) protein and *acdS* is the ACC deaminase structural gene (Fig. 5.21). The active form of the LRP protein is an octamer that can, when it is present in excess, bind to the LRP box on the DNA preventing additional transcription of the gene encoding this protein. Alternatively, this octamer can bind to the AcdB protein (another regulatory protein, shown as a green star) which is complexed with ACC, and as a result, in the presence of oxygen and a CRP (cAMP receptor) protein, bind to the CRP box and activate transcription from the *acdS* promoter. In the absence of oxygen, the LRP octamer–AcdB–ACC complex binds to the FNR (fumarate-nitrate reductase) protein before the entire complex binds to the FNR box, also activating an *acdS* promoter. In addition to all of the abovementioned proteins, the bacterial RNA polymerase is also required for transcription of both *acdR* and *acdS*. Following its synthesis, AcdS (i.e., ACC deaminase) cleaves ACC, producing ammonia and α-ketobutyrate. When ACC is degraded, the ammonia is assimilated and α-ketobutyrate is converted into branched chain amino acids including leucine. Finally, an excess of free leucine within the bacterial cell causes the LRP octamer to dissociate into an inactive dimeric form. Notwithstanding the fact that ACC deaminase is not needed for the survival of the bacteria in which it is encoded, this rather complex mode of gene expression suggests that the expression of this gene is highly important to the environmental functioning of the bacteria that contain this enzyme.

In some bacteria, the FNR and/or CRP proteins and boxes are not part of the regulatory mechanism of ACC deaminase gene expression. It is likely that these genetic elements facilitate protein expression with bacteria that are commonly found at the aerobic–anaerobic interphase within the soil. Bacteria that are found primarily at one or the other environment may not require this part of the regulatory mechanism. In addition, the direction of transcription of *acdR* and *acdS* can vary between different bacterial strains. Moreover, as indicated below, *Mesorhizobium* species transcriptionally regulate ACC deaminase gene expression in an entirely different manner, as do many *Actinobacteria*.

5.4.3 Ethylene and Nodulation

Bacteria that contain ACC deaminase activity are found relatively commonly in a wide range of soils. Genes coding for ACC deaminase have been found in Gram-negative and Gram-positive bacteria, rhizobia, methylobacteria, rhizosphere bacteria and endophytes, and fungi. In one study, ACC deaminase activity/genes were found in a wide range of different bacteria including *Azospirillum, Rhizobium, Agrobacterium, Achromobacter, Burkholderia, Ralstonia, Pseudomonas,* and *Enterobacter*. As an indication of the frequency of ACC deaminase genes in rhizobia in nature, in another study, 12% of the 233 *Rhizobium* spp. isolated and purified from nodules of wild legumes at more than 30 different sites in southern and central Saskatchewan, Canada (over an area of ~40,000 square kilometers; approximately half the size of Portugal or Ireland) encoded and expressed this enzyme. Thus, ACC deaminase may be considered to be commonly present in various rhizobia strains.

Ethylene is an inhibitor of rhizobial nodulation of legumes, and since the infection of plant roots by Rhizobia causes plants to locally produce ethylene, rhizobial infection is therefore a self-limiting process. On the other hand, many strains of Rhizobia produce either rhizobitoxine (Fig. 5.22), a naturally occurring chemical

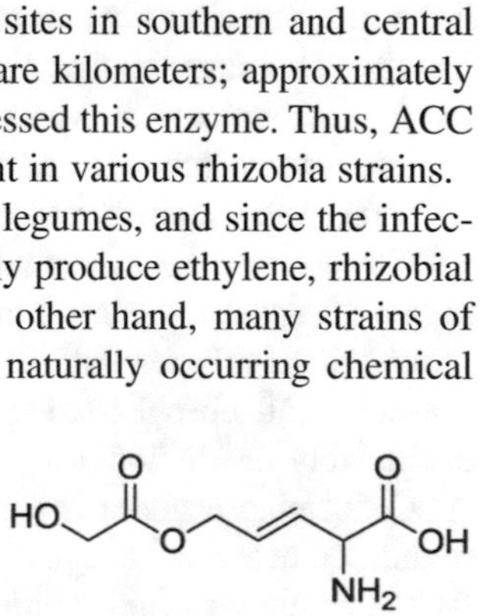

Fig. 5.22 Chemical structure of the compound rhizobitoxine

Rhizobitoxine

Fig. 5.23 Acquisition of ACC deaminase by *Sinorhizobium meliloti* increases nodulation and biomass of alfalfa. © 2010 American Society for Microbiology. Used with permission. No further reproduction or distribution is permitted without the prior written permission of American Society for Microbiology

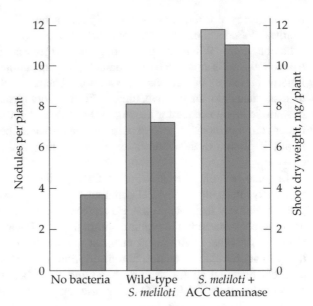

inhibitor of the enzyme ACC synthase (produced and secreted by some strains of *Bradyrhizobium elkanii*), or ACC deaminase, which allows these strains to locally lower the root ethylene levels (caused by rhizobial infection) and thereby increase both the extent of nodulation and the plant biomass, typically by ~30% (Fig. 5.23). When different strains of Rhizobia were assayed for the presence of ACC deaminase, it was found that some, but not all, of the strains tested contained this enzyme (Table 5.4). These strains were assayed for ACC deaminase in three different ways: (1) enzyme activity, (2) Southern hybridization (in case a strain had no enzyme activity but the gene for this enzyme was present), and (3) western blotting (in case a strain had no enzyme activity but the enzyme protein was synthesized). Interestingly, when they are assayed, a large number of commercially available rhizobial strains, but a much smaller number of newly isolated field strains, contain ACC deaminase activity. This suggests that the future screening of field bacterial isolates of Rhizobia for ACC deaminase activity may be one means of rapidly selecting for strains with superior commercial inoculant potential.

A few years after these experiments were reported, it was demonstrated that the ACC deaminase genes in a number of *Mesorhizobium* strains were transcriptionally regulated under the control of the *nifA* promoter. Recall that the *nifA* gene product (NifA) positively regulates the expression of all other *nif* genes and is universally present in all nitrogen-fixing bacteria. Importantly, the NifA protein is expressed exclusively inside nodules. Thus, the transcriptional regulation of *Mesorhizobium* ACC deaminase genes is significantly different from the more common mode of regulation described above. Some researchers have suggested that the expression of ACC deaminase genes within nitrogen-fixing nodules decreases the rate of nodule senescence—as nitrogen fixation with its high energy demand could potentially

Table 5.4 Assaying *Rhizobium* spp. for ACC deaminase

Strain	ACC deaminase activity	Southern hybridization	Western blots	Conclusion
R. leguminosarum bv. *viciae* 128C53K	+	++	+	Yes
R. leguminosarum bv. *viciae* 128C53	+	++	+	Yes
R. leguminosarum bv. *viciae* 99A1	+	++	+	Yes
R. leguminosarum bv. *phaseoli* 657	−	−	−	No
R. sp. Designati	−	−	−	No
R. Hedysari	+	+	+	Yes
M. Ciceri	−	−	+	Yes/no
R. leguminosarum bv. *trifolii*	−	+	−	No
R. radicicola	−	−	−	No
M. loti MAFF303099	−	++	−	Yes/no
S. meliloti 1021	−	−	−	No

M, *Mesorhizobium*; S, *Sinorhizobium;* bv., biovar.; −, no hybridization; +, moderate hybridization; ++, strong hybridization

activate stress ethylene synthesis—and by prolonging the lifetime of the nodule, effectively increases the amount of nitrogen that can be fixed.

However, an ACC deaminase gene from *Mesorhizobium* spp. that is under the control of a *nifA* promoter, does not facilitate nodulation of legume roots by those strains. With this in mind, one group of researchers transformed a strain of *Mesorhizobium ciceri* with an ACC deaminase gene from a free-living bacterium so that the transformants contained two separate ACC deaminase genes, each with a different mode of transcriptional regulation (Fig. 5.24). When the behavior of the wild-type *Mesorhizobium ciceri* strain was compared to the transformed strain, it was found that, in addition to having longer lived nodules, plants treated with the transformed strain displayed a significantly increased number of nodules, average weight per nodule, and total plant biomass (with two different chickpea cultivars). In addition, plants treated with the transformed *Mesorhizobium ciceri* strain were more resistant to *Fusarium* spp.-caused root rot disease. Unfortunately, despite the obvious benefits of this simple genetic manipulation under laboratory conditions, this approach remains to be tested under field conditions, a situation which requires regulatory approval.

In one series of experiments, researchers noted that *Rhizobium* sp. TAL1145 has an ACC deaminase gene that is regulated by *nifA* and is not induced by ACC but rather by mimosine. The alkaloid mimosine is β-3-hydroxy-4 pyridone amino acid (Fig. 5.25). This (otherwise) toxic compound occurs in several *Mimosa* spp. and *Leucaena* spp. plants. In this case, similar to what has been suggested with *Mesorhizobium* spp., this ACC deaminase gene is thought to decrease the rate of

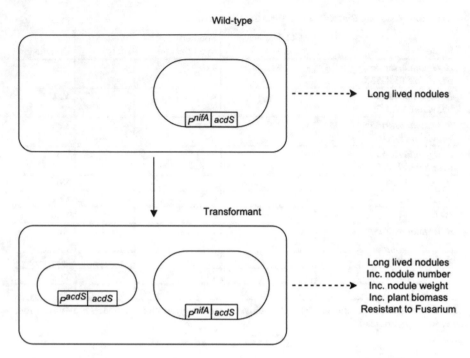

Fig. 5.24 Transformation of a wild-type strain of *Mesorhizobium ciceri* increases nodule number, nodule weight, plant biomass and resistance of treated to chickpea plants to *Fusarium* spp.-caused root rot while retaining the ability of the wild-type strain to induce long lived nodules. The wild-type contains a single ACC deaminase gene (*acdS*) under the control of a *nifA*. The transformant contains an ACC deaminase gene under the control of a *nifA* promoter (P^{nifA}) and an ACC deaminase gene under the control of an ACC deaminase gene promoter (P^{acdS}) from a free-living bacterium

Fig. 5.25 The chemical structure of the alkaloid mimosine

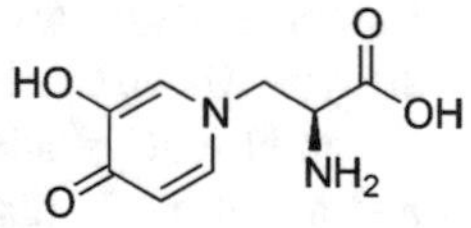

nodule senescence. In one series of experiments, this bacterium was genetically transformed with an ACC deaminase gene from *Sinorhizobium* sp. BL3 which, like most other rhizobial ACC deaminase genes is transcriptionally regulated by *acdR* and induced by ACC. In response to the introduction of the foreign *acdS* gene, when the transformed *Rhizobium* sp. TAL1145 strain was used in association with *Leucaena leucocephala* plants, a small species of Mimosoid tree native to Mexico and Central America, both nodule number and nodule size as well as root biomass were significantly increased. In this instance, with two different *acdS* genes, each transcriptionally regulated by a different mechanism, the transformants caused the formation of more and larger nodules that were more long lived than usual.

Table 5.5 Co-inoculation of legumes with rhizobial strains with either an ACC deaminase-containing PGPB or the same PGPB with a mutated ACC deaminase gene and subsequent assessment of nodule number and dry plant shoot and root biomass

Treatment	Number of nodules	Shoot dry weight (mg)	Root dry weight (mg)
Plant: Phaseolus vulgaris			
Negative control	0	105	40
Rhizobia only	7	160	40
Rhizobia + PGPB	32	210	65
Rhizobia + mutnat	10	165	40
Plant:Mimosa pudica			
Negative control	0	10	8
Rhizobia only	10	37	22
Rhizobia + PGPB	18	55	40
Rhizobia + mutant	14	38	18

Each treatment consisted of ten plants. Plants were harvested 20 days after inoculation

Unfortunately, at this time, regulatory agencies are unlikely to readily approve the field use of such a genetically manipulated bacterium.

Despite a similar mode of expression and regulation to PGPB that contain ACC deaminase, most Rhizobia usually express a low percentage of ACC deaminase enzyme activity (typically about 3–10%), compared to the amount expressed by free-living PGPB. The free-living bacteria that bind relatively nonspecifically to plant roots, with a high level of ACC deaminase activity, protect plants from a range of different abiotic and biotic stresses by lowering ethylene levels throughout the plant (see Chaps. 6, 8 and 10). On the other hand, various Rhizobia bind tightly only to the roots of specific plants and have a low level of enzyme activity which facilitates nodulation by locally lowering ethylene levels without affecting the overall ethylene concentration in the plant.

Recently, it was reported that co-inoculation of a free-living endophytic ACC deaminase-producing PGPB strain with either the alpha-rhizobia *Rhizobium tropici* or the beta-rhizobia *Cupriavidus taiwanensis* significantly increased the nodulation abilities of both Rhizobia strains (Table 5.5). The *Rhizobium tropici* strain was used to nodulate common bean plants (*Phaseolus vulgaris*) while *Cupriavidus taiwanensis* was used to nodulate touch-me-not (*Mimosa pudica*) plants. In these experiments, treatment with the wild-type PGPB, but not with the ACC deaminase minus mutant PGPB, significantly increased both the number of nodules per plant and the dry weights of roots and shoots. Although it was not tested in these experiments, co-inoculation with a free-living PGPB and an appropriate rhizobial strain should also protect the plants against inhibition by various environmental stresses potentially making this approach a highly effective commercial strategy; however, this remains to be proven to be effective in the field. In addition, it is important to bear in mind that using a co-inoculation strategy does not require the

Table 5.6 Nodulation, nitrogen fixation, and plant growth of soybean plants inoculated with various strains of *B. japonicum*

Strain	Number of nodules per plant	Nodule dry weight (mg)	Shoot dry weight (mg)	Nitrogenase activity, nmol ethylene·h^{-1} mg nodule dry weight $^{-1}$
Negative control	0.0	0.0	169.5	0.0
E109	24.7	12.3	223.2	31.7
163	24.0	12.2	230.0	32.8
366	34.5	14.3	231.0	32.5

genetic manipulation of any of the rhizobial strains thereby obviating potential regulatory concerns.

In Argentina, the most efficient strain of *Bradyrhizobium japonicum*, that has also been used commercially for many years in soybean production, is strain E109. Nevertheless, researchers recently isolated and characterized two *B. japonicum* strains, 163 and 366, that fix more nitrogen than strain E109 (Table 5.6). In particular, strain 366 yields more and bigger nodules than the commercial strain. Multiplying the number of nodules per plant times the nodule dry weight times the level of nitrogenase activity provides an estimate of the amount of fixed nitrogen produced per hour at a particular point in time in soybean plants by these bacterial strains. It was also noted that leghemoglobin levels in nodules induced by strain 163 lasted much longer than nodules induced by either of the other two strains; as observed at weeks 5–7 post inoculation. This means that nodules induced by strain 163 senesced at a much slower rate than nodules induced by either of the other two strains. In these experiments, the increased level of fixed nitrogen from soybean plants inoculated with strains 366 and 163 were attributed to an increase in nodule number and size and to the presence of ACC deaminase, respectively. In fact, the scientists who performed this study confirmed that the observed ACC deaminase expression, which reduced the amount of ethylene and delayed nodule senescence, was dependent upon the *nifA* promoter. This is not only the first time that delayed senescence has been shown to lead to higher nitrogen fixation in plants, but also it is the first time that regulation of ACC deaminase by the *nifA* promoter has been observed in *B. japonicum*.

5.4.4 Ethylene and Plant Transformation

Agrobacterium tumefaciens, a Gram-negative soil bacterium, is a phytopathogen that genetically transforms plant cells leading to the formation of crown gall tumors on the infected plant that interfere with the plant's normal growth. Crown gall formation occurs as a consequence of the transfer, integration, and expression of genes of a segment of bacterial plasmid DNA into the plant cell genome. The transferred DNA (T-DNA) is region of the bacterium's tumor-inducing (Ti) plasmid that is commonly found in most strains of *A. tumefaciens*. Most of the genes that are located within the T-DNA region become activate only after the T-DNA is inserted into the plant genome. This is because genes in this region are essentially plant genes that cannot be expressed in bacteria because of the differences in transcriptional and translational regulatory sequences between the two types of organisms. The products of these genes are responsible for crown gall formation. Although Ti plasmids are effective as natural vectors, it is necessary to remove most of the genes within the T-DNA region before they can be used as cloning vectors. Following extensive modification of the Ti plasmid and its T-DNA region, two types of Ti plasmid-based vectors have been created that enable scientists to introduce foreign DNA into the plant host genome.

Although the *A. tumefaciens*-mediated gene transfer systems were effective in many species of plants, monocotyledonous plants, including the world's major cereal crops (rice, wheat, and corn), were not as readily transformed by *A. tumefaciens*. However, by refining and modifying conditions, protocols were devised for the transformation of monocotyledonous plants by *A. tumefaciens* carrying Ti plasmid vectors. These altered conditions included the use of broad-host-range strains of *A. tumefaciens* that infect plants that previously appeared to be refractory to transformation by *A. tumefaciens*. In addition, it was discovered that the conditions that were used in *Agrobacterium*-mediated plant transformation produced moderate amounts of ethylene during plant transformation thereby significantly decreasing the frequency of transformation. To overcome this problem, a bacterial gene encoding the enzyme ACC deaminase was introduced into an *A. tumefaciens* strain and, when expressed, lowered plant ethylene levels. When melon cotyledon segments were genetically transformed using the *A. tumefaciens* strain expressing ACC deaminase, the transformation frequency of the plants increased significantly. Moreover, other researchers reported that an *A. tumefaciens* strain that had been genetically engineered to express the ACC deaminase gene was more efficient at transforming commercial canola cultivars than the nonengineered strain. More recently, the researchers who first engineered a strain of *A. tumefaciens* to produce ACC deaminase developed a strain of *A. tumefaciens* that synthesized both ACC deaminase and GABA transaminase thereby increasing the transformation rate of host plants by 3.6-fold compared to using the *A. tumefaciens* strain with ACC deaminase alone. This was done because gamma-aminobutyric acid (GABA), a non-protein amino acid, was another negative factor in *A. tumefaciens*-plant interactions. As a consequence of these manipulations, it is expected that the genetic transformation of a wide range of different plants simpler, faster, and more efficient.

5.5 Volatile Organic Compounds

In the past 10–15 years, it has been observed that some PGPB produce and subsequently release volatile organic compounds that are able to stimulate plant growth. The activity of these volatile organic compounds may be detected in several ways. (1) For example, to detect this mode of plant growth promotion, a number of germinated surface-sterilized seeds may be planted on one side of an agar petri plate

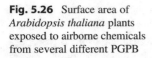 **Fig. 5.26** Surface area of *Arabidopsis thaliana* plants exposed to airborne chemicals from several different PGPB

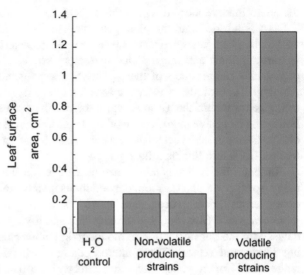

while a small aliquot of a bacterial inoculant is spotted in the middle of the other half of the plate. In this case, the two halves of the agar plate are physically separated by a center partition. (2) A bacterium growing on an agar petri plate is placed beneath growing seedlings with the petri plate and the plants separated by a wire mesh. (3) A bacterium growing on an agar petri plate or bacteria in an open container, previously grown in liquid medium, and plant seedlings placed adjacent to one another in a sealed container. The key to all of these approaches is that there is no possibility of physical contact between the bacteria (or their solutes) and the plant seedlings. Thus, any alteration to the growth of the plant seedlings may be attributed to volatile organic compounds produced by the bacteria.

In one study, the interaction between some Gram-positive bacteria (including *Bacillus subtilis* and *B. amyloliquefaciens*) and the model plant *Arabidopsis thaliana* using method #1 (above) was examined. In this case, following an incubation period of 14 days, the leaf surface area of the plants on each plate was measured (Fig. 5.26). Based on this assay, it was found that two of the PGPB strains (*B. subtilis* GB03 and *B. amyloliquefaciens* IN937a) significantly increased the plant leaf surface area. The volatile compounds produced by these bacteria were analyzed by gas chromatography and it was subsequently determined that two compounds acetoin and 2,3 butanediol (Fig. 5.27) were responsible for the bulk of the observed increase in leaf surface area. Moreover, in this system, PGPB mutants unable to synthesize these two compounds do not promote plant growth. It should, however, be noted that other (so-called minor) volatile compounds such as 2-pentylfuran and dimethylhexadecylamine (Fig. 5.25) that are produced by these bacteria may also play a significant role in signaling the plant to alter its gene expression and thereby its physiology.

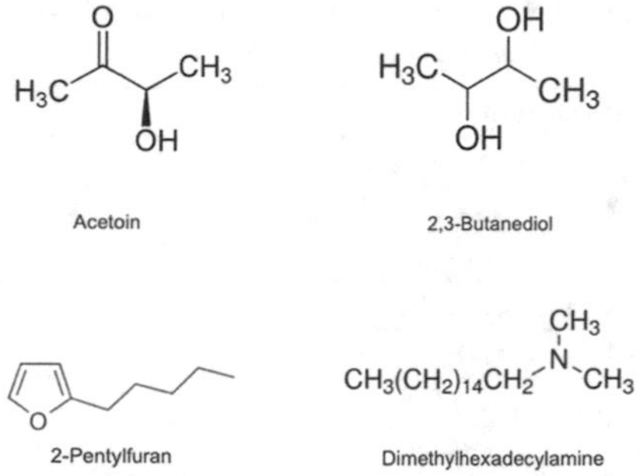

Fig. 5.27 Chemical structure of acetoin, 2,3-butanediol, 2-pentylfuran and dimethylhexadecylamine

Acetoin

2,3-Butanediol

2-Pentylfuran

Dimethylhexadecylamine

The abovementioned experiment was repeated with mutant strains of *A. thaliana* that were altered in their response to one or more plant hormones. The results of these experiments suggested that volatile compound plant growth promotion did not depend at all on brassinosteroids, gibberellin, or ethylene signaling. However, on the basis of these experiments, the participation of auxin cannot be excluded. In a more definitive experiment, *A. thaliana* plants treated with volatile organic compounds were subjected to microarray analysis at 48 and 72 h after treatment. At both times the expression of genes encoding flavonoid and auxin synthesis, and cell wall-loosening enzymes were increased. Also, the level of ribosomal protein synthesis was increased at 72 h after treatment indicating elevated protein synthesis. In addition, volatile organic compound-mediated growth promotion is inhibited by blocking auxin transport with the result that gene expression for auxin synthesis is upregulated in the aerial regions of GB03-exposed plants. That is, following treatment with volatile organic compounds, auxin accumulation decreases in leaves and increases in roots. All of this data is consistent with the notion that volatile organic compounds regulate auxin homeostasis in plants thereby promoting plant growth.

Arabidopsis seedlings exposed to *B. subtilis* GB03 or *B. amyloliquefaciens* IN937a for 4 days developed significantly fewer symptomatic leaves 24 h after inoculation with the soft rot-causing pathogen *Erwinia carotovora* subsp. *carotovora*. This effect was attributed to the development of induced systemic resistance or ISR (discussed in Chap. 6) in the more resistant plants. It was further deduced that the volatile compounds from strain IN937a triggered ISR through an ethylene-independent signaling pathway, whereas the volatile compounds from strain GB03 operate through an ethylene-dependent pathway.

In a different type of experiment, volatile organic compounds released by three different PGPB (*Pseudomonas fluorescens*, *Bacillus subtilis* and *Azospirillum brasilense*) were found to stimulate the growth parameters and the composition of essential oils in the aromatic plant *Mentha piperita* (peppermint). As in the experiments described above, the bacteria and plants were grown on the same petri plate, but were separated by a physical barrier so that the plants were exposed only to the volatile organic compounds but not to solutes from the bacteria. Based on the observation that each of the three strains affects the plants somewhat differently, these findings suggest that volatile organic compounds can both induce the

Table 5.7 Bacteria that produce volatile organic compounds and the plants whose growth these compounds can stimulate

Bacterium	Plant
Alcaligenes faecalis	*Arabidopsis thaliana*
Azospirillum brasilense	*Arabidopsis thaliana*
Azomonas agilis	*Medicago truncatula* (barrel medic)
Azomonas agilis	*Sorghum bicolor* (sorghum)
Azomonas agilis	*Medicago sativa* (alfalfa)
Bacillus ambifaria	*Arabidopsis thaliana*
Bacillus subtilis	*Arabidopsis thaliana*
Bacillus subtilis	*Ocimum basilicum* (basil)
Bacillus subtilis	*Mentha piperita* (peppermint)
Bacillus megaterium	*Arabidopsis thaliana*
Bacillus pyrrocinia	*Arabidopsis thaliana*
Bacillus sp.	*Nicotiana attenuata* (wild tobacco)
Bacillus vallismortis	*Nicotiana tabacum* (tobacco)
Bacillus amyloliquefaciens	*Arabidopsis thaliana*
Chromobacterium violaceum	*Arabidopsis thaliana*
Paenibacillus polymyxa	*Arabidopsis thaliana*
Pseudomonas chlororaphis	*Arabidopsis thaliana*
Pseudomonas fluorescens	*Arabidopsis thaliana*
Pseudomonas simiae	*Glycine max* (soybean)

biosynthesis of secondary metabolites and modulate specific steps of monoterpene metabolism.

Since 2003 when bacterial volatile organic compounds were first shown to promote plant growth, several dozen bacterial strains that synthesize these compounds and use them to facilitate plant growth have been identified (Table 5.7). The fact that most of the bacteria that have been shown to produce volatile organic compounds were identified using a test system that included *Arabidopsis thaliana* is a direct consequence of the small size of these plants so that several seedlings can conveniently be placed on one side of a petri plate with a physical separation in the middle, by far the simplest of the test systems. Moreover, while the all of the bacteria that were originally identified as being able to produce volatile organic compounds were Gram-positive, more recently several Gram-negative bacteria have also been shown to produce these compounds. In addition, some fungi have also been reported to emit bioactive compounds that can induce plant growth.

Questions

1. How would you prove that bacterial IAA contributes significantly to the promotion of plant growth by a PGPB strain?

2. Why does the addition of PGPB that can synthesize IAA to the roots of developing plants stimulate the growth of some plants and inhibit the growth of others?

3. What is rhizobitoxine and what does it do?

4. Describe some of the pathways involved in the biosynthesis of bacterial IAA.

5. How can IAA biosynthesis be regulated in bacteria?

6. Having isolated a *Pseudomonas* gene encoding the enzyme indole-3-pyruvic acid decarboxylase that is involved in the biosynthesis of IAA, suggest a strategy for creating a mutant of the starting bacterium that produces a decreased amount of IAA.

7. What are cytokinins and how do they affect plants?

8. How would you select bacterial mutants that either overproduce or underproduce cytokinins?

9. What are gibberellins and how do they affect plants?

10. What are the major steps in the synthesis of ethylene in plants?

11. What are some of the ways in which ethylene affects plant growth and development?

12. How does the enzyme ACC deaminase lower plant ethylene levels?

13. How is ACC deaminase expression regulated in free-living soil bacteria?

14. How does ACC deaminase affect nodulation of legumes by *Rhizobia* spp.?

15. Suggest an explanation as to why the *Mesorhizobium loti acdS* gene is detected by Southern hybridization but the protein is not found by Western blots.

16. How do volatile organic compounds promote plant growth?

Bibliography

Abeles FB, Morgan PW, Saltveit ME Jr (1992) Ethylene in plant biology. Academic Press, New York

Ahmed A, Hasnain S (2010) Auxin-producing *Bacillus* sp.: auxin quantification and effect on the growth of *solanum tuberosum*. Pure Appl Chem 82:313–319

Arkhipova TM, Veselo SU, Melentiev AI, Martynenko EV, Kudoyarova GR (2005) Ability of bacterium *Bacillus subtilis* to produce cytokinins and to influence the growth and endogenous hormone content of lettuce plants. Plant Soil 272:201–209

Arshad M, Frankenberger WT Jr (2002) Ethylene: agricultural sources and applications. Kluwer Academic/Plenum Publishers, New York

Baudoin E, Lerner A, Mirza MS, El Zemrany H, Prigent-Combaret C, Jurkevich E, Spaepen S, Vanderleyden J, Nazaret S, Okon Y, Moënne-Loccoz Y (2010) Effects of *Azospirillum brasilense* with genetically modified auxin biosynthesis gene ipdC upon the diversity of the indigenous microbiota of the wheat rhizosphere. Res Microbiol 161:219–226

Bianco C, Defez R (2010) Improvement of phosphate solubilization and *Medicago* plant yield by an indole-3-acetic acid-overproducing strain of *Sinorhizobium meliloti*. Appl Environ Microbiol 76:4626–4632

Bianco C, Imperlini E, Calogero R, Senatore B, Amoresano A, Carpentieri A, Pucci P, Defez R (2006) Indole-3-acetic acid improves *Escherichia coli*'s defences to stress. Arch Microbiol 185:373–382

Bianco C, Imperlini E, Defez R (2009) Legumes like more IAA. Plant Signal Behav 4:763–765

Bianco C, Senatore B, Arbucci S, Pieraccini G, Defez R (2014) Modulation of endogenous indole-3-acetic acid biosynthesis in bacteroids within *Medicago sativa* nodules. Appl Environ Microbiol 80:4286–4293

Chen S, Qiu C, Huang T, Zhou W, Qi Y, Gao Y, Shen J, Qiu L (2013) Effect of 1-aminocyclopropane-1-carboxylic acid deaminase producing bacteria on the hyphal growth and primordium initiation of Agaricus bisporus. Fungal Ecol 6:110–118

Costacurta A, Vanderleyden J (1995) Synthesis of phytohormones by plant-associated bacteria. Crit Rev Microbiol 21:1–18

Duan J, Muller KM, Charles TC, Vesley S, Glick BR (2009) 1-Aminocyclopropane-1-carboxylate (ACC) deaminase genes in rhizobia from southern Saskatchewan. Microb Ecol 57:423–436

Duca D, Lorv J, Patten CL, Rose D, Glick BR (2014) Microbial indole-3-acetic acid and plant growth. Anton Van Leeuwenhoek 106:85–125

Duca D, Rose DR, Glick BR (2018) Indoleacetic acid overproduction transformants of *Pseudomonas* sp. UW4. Anton Van Leeuwenhoek 111:1645–1660

Fincheira P, Quiroz A (2018) Microbial volatiles as plant growth inducers. Microbiol Res 208:63–75

Frankenberger WT, Arshad M (1995) Auxins. In: Frankenberger WT, Arshad M (eds) Phytohormones in soils microbial production & function. Marcel Dekker, New York, pp 17–136

Gamalero E, Glick BR (2011) Mechanisms used by plant growth-promoting bacteria. In: Maheshwari DK (ed) Bacteria in agrobiology: plant nutrient management. Springer, Berlin, pp 17–46

Gamalero E, Glick BR (2012) Ethylene and abiotic stress tolerance in plants. In: Ahmad P, MNV P (eds) Environmental adaptations and stress tolerance of plants in the era of climate change. Springer, Berlin, pp 395–412

Gamalero E, Glick BR (2015) Bacterial modulation of plant ethylene levels. Plant Physiol 169:13–22

Glick BR (2014) Bacteria with ACC deaminase can promote plant growth and help to feed the world. Microbiol Res 169:30–39

Grossmann K (2010) Auxin herbicides: current status of mechanism and mode of action. Pest Manag Sci 66:113–120

Guinel FC (2015) Ethylene, a hormone at the center-stage of nodulation. Front Plant Sci 6:1121

Hao Y, Charles TC, Glick BR (2010) ACC deaminase increases the *Agrobacterium tumefaciens*-mediated transformation frequency of commercial canola cultivars. FEMS Microbiol Lett 307:185–190

Hershey DM, Lu X, Zi J, Peters RJ (2014) Functional conservation of the capacity for *ent*-kaurene biosynthesis and an associated operon in certain rhizobia. J Bacteriol 196:100–106

Hontzeas N, Zoidakis J, Glick BR, Abu-Omar MM (2004) Expression and characterization of 1-aminocyclopropane-1-carboxylate deaminase from the rhizobacterium *Pseudomonas putida* UW4: a key enzyme in bacterial plant growth promotion. Biochim Biophys Acta 1703:11–19

Houben M, Van de Poel B (2019) 1-Aminocyclopropane-1-carboxylic acid oxidase (ACO): the enzyme that makes the plant hormone ethylene. Front Plant Sci 10:695

Hussain A, Hasnain S (2009) Cytokinin production by some bacteria: its impact on cell division in cucumber cotyledons. Afr J Microbiol Res 3:704–712

Ibort P, Molina S, Núñez R, Zamarreño AM, García-Mina JM, Ruiz-Lozano JM, Orozco-Mosqueda MC, Glick BR, Aroca R (2017) Tomato ethylene sensitivity determines interaction with plant growth-promoting bacteria. Ann Bot 120:101–122

Idris EE, Iglesias DJ, Talon M, Borriss R (2007) Tryptophan-dependent production of indole-3-acetic acid (IAA) affects level of plant growth promotion by *Bacillus amyloliquefaciens* FZB42. Mol Plant Microbe Interact 20:619–626

Imperlini E, Bianco C, Lonardo E, Camerini S, Cermola M, Moschetti G, Defez R (2009) Effects of indole-3-acetic acid on *Sinorhizoboium meliloti* survival and on symbiotic nitrogen fixation and stem dry weight production. Appl Microbiol Biotechnol 83:727–738

Kai M, Effmert U, Piechulla B (2016) Bacterial-plant-interactions: approaches to unravel the biological function of bacterial volatiles in the rhizosphere. Front Microbiol 7:108

Kazan K, Manners JM (2009) Linking development to defense: auxin in plant-pathogen interactions. Trends Plant Sci 14:373–382

Khandelwal A, Sindhu SS (2013) ACC deaminase containing rhizobacteria enhance nodulation and plant growth in Clusterbean (*Cyamopsis tetragonoloba* L.). J Microbiol Res 3:117–123

Lee K-E, Radhakrishnan R, Kand S-M, You Y-H, Joo G-J, Lee I-J, Ko J-H, Kim J-H (2015) *Enterobacter faecium* LKE12 cell-free extract accelerates host plant growth *via* gibberellin and indole-3-acetic acid secretion. J Microbiol Biotechnol 25:1467–1475

López SMY, Sánchez MDM, Pastorino GM, Franco MEE, García NT, Balatti PA (2018) Nodulation and delayed nodule senescence: strategies of two *Bradyrhizobium japonicum* isolates with high capacity to fix nitrogen. Curr Microbiol 75:997–1005

Malhotra M, Srivastava S (2008) An *ipdC* gene knock-out of *Azospirillum brasilense* strain SM and its implications on indole-3-acetic acid biosynthesis and plant growth promotion. Anton Van Leeuwenhoek 93:425–433

Nascimento FX, Brígido C, Glick BR, Oliveira AL (2012) *Mesorhizobium ciceri* LMS-1 expressing an exogenous 1-aminocyclopropane-1-carboxylate (ACC) deaminase increases its nodulation abilities and chickpea plant resistance to soil constraints. Lett Appl Microbiol 55:15–21

Nascimento FX, McConkey BJ, Glick BR (2014) New insights into ACC deaminase phylogeny, evolution and evolutionary significance. PLoS One 9(6):e99168

Nascimento FX, Rossi MJ, Glick BR (2018) Ethylene and 1-aminocyclopropane-1-carboxylate (ACC) in plant-bacterial interactions. Front Plant Sci 9:114

Nascimento FX, Tavares MJ, Franck J, Ali S, Glick BR, Rossi MJ (2019) ACC deaminase plays a major role in *Pseudomonas fluorescens* YsS6 ability to promote the nodulation of Alpha- and Betaproteobacteria rhizobial strains. Arch Microbiol. https://doi.org/10.1007/s00203-019-01649-5

Naveed M, Qureshi MA, Zahir ZA, Hussain MB, Sessitsch MB (2014) L-tryptophan-dependent biosynthesis of indole-3-acetic acid (IAA) improves plant growth and colonization of maize by *Burkholderia phytofirmans* PsJN. Ann Microbiol 65:1381–1389

Nonaka S, Sugawara M, Minamisawa K, Yuhashi K, Ezura H (2008) 1-Aminocyclopropane-1-carboxylate de3aminase enhances *Agrobacterium tumefaciens*-mediated gene transfer into plant cells. Appl Environ Microbiol 74:2525–2528

Nonaka S, Someya T, Kadota Y, Nakamura K, Ezura H (2019) Super-*Agrobacterium* ver. 4: improving the transformation frequencies and genetic engineering possibilities for crop plants. Front. Plant Sci 10:1204

Nukui N, Minamisawa K, Ayabe SI, Aoki T (2006) Expression of the 1-aminocyclopropane-1-carboxylic acid deaminase gene requires symbiotic nitrogen-fixing regulator gene *nifA2* in *Mesorhizobium loti* MAFF303099. Appl Environ Microbiol 72:4964–4969

Onofre-Lemus J, Hernandez-Lucas I, Girard L, Caballero-Mellado J (2009) ACC (1-aminocyclopropane-1-carboxylate) deaminase activity, a widespread trait in *Burkholderia* species, and its growth-promoting effect on tomato plants. Appl Environ Microbiol 75:6581–6590

Ouyang L, Pei H, Xu Z (2016) Low nitrogen stress stimulating the indole-3-acetic acid biosynthesis of *Serratia* sp. ZM is vital for the survival of the bacterium and its plant growth-promoting characteristic. Arch Microbiol 199:425–432

Patil C, Suryawanshi R, Koli S, Patil S (2016) Improved method for effective screening of ACC (1-aminocyclopropane-1-carboxylate) deaminase producing microorganisms. J Microbiol Methods 131:102–104

Patten C, Glick BR (1996) Bacterial biosynthesis of indole-3-acetic acid. Can J Microbiol 42:207–220

Patten CL, Glick BR (2002a) The role of bacterial indoleacetic acid in the development of the host plant root system. Appl Environ Microbiol 68:3795–3801

Patten CL, Glick BR (2002b) Regulation of indoleacetic acid production in *Pseudomonas putida* GR12-2 by tryptophan and the stationary phase sigma factor RpoS. Can J Microbiol 48:635–642

Patten CL, Kirchof MD, Schertzberg MR, Morton RA, Schellhorn HE (2004) Microarray analysis of RpoS-mediated gene expression in *Escherichia coli* K-12. Mol Gen Genet 272:580–591

Patten CL, Blakney AJC, Coulson TJD (2013) Activity, distribution and function of indole-3-acetic acid biosynthetic pathways in bacteria. Crit Rev Microbiol 39:395–415

Penrose DM, Glick BR (2003) Methods for isolating and characterizing ACC deaminase-containing plant growth-promoting rhizobacteria. Physiolog Plant 118:10–15

Polko JK, Kieber JJ (2019) 1-Aminocyclopropane 1-carboxylic acid and its emerging role as an ethylene-independent growth regulator. Front Plant Sci 10:1602

Poupin MJ, Greve M, Carmona V, Pinedo I (2016) A complex molecular interplay of auxin and ethylene signaling pathways is involved in *Arabidopsis* growth by *Burkholderia phytofirmans* PsJN. Front Plant Sci 7:492

Ryu RJ, Patten CL (2008) Aromatic amino acid-dependent expression of indole-3-pyruvate decarboxylase is regulated by tyrR in *Enterobacter cloacae* UW5. Bacteriology 190:7200–7208

Ryu C-M, Farag MA, Hu C-H, Reddy MS, Wei H-X, Pare PW, Kloepper JW (2003) Bacterial volatiles promote growth in *Arabidopsis*. Proc Natl Acad Sci USA 100:4927–4932

Ryu C-M, Farag MA, Hu C-H, Reddy MS, Kloepper JW, Pare PW (2004) Bacterial volatiles induce systemic resistance in *Arabidopsis*. Plant Physiol 134:1017–1026

Saleh SS, Glick BR (2001) Involvement of gacS and rpoS in enhancement of the plant growth-promoting capabilities of *Enterobacter cloacae* CAL2 and UW4. Can J Microbiol 47:698–705

Santoro MV, Zygadlo J, Giordano W, Banchio E (2011) Volatile organic compounds from rhizobacteria increase biosynthesis of essential oils and growth parameters in peppermint (*Mentha piperita*). Plant Physiol Biochem 49:1177–1182

Sugawara M, Okazaki S, Nukui N, Ezura H, Mitsui H, Minamisawa K (2006) Rhizobitoxine modulates plant–microbe interactions by ethylene inhibition. Biotechnol Adv 24:382–388

Tanimoto E (2012) Tall or short? Slender or thick? A plant strategy for regulating elongation growth of roots by low concentrations of gibberellin. Ann Bot 110:373–381

Tittabutr P, Awaya JD, Li QX, Borthakur D (2008) The cloned 1-aminocyclopropane-1-carboxylate (ACC) deaminase gene from *Sinorhizobium* sp. strain BL3 in *Rhizobium* sp. strain TAL1145 promotes nodulation and growth of *Leucaena leucocephala*. Syst Appl Microbiol 31:141–150

Tromas A, Perrot-Rechenmann C (2010) Recent progress in auxin biology. C R Biol 333:297–306

Vanderstraeten L, Van Der Straeten D (2017) Accumulation and transport of 1-aminocyclopropane-1-carboxylic acid (ACC) in plants: current status, considerations for future research and agronomic applications. Front Plant Sci 8:38

Vanderstraeten L, Depaepe T, Bertrand S, Van Der Straeten D (2019) The ethylene precursor ACC affects early vegetative development independently of ethylene signaling. Front Plant Sci 10:1591

Woodward AW, Bartel B (2005) Auxin: regulation, action, and interaction. Ann Bot 95:707–735

Yamaguchi S (2008) Gibberellin metabolism and its regulation. Annu Rev Plant Biol 59:225–251

Yang SF, Hoffman NE (1984) Ethylene biosynthesis and its regulation in higher plants. Annu Rev Plant Physiol 35:155–189

Zemlyanskaya EV, Omelyanchuk NS, Ubogoeva EV, Mironova VV (2018) Deciphering auxin-ethylene crosstalk at a systems level. Int J Mol Sci 19:4060

Zhang H, Kim MS, Krishnamachari V, Payton P, Sun Y, Grimson M, Farag MA, Ryu C-M, Allen R, Melo IS, Pare PW (2007) Rhizobacterial volatile emissions regulate auxin homeostasis and cell expansion in *Arabidopsis*. Planta 226:839–851

Zhang S, Gan Y, Xu B (2019) Mechanisms of the IAA and ACC-deminasee producing strain of *Trichoderma longbrachiatum* T6 in enchancing wheat seedling tolerance to NaCl stress. BMC Plant Biol 19:22

Zhao Y (2010) Auxin biosynthesis and its role in plant development. Annu Rev Plant Biol 61:49–64

Biocontrol of Bacteria and Fungi

6

Abstract

In this chapter the indirect promotion of plant growth is discussed in some detail (i.e., their use as biocontrol bacteria). That is, the many mechanisms that plant growth-promoting bacteria use to prevent the proliferation of bacterial and fungal plant pathogens are considered. In particular, the bacterial synthesis of antibiotics, hydrogen cyanide, siderophores, cell wall degrading enzymes, volatile compounds, and quorum quenching compounds is addressed. Moreover, the use and efficacy of additional biocontrol mechanisms by plant growth-promoting bacteria is considered including competition with pathogens, lowering plant ethylene levels, and inducing systemic resistance within plants is discussed. Finally, although they are not plant growth-promoting bacteria, the use of certain bacteriophages as biocontrol tools is discussed.

6.1 Phytopathogens

It has been estimated that various plant diseases typically reduce plant yields by ~10% per year in the more developed countries and by an average of >20% per year in the less developed countries of the world. From the perspective of many farmers in the developed world, the control of plant diseases is often not considered to be especially problematic. Plant disease control may be achieved by the use of plants that have been bred for good resistance to many diseases, and by various plant cultivation approaches such as crop rotation, the use of pathogen-free seed, appropriate planting date and plant density, control of field moisture, and, perhaps most importantly, the use of chemical pesticides. Independent of human intervention, plants typically synthesize a variety of phytochemicals as secondary metabolites and many of these chemicals have antimicrobial properties and can, to some extent, protect the plants against damage from diseases caused by different pathogens.

© Springer Nature Switzerland AG 2020
B. R. Glick, *Beneficial Plant-Bacterial Interactions*,
https://doi.org/10.1007/978-3-030-44368-9_6

Interestingly, some of these plant secondary metabolites are also useful in treating a wide range of human diseases.

A wide variety of infectious organisms can cause plant diseases including fungi, oomycetes (water molds), bacteria, viruses, phytoplasmas (obligate bacterial parasites of plant phloem tissue), protozoa, and nematodes (also called roundworms). However, fungi and bacteria cause the majority of common plant diseases. Most of the phytopathogenic fungi belong to the Ascomycetes and the Basidiomycetes. Some important disease-causing Ascomycetes include *Fusarium* spp., *Thielaviopsis* spp., *Verticillium* spp., *Magnaporthe grisea*, and *Sclerotinia sclerotiorum*. Some common disease-causing Basidiomycetes include *Ustilago* spp., *Rhizoctonia* spp., *Phakospora pachyrhizi*, and *Puccinia* spp. Fungal phytopathogens that are biotrophic colonize living plant tissue and obtain nutrients from living plant cells. By definition, biotrophic pathogenic fungi establish a long-term feeding relationship with their hosts' cells, rather than killing the host cells as part of the infection process. However, fungal pathogens that are necrotrophic infect and subsequently kill plant tissue, and then extract nutrients from the dead plant cells. Most common fungal diseases of plants are generally controlled through the use of chemical fungicides. A smaller number of bacteria than fungi can cause disease in plants. Pathogenic bacteria include some species of *Burkholderia*, *Pseudomonas*, and *Xanthomonas*. Unfortunately, many of the chemicals that are used to control phytopathogens are hazardous to animals and humans, and some of them persist and accumulate in natural ecosystems. It is therefore desirable to replace these chemical agents with biological control agents that are "friendlier" to the environment and to humans.

To decrease our dependence upon use of chemical pesticides or, ideally, to avoid their use altogether, scientists have been studying PGPB that are able to act as biocontrol agents utilizing a wide range of strategies to limit the growth and pathogenicity of fungal and bacterial phytopathogens. This chapter presents an overview of some of the most prominent of the strategies that are used by some PGPB strains that act as biocontrol agents. While some researchers consider biocontrol bacteria to be separate and distinct from PGPB that directly promote plant growth, here, biocontrol bacteria are considered to be PGPB that employ mechanisms that are intended to indirectly promote plant growth, by limiting pathogen inhibition of plant growth and proliferation. Moreover, many biocontrol PGPB, in addition to limiting pathogen proliferation, can also directly promote plant growth.

With the enormous interest in isolating, characterizing, and developing new biocontrol PGPB, it is perhaps surprising that the procedures for finding these important bacterial strains are generally rather rudimentary. That is, researchers typically collect a large number of samples (sometimes hundreds or even thousands), from rhizosphere soil, bulk soil, plant endospheres, and compost from different locations with different types of soil and then isolate the bacteria contained in these samples before testing them for antagonism to specific pathogens. Pathogen antagonism may be assessed by the inhibition of pathogen proliferation by the new bacterial isolates on either solid media in Petri dishes or in liquid media by monitoring pathogen growth. In addition, some researchers have monitored plant proliferation in the presence of both a pathogen and a potential biocontrol PGPB, selecting for PGPB strains that allow the plant to grow as if little or no pathogen were present. Unfortunately, these screening procedures are quite tedious and time consuming and require a large number of samples to be tested. More recently, some researchers have sequenced the complete genomes of prospective biocontrol PGPB strains in an effort to understand what antipathogenic agents are synthesized by selected bacterial strains.

The practice of growing the same crop year after year on the same land (monocropping) often leads to the enrichment of fungal and bacterial pathogens in the soil. This problem is generally addressed by crop rotation. In one series of experiments, scientists evaluated the effect of crop rotations on cucumber (*Cucumis sativus*) infected with *Fusarium* wilt disease using Indian mustard (*Brassica juncea*) and wild rocket or arugula (*Diplotaxis tenuifolia*). Importantly, at the same time that they monitored cucumber disease, scientists characterized in detail the cucumber rhizosphere bacterial composition. They observed that crop rotations significantly suppressed *Fusarium* wilt disease and, at the same time, significantly increased rhizosphere bacterial diversity. While additional work needs to be done to better understand this result, it is speculated that within the increased bacterial diversity there is an increase in biocontrol PGPB that prevent the pathogen damage.

The vast majority of the experiments discussed below utilized a single PGPB strain in an effort to control the damage from a specific pathogen. In a very few experiments, scientists have employed a consortium of several bacterial strains to make plants more resistant to disease. For example, one group of researchers simultaneously applied three separate bacterial strains, *Bacillus cereus*, *Bacillus subtilis*, and *Serratia* sp. in order to reduce phytophthora blight of sweet peppers. This combination of bacterial strains was successful in decreasing the pathogen damage compared to untreated plants. Unlike other biocontrol experiments, this effect was not attributed to any particular traits possessed by the three members of this consortium. Rather, it was suggested that treatment with this consortium altered the composition of the rhizospheric microbial community to one that suppressed soilborne disease. In addition to a poor understanding of how this bacterial consortium actually functioned, it is unclear whether this consortium would be sufficiently stable to provide this sort of protection in various soils and against different pathogens.

Given the very large number of microorganisms that are present in the soil, an interesting question that arises is: how do plant roots distinguish beneficial microbes from pathogens. A very recent study by an international group of scientists has now shown that for two different legumes, the root cells possess receptors that can distinguish between chitin oligomeric microbe-associated molecular patterns and lipochitin oligosaccharides. In this case, the perception of chitin triggers plant responses to pathogens while the perception of lipochitin oligosaccharides (also known as Nod factors) is an early step in nodule formation. Whether this observation can be extended to a wide variety of different plants and whether it can be used by scientists to develop strategies that are effective at protecting plants against certain pathogens remains to be seen; however, the possibilities of this observation are intriguing.

6.2 Allelochemicals Including Antibiotics and HCN

PGPB indirectly promote plant growth by suppressing the functioning of phytopathogens, primarily by producing allelochemicals that inhibit pathogen growth; these PGPB are sometimes called biocontrol PGPB or biocontrol bacteria. Allelochemicals may be defined as chemicals that are produced by living organisms that are released into the environment and affect the growth and development of other organisms. Allelochemicals include antibiotics, hydrogen cyanide (HCN), lipopeptides and a variety of compounds discussed in separate subsections of this chapter including siderophores, various enzymes, volatile organic compounds, and plant hormones.

One of the major mechanisms that control PGPB use to kill plant pathogens is the ability of the biocontrol strain to synthesize one or more antibiotics. The antibiotics synthesized by biocontrol PGPB include, but are not limited to, agrocin 84, amphisin, bacillomycin, 2,4-diacetylphloroglucinol, fengycin, herbicolin, iturin, oomycin, phenazine-1-carboxylic acid, pyoluteorin, pyrrolnitrin, surfactin, tensin, and viscosinamide (Fig. 6.1). In fact, the biocontrol activity of a number of PGPB strains is directly related to the ability of the bacterium to produce one or more of these antibiotics. However, an antibiotic that is effective in the laboratory against one strain of a pathogenic agent may not prevent the damage to the plant that occurs as a consequence of other strains of the same pathogen, and may behave differently than expected under more variable field conditions. For example, the expression of bacterial antibiotic biosynthesis genes is often dependent upon the nutritional environment provided by the host plant. In addition, the effectiveness of a biocontrol organism can be significantly affected by the way in which the biocontrol PGPB is grown in culture as well as how it is formulated for application.

By limiting the proliferation of other soil microorganisms, antibiotic-secreting biocontrol plant growth-promoting bacteria facilitate their own proliferation since they will have fewer competitors for the limited nutritional resources found in most soils. Thus, by inhibiting the growth of other soil microorganisms, biocontrol PGPB effectively create a niche for themselves to grow and function and stimulate plant growth directly as well as indirectly.

Evidence for the involvement of antibiotics in disease suppression by biocontrol PGPB comes from experiments in which mutant strains of disease-suppressive bacteria that were no longer able to produce antibiotics concomitantly lost either all or a substantial portion of the ability to prevent the damage to plants caused by phytopathogens. Moreover, in a number of instances it has been possible to isolate and purify specific antibiotics from biocontrol PGPB and subsequently demonstrate that the purified antibiotics by themselves were inhibitory to the same spectrum of fungal phytopathogens as the biocontrol bacteria had the ability to inhibit the growth of both *Fusarium* spp. and *Alternaria* spp.

Fig. 6.1 The structures of some of the antibiotics that are produced by biocontrol PGPB

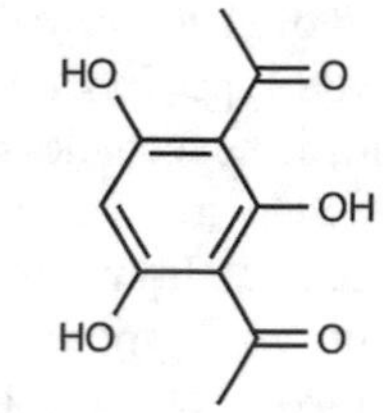

2, 4-diacetyl-phloroglucinol

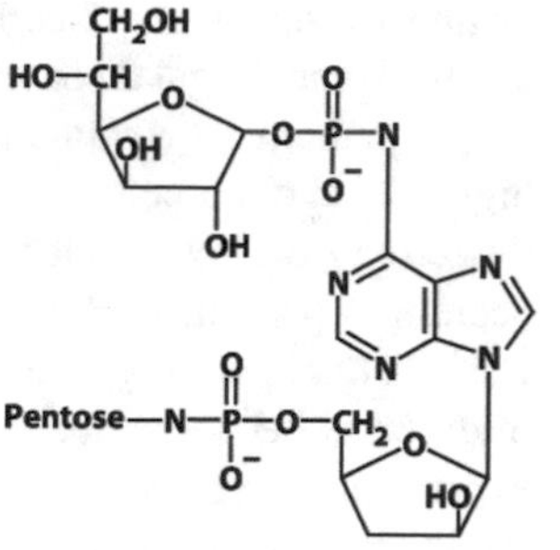

Agrocin 84

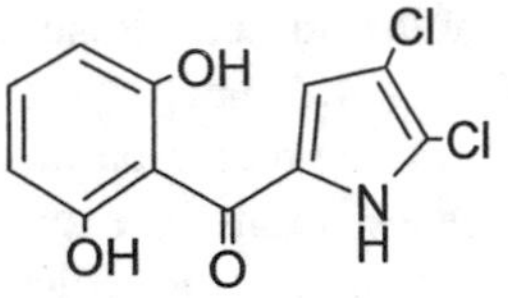

Pyluteorin

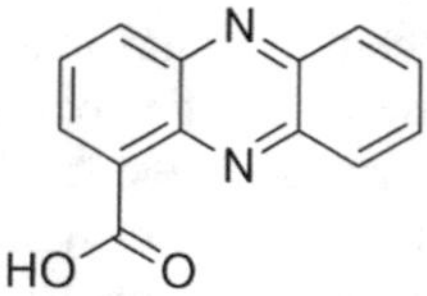

Phenazine-1-carboxylate

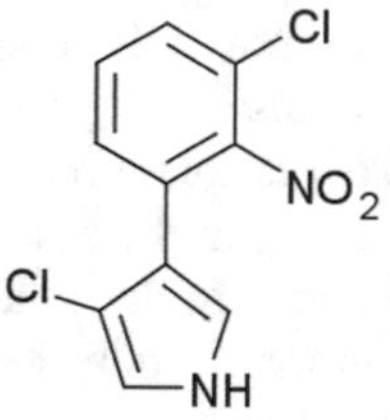

Pyrrolnitrin

Most of the biocontrol PGPB that have been isolated and characterized are either rhizospheric or endophytic in nature. However, some researchers have shown that it is also possible to isolate biocontrol PGPB from the plant phyllosphere (the surface of the above ground portions of the plant) as well. For example, in one study, researchers found that 40 out of 175 fluorescent pseudomonads isolated from wheat leaves were able to inhibit fungal growth. Moreover, 22 of these newly isolated biocontrol strains could suppress the growth of both *Fusarium* spp. and *Alternaria* spp., pathogenic fungi that are commonly found on wheat leaves. These workers suggest that it may be possible to spray one or more of these biocontrol PGPB onto wheat leaves as a means of protecting the plants against these ubiquitous fungal pathogens.

One way to improve the efficacy of biocontrol PGPB strains is to genetically combine several biocontrol activities into a single bacterial strain. Thus, in one experiment the operon carrying all seven of the genes that encode the biosynthesis of the antibiotic phenazine-1-carboxylic acid (i.e., *phzABCDEFG*) were inserted into the chromosomal DNA of a PGPB strain of *P. fluorescens* using the transposon Tn*5* (Fig. 6.2a). Despite the fact that the wild-type version of this bacterium does not synthesize phenazine-1-carboxylic acid, it nevertheless acts as a biocontrol agent against some fungal diseases. As indicated by a much larger zone of clearance of the fungal pathogen *Pythium ultimum* on solid medium, the engineered bacterium has a higher level of biocontrol activity than the wild-type strain (Fig. 6.2b). In addition to its activity on Petri plates, the engineered phenazine-1-carboxylic acid-producing bacterium prevented *P. ultimum*-caused damping-off disease in pea plants in soil. Besides manipulating antibiotic biosynthesis genes per se, antibiotic production can often be either increased or decreased to a significant extent by manipulating the expression of either the vegetative- or the stationary-phase sigma factor thereby modulating antibiotic expression at the transcriptional level. While most antibiotics are preferentially synthesized during the stationary phase, some are synthesized during the vegetative phase so that overproduction of the appropriate sigma factor can increase the expression of genes normally expressed at a specific phase of bacterial growth. This work demonstrates the efficacy of this approach under greenhouse conditions; that is, by judicious manipulation, it is possible to engineer biocontrol strains that are more effective at producing antibiotics. Nevertheless, it remains to be demonstrated whether these altered bacteria are as effective in the field as they are in the lab. Thus, for example, altering the ratio of the vegetative to the stationary phase sigma factor will not only change the amount of antibiotic synthesized by the bacterium, it will also change the overall metabolic functioning of the bacterium. This may ultimately decrease the environmental persistence of the bacterium possibly making it ineffective in the field. Moreover, it is likely that such manipulated bacteria may not be tested in the field for some time because of the reluctance of regulatory agencies in many countries to agree to the field release of genetically engineered bacterial strains.

Bacillus amyloliquefaciens subsp. *plantarum* FBZ42 is a Gram-positive bacterium that is used commercially as both a biofertilizer and a biocontrol agent. This bacterium has the ability to inhibit the growth and proliferation of a number of different bacterial and fungal pathogens. In addition, strain FBZ42 has both antiviral and nematicidal activities. To elaborate the range of mechanisms that this biocontrol bacterium employed, its genome was sequenced, and it was found that nearly 10% of the genome encodes various antimicrobial metabolites and their corresponding immunity genes (i.e., genes encoding determinants that protect this strain from the antimicrobial metabolites that it produces). The wide range and the large number of antimicrobial metabolites that are synthesized by this bacterium makes it a potent biocontrol strain, obviating the need to genetically engineer this bacterium to synthesize additional antimicrobial metabolites (Table 6.1). In another study, scientists isolated 267 different *Bacillus* strains from the rhizosphere of field-grown tomato plants. Following phenotypic characterization of these strains, it was observed that the seven different strains of *Bacillus thuringiensis* (a bacterium better known for its activity against various insects; see Chap. 7) all displayed antagonistic activity against the tomato pathogenic fungi *Verticillium dahlia*. When the genomes of these bacteria were sequenced, similar to what was observed

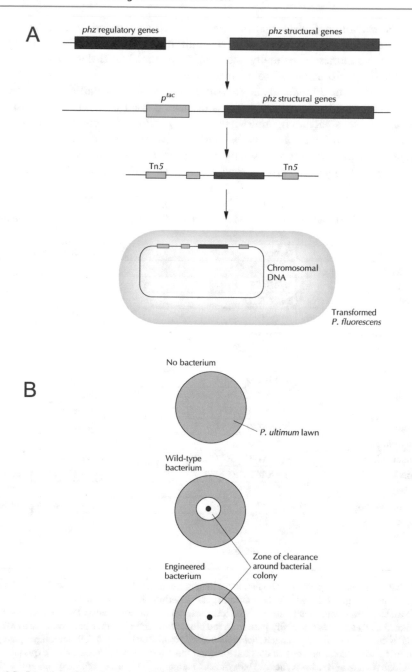

Fig. 6.2 Insertion of the phenazine biosynthesis operon into the chromosomal DNA of a biocontrol strain of *Pseudomonas fluorescens* (**a**). The regulatory genes that normally control the expression of the seven biosynthetic genes were removed, and the entire operon was placed under the control of the *tac* promoter (p^{tac}). Since *P. fluorescens* does not utilize lactose as a carbon source, it does not encode the *lac* repressor, and in the absence of the *lac* repressor, any genes under the control of the *tac* promoter are expressed constitutively. The operon, under the control of the *tac* promoter, was inserted into a derivative of transposon Tn*5* adjacent to a kanamycin resistance gene (not shown) on

with *B. amyloliquefaciens* subsp. *plantarum*, it was found that these strains synthesized a large number of different antifungal metabolites. Thus, it appears that there are numerous naturally occurring biocontrol PGPB that have the ability to synthesize multiple antimicrobial metabolites.

The bacterium *Pseudomonas chlororaphis* 30–84 is a particularly effective biocontrol strain that is able to suppress diseases caused by a number of different fungal pathogens including *Alternaria brassicae*, *Botrytis cinerea*, *Cochliobolus heterotrophs*, *Colletotrichum graminicola*, *Fusarium graminearum*, *Fusarium oxysporum*, *Gaeumannomyces graminis* var. *graminis*, *Gaeumannomyces graminis* var. *tritici*, *Phytophthora capsica*, *Pythium ultimum*, *Rhizoctonia solani*, *Sclerotinia*

Table 6.1 Genes encoding biocontrol metabolites in *Bacillus amyloliquefaciens* subsp. *plantarum* FZB-42

Metabolite	Genes	Directed against
Lipopeptides		
Surfactin	*srfABCD*	Viruses
Bacillomycin D	*bmyCBAD*	Fungi
Fengycin	*fenABCDE*	Fungi
Bacillibactin	*dhbABCDEF*	Bacteria
Unknown metabolite	*nrsABCDEF*	Unknown
Antibiotics		
Macrolactin	*mlnABCDEFGHI*	Bacteria
Bacillaene	*baeBCDE, acpK, baeGHIJLMNRS*	Bacteria
Difficidin	*dfnAYXBCDEFGHIJKLM*	Bacteria
Bacilysin	*bacABCD, ywfG*	Bacteria, cyanobacteria
Bacteriocins		
Plantazolicin	*pznFKGHIAJC DBEL*	Nematodes
Amylocyclicin	*acnBACDEF*	Closely related bacteria
Volatiles[a]		
2,3-Butanediol	*alsDRS*	Various plant pathogens

[a]See Sect. 6.7

Fig. 6.2 (continued) a plasmid. Tn*5* facilitates integration of DNA into the chromosome of the host cell. Transconjugants in which the chromosomal insertion had not inactivated any important bacterial functions were tested for their effectiveness as biocontrol strains. The Tn*5* derivative is designed so that it does not easily pass from the biocontrol strain to other bacteria in the environment. Killing of the fungal pathogen *Pythium ultimum* around the bacterial cells on an agar plate (**b**). The increased size of the zone of clearance around the engineered *P. fluorescens* (expressing the antibiotic phenazine-1-carboxylic acid biosynthetic pathway) indicates increased biocontrol activity preventing growth of the fungus on the medium. The antifungal activity of the bacterium is approximately proportional to the area of the zone of clearance around the center of the petri plate to which the bacterium is added. © 2010 American Society for Microbiology. Used with permission. No further reproduction or distribution is permitted without the prior written permission of American Society for Microbiology

minor, and *Sclerotinia sclerotiorum.* This biocontrol PGPB produces three different phenazine antibiotics including phenazine-1-carboxylic acid, 2-hydroxy-phenazine-1-carboxylic acid, and 2-hydroxy-phenazine. From a biocontrol perspective, the type of phenazines produced may affect the capacity of a producing bacterium to suppress disease. In most of the literature studies on the specificity of various phenazines, the antibiotics were produced in different bacterial strains so that it is not possible to sort out what activities are dictated by the phenazine produced by a particular bacterium and any other activities that the bacterium encodes. In one study, the three abovementioned phenazines plus several derivatives of these phenazines were produced in the same genetic background (i.e., strain 30–84) and then the ability of these strains/antibiotics to inhibit the proliferation of the above mentioned fungal pathogens was assessed. Briefly, the data are consistent with the notion that not all phenazines are equally effective in their ability to suppress the proliferation of these fungi. While a few of the phenazine derivatives that were constructed were somewhat more effective at inhibiting the growth of one or two fungal pathogens, overall, the wild-type bacterium that produced three different phenazines was most effective at inhibiting a wide range of fungal pathogens.

In yet another study, scientists closely examined a landrace (domesticated, traditional variety of a plant that has developed naturally over time) of corn that was estimated to have been grown continuously by farmers in Mexico for between 2000 and 3000 years. This particular strain of corn, which is grown without any chemical fungicides was found to contain a bacterial endophyte, *Burkholderia gladioli*, that protects the plant against a range of fungal pathogens. Detailed examination of this bacterium revealed that the mechanisms of antifungal activity include chitinase activity, phenazine antibiotics, bacteriocin synthesis, and the ability to synthesize volatile organic compounds. Moreover, it has been shown that this bacterium can recognize and then swarm toward a pathogen of its host plant, attach to the pathogen, form microcolonies, and eventually kill the pathogen. Clearly, nature has engineered some very powerful biocontrol PGPB.

Botrytis cinerea is a pathogenic fungus that causes gray mold disease which detrimentally affects more than 200 different plant species so that this fungus is considered to be the second most important/problematic plant pathogen worldwide. As a consequence of forming conidia (nonmotile fungal spores), this fungus may lie dormant within a plant for long periods of time until conditions are favorable to produce an infection. The current treatment for a gray mold infection is the use of chemical pesticides. Until recently, the application of some of the more well-characterized biocontrol PGPB was not particularly effective at inhibiting the functioning of this pathogen. However, scientists have recently shown that some strains of *Bacillus* spp. that can synthesize certain lipopeptides (short peptides, typically 6–10 amino acids with covalently attached lipids chains) are effective inhibitors of this pathogen. Currently, a *Bacillus methylotrophicus* PGPB strain that is highly effective in killing the *Botrytis cinerea* pathogen is being tested and developed by a company in Spain for possible eventual commercial use.

In addition to *Bacillus* spp., several strains of *Pseudomonas* spp. have been shown to produce cyclic lipopeptides that are inhibitory to a variety of fungal pathogens. For example, some of these strains have been shown to be able to suppress the rice blast fungus *Magnaporthe oryzae* both directly and by induced systemic resistance. The strains that effectively inhibited the rice blast fungus produced one of the cyclic lipopeptides lokisin, WLIP (White-Line-Inducing Principle) or entolysin. In addition, extracts from several other cyclic lipopeptide-producing pseudomonads significantly reduced the disease caused by the rice blast fungus although they were not as effective as the abovementioned cyclic lipopeptides.

Soilborne diseases such as soft rot of a wide range of vegetables and flowers are particularly difficult to overcome. In one study, in an effort to avoid having to use chemical fungicides, researchers isolated an endophytic strain of *Bacillus amyloliquefaciens* and showed that it was able to suppress soft rot of Chinese cabbage which is caused by the pathogenic bacterium *Pectobacterium carotovorum* subsp. *caraotovorum*. The isolated biocontrol strain was shown to synthesize the polyketide antibiotics difficidin, bacillaene, and macrolactin as well as the antibiotic dipeptide bacilysin. As a consequence of the action of these antibiotics, the biocontrol strain decreased the growth of the pathogenic strain by approximately three orders of magnitude thereby decreasing the damage to plants by a factor of around ten. Thus, treatment with this biocontrol strain significantly reduced both the extent of the soft rot and the subsequent transmission of this disease to other plants.

Sometimes, biocontrol PGPB strains lose their effectiveness in the laboratory or following long-term storage of bacterial cultures. This is often attributed to mutations in *lemA* or *gacA* genes, mutations that often occur at a high frequency and subsequently negatively affect a whole range of biological activities including antibiotic biosynthesis.

Interestingly, it has been suggested that in antibiotic-producing biocontrol PGPB the antibiotic biosynthesis genes and their regulatory regions often include several rare codons. These rare codons act to decrease the rate of synthesis of these genes and hence limit the amount of antibiotic that can be synthesized. In one laboratory study, researchers modified several of these rare codons, replacing them with more commonly used codons. In the case of pyoluteorin, when several rare codons in a gene that regulates the biosynthesis of this antibiotic were replaced (leaving the encoded amino acid exactly the same), the resultant mutant bacterium increased its synthesis of pyoluteorin by 15-fold (Fig. 6.3). However, whether dramatically increasing the level of antibiotic synthesis of a biocontrol PGPB results in a similar change in the efficacy of modified bacterium in the field remains to be seen. Also, producing constitutively high levels of an antibiotic during periods when the antibiotic is not needed might make the biocontrol PGPB less competitive in the environment.

At the present time, there is only one commercially available genetically engineered biocontrol bacterial strain. A modified version of the bacterium *Agrobacterium radiobacter* K84 has been marketed, first in Australia in 1989, and more recently all over the world, as a means of controlling crown gall disease. This disease, which is caused by the bacterium *Agrobacterium tumefaciens*, deleteriously affects almond trees and stone fruit trees, such as peach trees as well as grapevines

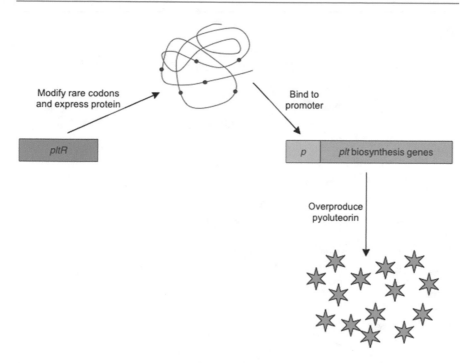

Fig. 6.3 Modifying a biocontrol PGPB to overproduce the antibiotic pyoluteorin. Several rare codons in the gene *pltR* encoding a regulatory protein were changed to more commonly used codons that encoded the same amino acid (so that the amino acid sequence of the protein synthesized by this gene did not change at all). This resulted in a much higher level of synthesis of protein PltR which is directly involved in modulating the transcription of the *plt* biosynthesis genes with the net result that the antibiotic pyoluteorin is overproduced many times

(Fig. 6.4). The antibiotic agrocin 84, which is produced by *A. radiobacter* K84, is toxic to *A. tumefaciens*. Unfortunately, agrocin 84-resistant strains of *A. tumefaciens* can develop following treatment with the *A. radiobacter* strain. This may occur if the plasmid that carries the genes encoding the biosynthesis of agrocin 84 is accidentally transferred from *A. radiobacter* to the pathogenic *A. tumefaciens*, a not unlikely occurrence. To avoid this possibility, the region of DNA responsible for plasmid transfer was removed from the agrocin 84 plasmid, pAgK84 (Fig. 6.5). As a result of the deletion of the *tra* (transfer) gene, the *A. radiobacter* K84 strain retains the capacity to act as a biocontrol agent, but it can no longer transfer the agrocin 84 plasmid to other bacteria in the environment including pathogenic agrobacteria.

Perhaps surprisingly, many biocontrol PGPB strains have the ability to synthesize hydrogen cyanide (HCN), a colorless and extremely poisonous liquid with a boiling temperature of 25.6 °C. In fact, many antibiotic-producing PGPB strains also synthesize HCN. Moreover, it was found in several studies that a high percentage (30–90%) of *Pseudomonas* and *Bacillus* PGPB strains are positive for HCN production. By itself, the low level of HCN that is synthesized by most biocontrol PGPB is unlikely to have much biocontrol activity. The one notable exception to this is that

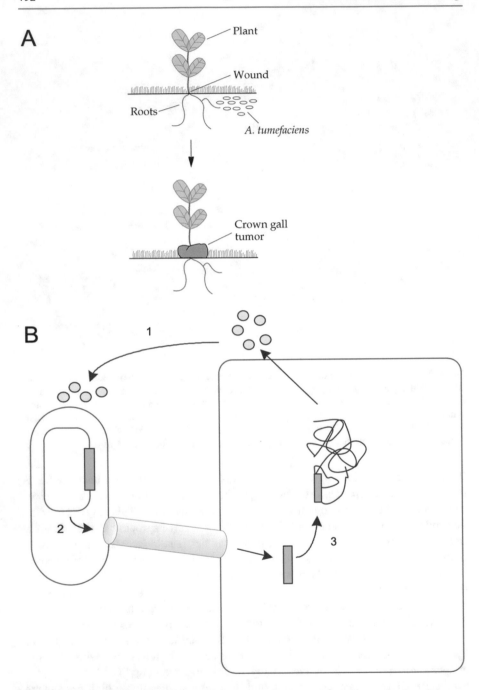

Fig. 6.4 Infection of a plant with *Agrobacterium tumefaciens* results in the production of a crown gall tumor (**a**) © 2010 American Society for Microbiology. Used with permission. No further reproduction or distribution is permitted without the prior written permission of American Society for Microbiology. Insertion of T-DNA from the Ti plasmid of *Agrobacterium tumefaciens* into plant chromosomal DNA (**b**). Following attachment of *A. tumefaciens* to a plant cell at the site of an open wound, the bacterium produces a network of cellulose fibrils that bind it to the plant surface. (1) The

the HCN that is produced by some strains of *Pseudomonas* spp. has been reported to suppress black root rot of tobacco caused by the fungus *Thielaviopsis basicola*. HCN is believed to act by inhibiting the mitochondrial enzyme cytochrome c oxidase. However, in most instances, it is thought that HCN probably acts synergistically with bacterially encoded antibiotics. In fact, the transfer of genes that encode HCN biosynthesis has been shown to increase the biocontrol activity of some bacteria. The synergism between antibiotics and HCN would likely help ensure that various pathogens do not develop resistance to specific antibiotics as such resistance would negate the effectiveness of an antibiotic-producing PGPB strain. In addition, although not discussed in this chapter, there is some evidence that HCN facilitates the insecticidal and nematicidal activity of some bacterial biocontrol strains.

A major advantage of employing endophytic biocontrol strains is that the added bacterial strains are much more likely to persist in the plant for a relatively long period of time. This is in comparison to rhizospheric biocontrol bacteria that are more likely to be negatively affected by soil biotic and abiotic conditions. With this in mind, one group of researchers isolated 100 endophytic bacterial strains from wheat plants (primarily from the roots). Following the identification of those bacterial strains by the sequencing of their 16S ribosomal DNA, all strains were assayed for the ability to synthesize IAA, solubilize phosphate, produce siderophores, and synthesize HCN. Based on those results, a number of strains were then tested for their ability to inhibit wheat rust spore (*Puccinia striformis*) germination. In fact, two of the strains significantly inhibited the germination of this wheat pathogen, making them potential biocontrol strains in dealing with this phytopathogen.

6.3 Siderophores

Iron is one of the most abundant minerals on Earth; however, it is not readily available in soil for direct assimilation by bacteria. This is because ferric ion, Fe^{+3}, the predominant form of iron in nature, is only sparingly soluble, i.e., about 10^{-18} M at pH 7.4. In order to find sufficient iron for their growth and metabolism, soil bacteria synthesize and secrete low molecular mass (~400–1000 dalton) iron-binding molecules known as siderophores. These molecules bind Fe^{+3} with a very high affinity ($Kd = \sim 10^{-20}$ to $\sim 10^{-50}$), transport the iron–siderophore complex back to the bacterial cell where it is taken up by means of an outer cell membrane receptor, and then made available for bacterial growth.

PGPB have been shown to prevent the proliferation of phytopathogens, and thereby facilitate plant growth, by synthesizing siderophores that bind a large part

Fig. 6.4 (continued) plant cells synthesize phenolic compounds such as acetosyringone that are taken up by the bacterium. (2) The phenolic compounds activate the virulence (*vir*) genes that are part of the Ti plasmid leading to the synthesis of a protein pilus connecting the bacterium and plant cell. (3) Subsequently, the T-DNA from the Ti plasmid is transferred through the pilus and is inserted into the plant chromosomal DNA

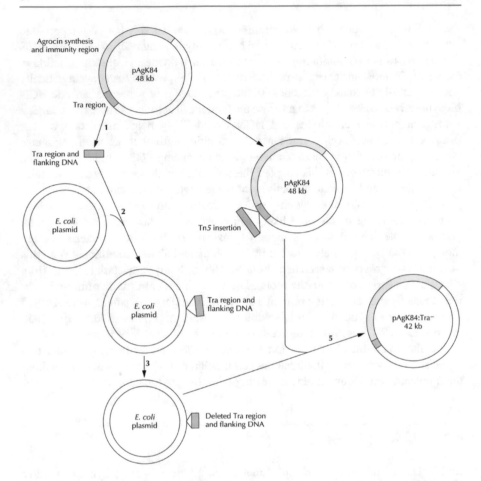

Fig. 6.5 Construction of a transfer-deficient (Tra⁻) derivative of plasmid pAgK84 from *A. radiobacter,* which encodes both synthesis of and immunity to the antibiotic agrocin 84. Based on knowledge of the restriction enzyme map of pAgK84, a DNA fragment containing the transfer (Tra) region, together with some of the flanking DNA, is isolated (1) and spliced into an *E. coli* plasmid vector (2). By restriction enzyme digestion, approximately 80% of the Tra region and some of the flanking DNA (a total of about 6 kb) is deleted from the cloned DNA containing the Tra sequence (3). Homologous recombination of the *E. coli* plasmid containing the deleted Tra region with plasmid pAgK84, in which transposon Tn5, which carries a kanamycin resistance gene, has been inserted into the Tra region, is performed (4). This results in some derivatives of pAgK84 in which a portion of the Tra region of the plasmid has been deleted (5). The resultant Tra⁻ mutant of pAgK84 can no longer be conjugationally transferred to other agrobacteria, although it is still able to synthesize and provide immunity to agrocin 84. None of the DNA fragments is shown to scale. © 2010 American Society for Microbiology. Used with permission. No further reproduction or distribution is permitted without the prior written permission of American Society for Microbiology

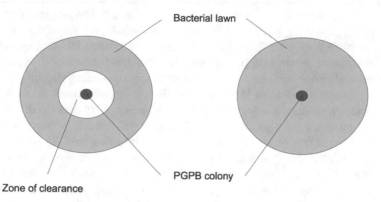

Fig. 6.6 Schematic representation of a biocontrol PGPB preventing the proliferation of a phyto-pathogen. On the Petri plates, the phytopathogen produces a lawn of growth over the entire plate. A colony of the same biocontrol PGPB strain is growing in the center of each plate. The plate on the right contains a sufficient amount of FeCl$_3$ so that the biocontrol PGPB does not need to sequester iron in order to grow. Therefore, on the plate on the right, siderophore biosynthesis is not induced. In the absence of added iron, as seen on the plate on the left, the biocontrol PGPB synthesizes and secretes siderophores that sequesters the iron around the PGPB colony thereby thwarting the growth of the pathogen and creating a zone of clearance

of the Fe^{+3} that is present in the rhizosphere of the host plant. These biocontrol PGPB prevent fungal and bacterial pathogens in the immediate vicinity of the roots of the host plant, where the biocontrol PGPB is bound, from obtaining sufficient iron for their growth so that they are unable to proliferate and act as pathogens because of a lack of iron (Fig. 6.6). While fungal and bacterial pathogens produce their own siderophores, the siderophores that they produce typically have a much lower affinity for iron than the biocontrol PGPB siderophores. In this way, the biocontrol PGPB effectively out-compete fungal and bacterial phytopathogens for available iron.

Unlike phytopathogens, plants are not generally affected by the localized deple-tion of iron in the soil caused by the siderophores that are produced by biocontrol PGPB. This is because most plants are able to grow at much lower iron concentrations than most microorganisms. In addition, many plants can bind and take up the biocontrol PGPB iron–siderophore complex, transport it through the plant, and then reductively release the iron from the bacterial siderophore so that the iron can be used by the plant.

The ability of siderophores to act as effective "disease-suppressive" agents is affected by the particular crop plant, the specific phytopathogen being suppressed, the soil composition (including the nutrients and the microorganisms), the bacterium that synthesizes the siderophore, and the affinity of the specific siderophore in question for iron. All of these possible variables suggest that even though a particular biocontrol PGPB is an effective disease-suppressive agent in the laboratory, its behavior in the field may be much more difficult to predict.

In addition, to the type of experiment shown in Fig. 6.6, several genetic studies of siderophore-producing PGPB are consistent with the involvement of siderophores in the suppression of disease caused by fungal and bacterial pathogens. Thus, researchers have shown that siderophore overproducing mutant PGPB strains are often more effective than the wild-type bacterium in controlling specific pathogens. Moreover, mutant strains of biocontrol PGPB that were selected for their loss of siderophore production concomitantly lost the ability, displayed by the wild-type strain, to protect tomato plants against damage from specific phytopathogens. In another study, it was observed that a single Tn5 insertion into the genome of the biocontrol PGPB *Alcaligenes* sp. strain MFA1 resulted in the simultaneous loss of the ability of the bacterium to grow in the absence of added iron, and to inhibit growth and development of the fungal pathogen *Fusarium oxysporum*.

It may be possible to improve biocontrol PGPB by extending the range of iron– siderophore complexes that a particular PGPB strain can utilize. In this case, a genetically altered biocontrol PGPB strain could take up and use siderophores that are synthesized by other soil microorganisms thereby giving it a competitive advantage in the soil environment. It has been observed that some *Pseudomonas* spp. strains naturally possess from 5 to 20 outer membrane receptors, each of which can recognize and take up a single siderophore, or a set of siderophores with very similar structures. These siderophore receptors allow these strains to proliferate in the soil, and especially in the rhizosphere, at the expense of other soil bacteria.

6.4   Cell Wall-Degrading Enzymes

Plants often respond to infection by phytopathogenic bacteria and fungi by synthesizing pathogenesis-related (PR) proteins that are able to hydrolyze the cell walls of various fungal pathogens. Moreover, some biocontrol PGPB produce enzymes that are very similar to the plants' PR proteins. These enzymes include chitinase which degrades chitin, a polymer of β-(1,4)-N-acetyl glucosamine residues and one of the main components of the cell wall of many phytopathogenic fungi (Fig. 6.7); β-1,3-glucanase; protease; and lipase, all of which can (at least partially) individually lyse fungal cells (Fig. 6.8). In one example of the effectiveness of these enzymes, researchers isolated a strain of *Pseudomonas stutzeri* that produced both extracellular chitinase and laminarinase (a β-glucanase), and found that these

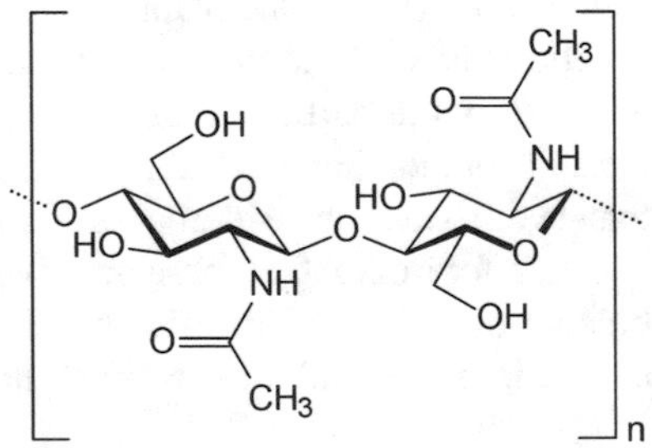

Fig. 6.7 The structure of chitin, a long-chain polymer of a *N*-acetylglucosamine

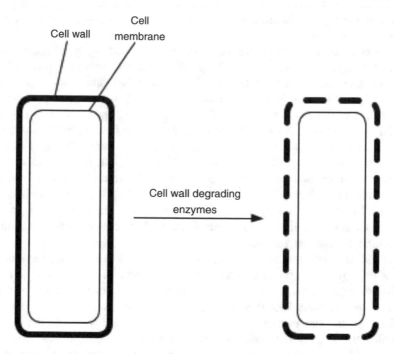

Fig. 6.8 Schematic representation of a fungal cell wall being degraded by one or more enzymes produced by a biocontrol PGPB. Following the breakdown of the fungal cell wall, the cell is readily lysed

enzymes could digest and lyse *Fusarium solani* mycelia and thereby prevent the fungus from causing crop loss due to root rot. In another experiment, a strain of the biocontrol PGPB *Enterobacter agglomerans* that possessed a complex of four separate enzymes responsible for the chitinolytic activity of the bacterium, and is antagonistic to various fungal phytopathogens, lost both chitinase and biocontrol activities at the same time when the bacterium was mutagenized with Tn5. When the wild-type *E. agglomerans* was used as a biocontrol agent, it significantly decreased the damage to cotton plants following infection with *Rhizoctonia solani*; however, the mutants deficient in chitinase activity were unable to protect the plant against damage that was caused by the fungal pathogen.

It may seem a bit surprising that proteases can act on fungal pathogen cell walls. This is because most fungi have a cell wall consisting largely of chitin and other polysaccharides. However, fungal cell walls also contain a number of glycoproteins, and the chitin, the glucans and the glycoproteins typically are covalently cross-linked. Some of the cell wall proteins contain transmembrane domains that anchor these proteins in the cell membrane (which is beneath the cell wall). Thus, researchers have isolated a strain of *Chryseobacterium aquaticum*, a Gram-negative, nonmotile, nonspore-forming rod that produces a novel antifungal protease. As a result of synthesizing this protease, this bacterium possesses a broad-spectrum antifungal activity that is effective against the pathogens *Rhizoctonia solani* and *Pseudocercospora theae*.

Since many of the enzymes from biocontrol PGPB that can lyse fungal cells are encoded by a single gene, it is relatively straightforward to isolate some of these genes and then transfer them to strains of biocontrol PGPB that do not express these enzymes. In one series of experiments, a chitinase gene was isolated from the bacterium *Serratia marcescens* and then introduced (separately) and expressed into *Trichoderma harzianum* (a plant growth-promoting fungus) and *Sinorhizobium meliloti*. In both instances the transformed microorganisms expressed chitinase activity and displayed a significant increase in antifungal activity. In a similar experiment, the *S. marcescens* chitinase gene was introduced into a biocontrol PGPB strain of *P. fluorescens*. As a consequence of this simple genetic manipulation, the transformant stably expressed and secreted active chitinase and was converted into a more effective biocontrol strain with increased activity against the phytopathogen *Rhizoctonia solani*.

In addition to using root-colonizing biocontrol PGPB to thwart the development of root pathogens, it is possible to use leaf-colonizing biocontrol PGPB to control phyllosphere pathogens. To this end, a chitinase gene from the marine bacterium *Alteromonas* sp. strain 79,401 was introduced into the leaf-colonizing bacterium *Erwinia ananas*. The transformed *E. ananas* cells were sprayed onto the leaves of cucumber plants that were then inoculated with the fungal pathogen *Botrytis cinerea* (the cause of cucumber gray mold) in a greenhouse. Compared to untreated plants, plants treated with the transformed *E. ananas* cells were significantly protected against leaf damage by the fungal pathogen.

It has been observed that in the laboratory, plasmids carrying cloned chitinase genes are often rapidly lost from biocontrol PGPB in the absence of an antibiotic that selects for the presence of the plasmid. Thus, to avoid the problem of plasmid loss from engineered biocontrol PGPB in the environment, the genes for chitinase or other fungal cell wall degrading enzymes should be integrated directly into the chromosomal DNA of the relevant biocontrol PGPB.

6.5   Phytohormones

As discussed previously (see Chap. 5), many PGPB actively synthesize various plant hormones. Moreover, by PGPB providing plants with more optimum levels of phytohormones, plant growth may be increased so that PGPB that do not contain any specific biocontrol mechanisms may nevertheless act as biocontrol agents by strengthening the host plant. In one series of experiments, *Arabidopsis* plants were infected with the pathogen *Pseudomonas syringae* pv. tomato in the presence or absence of various derivatives of a cytokinin-producing PGPB (a strain of *Pseudomonas fluorescens*). The wild-type PGPB strain that produced cytokinin but not the cytokinin minus mutants of this strain provided plants with a significant level of protection against damage by the pathogen (Table 6.2). Moreover, when the cytokinin minus mutants were transformed with a functional cytokinin biosynthesis gene,

Table 6.2 The effect of *Pseudomonas fluorescens* G20-18 (a cytokinin-producing PGPB) on the damage to *Arabidopsis thaliana* plants infected by the pathogen *Pseudomonas syringae*

Treatment (all include the pathogen)	Symptom score
No PGPB	4.8
Wild-type PGPB	1.1
Mutant 1	3.8
Mutant 2	3.9
Mutant 1 + cytokinin biosynthesis gene	2.0
Mutant 2 + cytokinin biosynthesis gene	2.0

The two PGPB mutants were created by the inactivation of the endogenous cytokinin biosynthesis gene by transposon mutagenesis. The complemented mutants included the transformation of the two mutant strains with an exogenous functional cytokinin biosynthesis gene. Pathogen damage was determined visually on a scale of 0–7 with a higher score indicating increasing pathogen damage. The symptom score data represent the means of ≥ 300 measurements

plants were again protected against pathogen damage. This experiment provides a clear demonstration of the effective protection against pathogen damage that is provided by a cytokinin-producing PGPB. Based on these laboratory results, it will be interesting to test this (and other cytokinin-producing PGPB) in the field. Moreover, there are likely to be synergistic effects in terms of pathogen biocontrol when combining cytokinin biosynthesis with other methods of pathogen biocontrol.

In addition to the abovementioned effect of cytokinin in decreasing the damage from a bacterial phytopathogen, some recent reports have suggested that PGPB that produce moderate to high levels of IAA can inhibit the proliferation of some fungal phytopathogens. For example, it was recently shown that some *Rhizobium* strains could control *Sclerotium rolfsii*, a pathogenic fungus that damages common beans. Moreover, in this study, it was shown that there was a direct correlation between IAA production by the tested *Rhizobia* and the observed inhibition of fungal pathogen mycelium growth. Notwithstanding the fact that some of these *Rhizobia* strains exhibited other activities that could be inhibitory to fungal growth, these results are consistent with the involvement of IAA as a significant component of fungal biocontrol. In fact, in another study, scientists identified eight PGPB strains that synthesized high levels of IAA that decreased crown rot in cucumber plants caused by the fungal pathogen *Phytophthora capsici*. Again, these strains contained a number of biocontrol traits; thus, these researchers suggested that IAA biosynthesis works together with other mechanisms to limit fungal proliferation.

The pathogenic fungus *Pseudocercospora fijiensis* is the causative agent of black Sigatoka disease that is a major problem for banana cultivation, worldwide, and especially in warmer climates where it can reduce fruit yields by more than 50%. In a recent study, researchers isolated approximately 150 unique bacterial endophytes from banana tree seedlings (Fig. 6.9) and then tested these bacterial strains for the ability to prevent or diminish black Sigatoka disease. One of the isolated bacterial strains, *Enterobacter cloacae*, that was positive for nitrogen fixation, IAA synthesis, and phosphate solubilization, facilitated the growth of banana plants in the presence of the fungal pathogen. While this study was not clear regarding the mechanism used

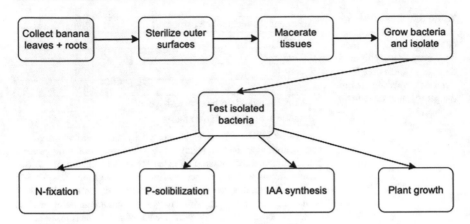

Fig. 6.9 Flow chart depicting the steps in the isolation and characterization of new bacterial endophytes from banana plant seedlings. The isolated bacteria were first tested for the ability to fix nitrogen, then for the ability to solubilize phosphate, then for IAA biosynthesis and finally to promote plant growth in the presence of the pathogenic fungus that causes black Sigatoka disease

by the newly isolated bacterial strain to suppress the fungal pathogen and allow the seedlings to grow, it was suggested that pathogen suppression might have been a consequence of the improved growth and health catalyzed by this PGPB. Clearly additional studies are needed to better understand precisely what is going on here. Unfortunately, *E. cloacae* is classified as a potential human pathogen so that this strain must be shown to nonpathogenic to humans before any field trials can proceed.

6.6   Competition

In addition to mechanisms where a biocontrol PGPB produces a substance that kills or inhibits the functioning of pathogenic agents, some biocontrol PGPB may out compete phytopathogens for nutrients and for suitable niches on the root surface. That is, competition between pathogens and PGPB can limit plant disease incidence and severity. In some instances, the more abundant PGPB may rapidly colonize plant surfaces and use most of the available nutrients, making it difficult for the pathogens to grow.

When several leaf surface bacteria were tested as potential biocontrol (protective) agents against the bacterial pathogen *Pseudomonas syringae* pv. tomato, all of the *Sphingomonas* spp. strains, that were isolated by one group of researchers, effectively protected plants against the pathogen. *Sphingomonas* spp. are Gram-negative, rod-shaped aerobic bacteria that are highly competitive plant leaf colonizers. Notwithstanding the suggestion that it may be common for competitive phyllosphere colonizers to act as biocontrol strains, this particular study did not rule out the possibility that other mechanisms of biocontrol might also be operative. On the other hand, the *Sphingomonas* spp. strains that were isolated from air, dust, or water did not act as biocontrol strains even though they were able to attain a high cell density on plant leaves. Thus, biocontrol activity is not necessarily a general trait amongst all *Sphingomonas* spp. strains.

One group of researchers endeavored to identify the *Sphingomonas* sp. traits that are essential components of the bacterium's ability to protect plants against the bacterial pathogen *P. syringae* (Fig. 6.10). Initially, a suspension of *Sphingomonas* sp. was mutagenized with a mini-Tn5 transposon (which is randomly inserted into the bacterial chromosomal DNA). Bacteria with a mini-Tn5 insertion were selected from a background of wild-type bacteria by their resistance to the antibiotic kanamycin (carried by the mini-Tn5). The binding of the pathogen *P. syringae* to leaf surfaces was readily visualized as a consequence of the expression of *lux* genes that had been added to this bacterium, causing the bacterium to be luminescent. In the absence of any added biocontrol strain, all of the leaf surfaces examined are luminescent (i.e., they are infected with the *lux*-labeled pathogen). In the presence of the biocontrol strain none of the leaf surfaces will be luminescent (i.e., no *lux*-labeled pathogen is present). In the presence of the mini-Tn5-mutagenized biocontrol strain, some of the leaf surfaces will be luminescent (i.e., the *lux*-labeled pathogen can bind when the traits that make the biocontrol strain effective are disrupted). In the latter case, the mini-Tn5 mutants that have lost biocontrol activity are isolated and their genomic DNA is purified and used to prepare a clone bank of genomic DNA in *E. coli*. From clones that contain the mini-Tn5 (which encodes resistance to the antibiotic kanamycin), the DNA regions that are adjacent to the mini-Tn5 are sequenced. These DNA sequences are used to identify proteins encoded by *Sphingomonas* sp. that are responsible for plant protective traits. In this way, ten separate *Sphingomonas* sp. genes that mediate plant protection were identified. This clever approach is a potential first step in either selecting or constructing strains of *Sphingomonas* sp. that are more effective than the wild-type biocontrol strain.

In another study, it was observed that a saprophytic (nonpathogenic) *P. syringae* strain completely protected pears against both gray mold and blue mold disease caused by the pathogenic fungi *Botrytis cinerea* and *Penicillium expansum*, respectively. In this case, it was assumed that the *P. syringae* strain protected plants by outcompeting the fungal pathogens. However, because this protective effect depended on applying an extremely high inoculum of the biocontrol agent compared to the phytopathogen, the use of this bacterium to combat these fungal diseases is impractical.

It has been shown that some biocontrol agents can outcompete the fungal pathogen *Pythium ultimum*, on sugar beets in soil, for the ability to proliferate on plant roots by metabolizing the constituents of the seed and root exudates. Moreover, some of these biocontrol PGPB subsequently produce compounds that are inhibitory to *P. ultimum*.

One approach to developing biocontrol strains includes using nonpathogenic variants of a phytopathogenic organism that can compete with the pathogen for the same ecological niche. This was done with the bacterial pathogen *Xanthomonas euvesicatoria*, one of the causative agents of Black Spot disease on plant leaves

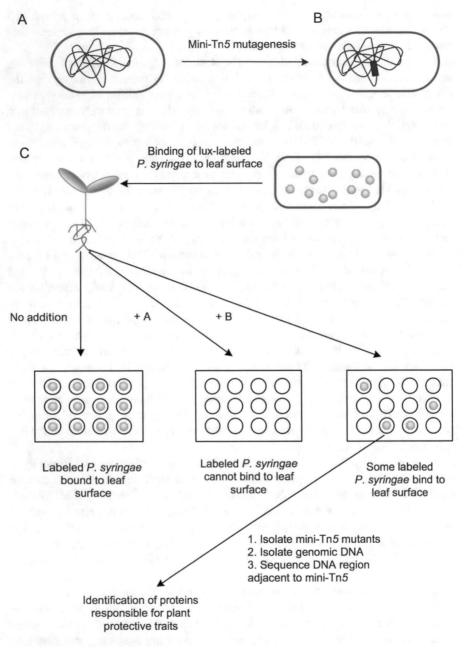

Fig. 6.10 Use of mini-Tn5 mutagenesis to identify regions of the *Sphingomonas* sp. genome that are responsible for protecting *Arabidopsis thaliana* plants against a pathogenic strain of *Pseudomonas syringae* that is labeled with *lux* genes

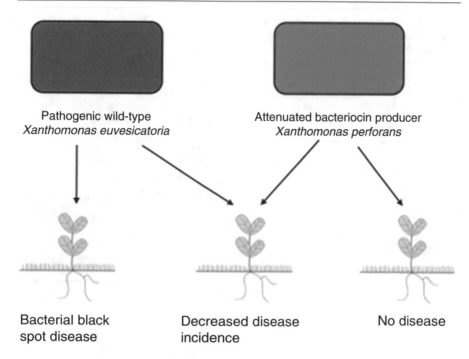

Fig. 6.11 Schematic representation of pathogen suppression by an attenuated *Xanthomonas* strain

(Fig. 6.11). In this case, the competitive (biocontrol) bacterium was an attenuated bacteriocin producer of *Xanthomonas perforans*. Bacteriocins are substances (peptide antibiotics) released by some bacteria that kill other bacteria. While the *X. perforans* strain was effective at limiting the damage caused by the pathogen *X. euvesicatoria*, it showed its greatest effectiveness on the leaf surface, with less of an effect observed inside of the leaf. This observation suggests that competition is the major mechanism of pathogen suppression and the bacteriocin merely enhances the competition since competition is known to be less effective inside the leaf.

In one recent experiment, researchers began with the bacterium *Pseudomonas fluorescens* F113rif, a biocontrol PGPB that was originally isolated from the sugar beet rhizosphere. This strain can suppress take-all disease caused by the fungal pathogen *Pythium ultimum*. The biocontrol ability of this strain is thought to be a consequence of several activities that the biocontrol strain possesses including the production of siderophores, the antibiotic diacetyl-phloroglucinol, hydrogen cyanide, and an extracellular protease. A triple mutant of this biocontrol strain that was affected in the genes *kinB*, *sadB*, and *wspR* (KSW) is hypermotile and is more competitive for rhizosphere colonization than the wild-type strain. In one set of experiments, the biocontrol activity of the wild-type strain (F113) was compared with a hypermotile strain derived from strain F113 (i.e., V35) and the abovementioned triple mutant strain (i.e. KSW). In competition experiments, the mutant strains were consistently more effective than the wild-type at colonizing plant

Fig. 6.12 Biocontrol activity of a mutagenized PGPB strain of *Pseudomonas fluorescens*. The pathogen in (**a**) is *Fusarium oxysporum* and the plant is tomato. The pathogen in (**b**) is *Phytophthora cactorum* and the plant is strawberry. Abbreviations: F113, a wild-type *Pseudomonas fluorescens* biocontrol strain; V35, a hypermotile phenotypic variant of strain F113; KSW, a triple mutant of strain F113 that is both hypermotile and more competitive at binding to plant roots than the wild-type

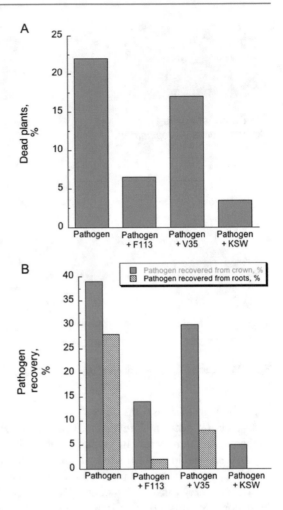

roots. In addition, the hypermotile mutant strain (KSW) was much more effective than either the hypermotile (V35) or wild-type strain (F113) in preventing damage to tomato plants by the fungal pathogen *Fusarium oxysporum* and to strawberry plants by the fungal pathogen *Phytophthora cactorum* (Fig. 6.12). This result indicates that the ability of a biocontrol PGPB to be an effective competitor for binding to plant roots can significantly augment other biocontrol activities that the bacterium may possess. Moreover, while hypermotility of a biocontrol strain may enhance the effectiveness of a biocontrol strain to a limited extent, the competitiveness of the bacterium in binding to plant roots is a more important trait.

To limit root colonization by phytopathogens, plants sometimes secrete active oxygen species, such as the hydroxyl radical, the superoxide anion and hydrogen peroxide, all of which can inhibit phytopathogen cell processes (Fig. 6.13). Plant roots also sometimes respond to colonization by PGPB by producing reactive oxygen species. Some more effective phytopathogens (that are effective at

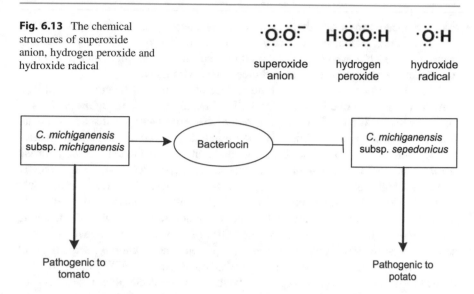

Fig. 6.13 The chemical structures of superoxide anion, hydrogen peroxide and hydroxide radical

Fig. 6.14 Schematic representation of strain *C. michiganensis* subsp. *michiganensis*, a pathogen of tomato plants, producing bacteriocin CmmAMP-1 which in turn inhibits the functioning of strain *C. michiganensis* subsp. *sepedonicus*, a pathogen of potato plants

overcoming the plant's defenses) contain elevated levels of enzymes such as superoxide dismutase, catalase, and peroxidase that can reduce the amount of active oxygen species. Similarly, it is possible by genetic manipulation of biocontrol PGPB, to increase the levels of one or more of the enzymes that reduce the amount of active oxygen species. This genetic manipulation results in the biocontrol PGPB being more effective at colonizing plant roots (i.e., more competitive) and more persistent in the rhizosphere with the result that these biocontrol PGPB are more effective antifungal pathogen agents.

In the soil, bacteria and fungi often compete with one another to occupy favorable ecological niches such as plant rhizospheres. To outcompete other similar bacteria, some soil bacteria synthesize antimicrobial compounds called bacteriocins, ribosomally encoded peptides that typically are active against species that are closely related to the producer strain. In one recent study, scientists identified, purified, and characterized a 14 kDa bacteriocin (*Cmm*AMP-1) synthesized by the tomato plant pathogen *Clavibacter michiganensis* subsp. *michiganensis*. The researchers subsequently showed that both the bacteriocin-producing pathogen and the purified bacteriocin were highly active at inhibiting the growth of the related pathogen *Clavibacter michiganensis* subsp. *sepedonicus* which is the cause of bacterial ring rot of potato (Fig. 6.14) (Tomato and potato both belong to an extremely diverse and large plant family, the Solanaceae). Following the isolation of the gene for this bacteriocin, the recombinant protein was produced and found to be highly active at a very low concentration, with 50% growth inhibition (IC_{50}) of ~10 pmol, but was not toxic to potato leaves or tubers. This demonstration opens up the possibility of identifying and characterizing a whole class of narrow inhibitory spectrum bacteriocins, each of which is synthesized by a specific bacterium as part of its effort to outcompete similar bacteria in its immediate environment.

It is often difficult to understand the precise mechanism of biocontrol used by a particular PGPB. In one instance, the PGPB *Paenibacillus dendritiformis* C454, a Gram-positive soil bacterium that can reduce bacterial pathogen disease in potato was studied in some detail. This PGPB significantly reduced the amount of maceration of potato slices that had been infected with the pathogen *Pectobacterium carotovorum* subsp. *carotovorum* in laboratory experiments and subsequently protected potato tubers in the field against this pathogen. When this PGPB was tested in plate assays (see Sects. 6.2 and 6.4) for any antagonism to the pathogen, none was found. Moreover, the PGPB did not demonstrate any quorum quenching activity (see Sect. 6.11). Nevertheless, the PGPB demonstrated significant biocontrol activity against the potato pathogen. When the genome sequence of *P. dendritiformis* C454 was examined, it was found that this PGPB strain encoded genes that made it an effective competitor over other soil bacteria for resources such as iron, amino acids, and sugars. Moreover, the DNA sequence of the genome indicated that this PGPB strain had the ability to synthesize both antibiotics and cell lytic enzymes. Thus, it was surmised the enhanced metabolic competitiveness of this PGPB strain together with a low level of other biocontrol activity enabled this bacterium to out-compete the otherwise plant-destroying pathogen.

Several species of the fungal phytopathogen *Fusarium* produce mycotoxins that may contaminate grain and therefore become a threat to both animals and humans. Fortunately, at least one strain of the PGPB *Paenibacillus polymyxa* are effective antagonists of *Fusarium* infection. Until recently, it was thought that this antagonism was largely the consequence of antibiotic lipopeptides synthesized by this bacterial strain. However, by constructing a deletion mutant of this PGPB strain that was unable to synthesize lipopeptides, researchers observed that while low levels of the mutant strain were not as effective as the wild-type strain, high levels of the mutant were as effective as the wild-type in antagonizing/inhibiting the pathogen (Fig. 6.15). These results were interpreted as indicating that while the antibiotic lipopeptide had a small effect on antagonizing the pathogen, the major determinant of the PGPR's effectiveness as a biocontrol strain was its ability to synthesize extracellular polysaccharides with a high D-glucuronic acid content.

6.7   Volatile Compounds

In a limited number of instances, biocontrol microorganisms that produce volatile compounds that are toxic to a variety of pathogens have been identified. One of the best examples of this mechanism is the plant growth-promoting endophytic fungus called *Muscodor albus*. This plant-dwelling fungus was first discovered, in the late 1990s, in the bark of a cinnamon tree in a rain forest in Honduras. *M. albus* is a highly effective antipathogenic agent against a wide range of pathogenic fungi and

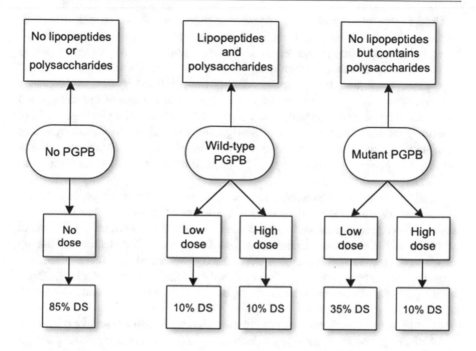

Fig. 6.15 Overview of the protocol used to assess the disease severity (DS) of *Fusarium graminearum* on wheat plants in the absence of added PGPB (*Paenibacillus polymyxa*) or in the presence of either wild-type or a mutant strain of *P. polymyxa* that is unable to synthesize lipopeptides. In the absence of the PGPB, the disease severity is 85% while either low or high doses of the PGPB or a high dose of the mutant lowers the disease severity to 10%

bacteria including, but not limited to, *Rhizoctonia solani, Pythium ultimum, Pythium aphanadermatum, Verticillium dahliae, Fusarium aveneaceum, Phytophthora capsici, Phytophthora palmivora, Sclerotinia minor,* and *Sclerotinia rolfsii.* In addition, *M. albus* can also act as an insecticide and nematocide.

Interestingly, *Muscodor albus* and other *Muscodor* species (found in various tropical forests around the world) do not have to come into contact with a pathogenic agent in order to inhibit or kill it. This is because *Muscodor* spp. synthesize a large number of different volatile organic compounds that are lethal to a wide range of bacteria and fungi. These volatile organic compounds include (among others) acetone, ethyl propionate, ethyl isobutyrate, methyl 2-methylbutyrate, ethanol, ethyl butyrate, isobutyl alcohol, 2-methylbutyl acetate, 2-methyl-1-butanol, isobutyric acid, octane and phenethyl alcohol. It should be pointed out that researchers have been able to duplicate some, but not all, of the anti-phytopathogenic effects observed with *M. albus* using mixtures of the abovementioned organic compounds. While the precise mode of action of volatile organic compounds produced by *Muscodor* species has not been established, it has been suggested that many of these compounds may induce the alkylation of pathogen DNA.

Unlike other biocontrol agents, *Muscodor* cannot readily be used in the field. However, it may be used as a fumigant in a greenhouse, or in a warehouse of harvested fruit. *M. albus* is also being developed for fumigation of greenhouse soil to control diseases that attack plant roots, post harvest disease control of fruit, nuts, vegetables, seeds and bulbs; insect control; pre-plant fumigation of tomatoes, peppers, strawberries, nursery and other crops; re-plant disease of nut trees, stone fruit trees and vines; and greenhouse soil diseases. Notwithstanding its suggested use as a fumigant, given that *Muscodor* was originally isolated as a tree endophyte, it is possible that some strains of *Muscodor* can be directly inoculated into agricultural and forest species in order to provide protection against invading pests and pathogens.

To utilize *M. albus* effectively, it has been necessary to develop a unique method of delivery for this biocontrol agent. Thus, a formulation of *M. albus* consists of desiccated rye grain colonized by the fungus. The delivery system includes a packet or sachet of the desiccated grain that has to be activated for postharvest use by rehydration (i.e., the addition of water). Once rehydration has taken place, the antipathogen treatment can be applied by simply placing activated *M. albus* sachets within packages (boxes or cartons) of grapes or peaches as is currently done with sulfur dioxide generator pads.

In addition to its antifungal activity, *M. albus* has been found to act as a biocontrol agent of the potato tuber moth, *Phthorimaea operculella,* a serious insect pest of stored potato as well as of a number of plant parasitic nematodes that, in aggregate, cause tens of billions of dollars of economic loss to US agriculture every year.

More recently, numerous biocontrol PGPB have been shown to produce volatile organic compounds that can decrease the growth of pathogenic fungi. For example, the biocontrol PGPB *Bacillus amyloliquefaciens* T-5 was recently shown to produce several volatile organic compounds that significantly inhibited the growth of *Ralstonia solanacearum*, the causative agent of tomato bacterial wilt disease. This biocontrol PGPB produced 13 different volatile organic compounds including benzenes, ketones, aldehydes, alkanes, acids, and one furan and naphthalene compound. Interestingly, individually these compounds demonstrated between 1% and 10% antibacterial activity against *R. solanacearum;* however, when all 13 compounds were tested together, they inhibited the growth of the pathogen by 62–85%, depending upon the medium being used in the test. While the concentration of the volatile organic compounds produced by this biocontrol PGPB are relatively low, they can nevertheless spread over a relatively long distance and they should be effective in making various pathogens more susceptible to other biocontrol approaches such as the use of antibiotics. In addition to the biocontrol activity observed with *Bacillus amyloliquefaciens* T-5, several researchers have reported that the volatile organic compounds produced by PGPB including: (1) *Paenibacillus polymyxa* caused the death of the canola pathogen *Verticillium longisporum*; (2) *Pseudomonas chlororaphis* caused the death of the potato pathogen *Rhizoctonia solani*; and (3) *Delftia lacustris* caused the death of the tomato pathogen *Fusarium oxysporum*. In addition to their direct effects on phytopathogens, there is some evidence that a number of volatile organic compounds can actuate induced systemic resistance in some plants.

Following the isolation of 114 different endophytic bacterial strains from wild licorice plants (*Glycyrrhiza uralensis*) the endophytes were tested for their antifungal activity against several different fungal pathogens. Of these, one particular bacterial strain, *Bacillus atrophaeus*, produced at least 13 compounds with antimicrobial activity including, but not limited to, 1,2-benzenedicarboxylic acid, bis (2-methylpropyl) ester; 9,12-octadecadienoic acid (Z,Z)-, methyl ester; 9-octadecenoic acid, methyl ester, (E)-; and decanedioic acid, bis(2-ethylhexyl) ester. Interestingly, these compounds were only expressed by the endophytic biocontrol strain in the presence of the pathogen *Verticillium dahlia*, a fungal pathogen that is usually extremely difficult to manage. It remains to be seen whether this bacterium will be an effective biocontrol strain against *Verticillium dahlia* infection in the field.

6.8 Lowering Ethylene

Plants respond to a wide variety of different stresses including fungal phytopathogen infection by synthesizing "stress" ethylene. In turn, the ethylene can trigger a stress/senescence response in the plant that may lead to the death of those cells that are at or near the site of the fungal infection. Stress ethylene is thought to act as a secondary messenger and can (in different tissues in different plants) stimulate senescence, leaf or fruit abscission, disease development, inhibition of growth, and/or antibiotic enzyme (e.g., chitinase or β-glucanase) synthesis.

The synthesis of stress ethylene in response to phytopathogen infection occurs in two phases (Fig. 6.16a). The first (and much smaller) peak of ethylene synthesis

Fig. 6.16 The synthesis of stress ethylene in response to various biotic and abiotic stresses including phytopathogen infection. (**a**) Represents the absence of an added PGPB while (**b**) represents the presence of an ACC deaminase-containing PGPB

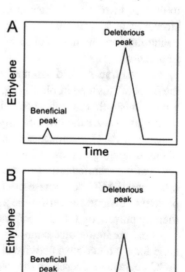

occurs very soon after the onset of the microbial infection (i.e., typically within a few hours). This short burst of ethylene is synthesized from the plant's pool of preexisting 1-aminocyclopropane-1-carboxylate (ACC), and appears to be one of the triggers that activates the plant's synthesis of defensive proteins, a response that is designed to limit the proliferation of the microbial pathogen within the plant. The second (and much larger) peak of ethylene synthesis usually occurs 2 or 3 days after the initial microbial phytopathogen infection. This ethylene peak is dependent upon newly synthesized ACC subsequent to the initial infection. It is believed that it is this second peak of ethylene synthesis that contributes to a large portion of the damage to a plant that follows microbial phytopathogen infection.

Many of the disease symptoms of a plant that is infected with a microbial phytopathogen may arise as a direct result of the stress that is imposed by the infection. That is, a significant portion of the damage to plants infected with microbial phytopathogens occurs as a result of the response of the plant to the increased level of stress ethylene that forms in response to the infection rather than as a consequence of the direct action of the pathogen itself. In this regard, it has been observed in laboratory experiments that the presence of exogenous ethylene generally increases the severity of a phytopathogenic microbial infection, and, conversely, inhibitors of ethylene synthesis usually significantly decrease the severity of a microbial infection. Thus for example, the addition of exogenous ethylene has been found to increase disease severity in tomato plants infected with *Verticillium* wilt. On the other hand, cotton plants that were treated with L-α-(aminoethoxyvinyl)-glycine, (AVG), an inhibitor of the enzyme ACC synthase and a chemical analog of the naturally occurring ethylene inhibitor rhizobitoxine, were much less damaged by the fungal phytopathogen *Alternaria* than were plants not treated with any ethylene inhibitors. There is little question of the laboratory effectiveness of adding chemical ethylene inhibitors in terms of their ability to reduce pathogen-associated plant damage. However, as a consequence of both the high cost and the potential environmental contamination, the use of chemical ethylene inhibitors on a large scale in the field is impractical.

The enzyme ACC deaminase (see Sect. 5.5) when present in PGPB, modulates the level of ethylene in plants (Fig. 6.16b). The functioning of this enzyme is part of a mechanism that is widespread in PGPB to directly stimulate plant growth. ACC deaminase lowers the level of ethylene in both developing and stressed plants. In the absence of ACC deaminase, pathogen-induced stress ethylene is produced in most plants following pathogen infection. The stress ethylene mediates and exacerbates some of the manifestations of the damage to plants caused by the pathogen. On the other hand, ACC deaminase-containing biocontrol PGPB act to modulate the level of stress ethylene (lowering the second or deleterious ethylene peak) in a plant and thereby prevent or at least decrease the damage normally caused by phytopathogens.

There are numerous examples of the presence of ACC deaminase being responsible for the biocontrol activity of various PGPB. For a start, it has been shown that ACC deaminase decreases the damage to cucumber plants infected with the fungal pathogen *Pythium ultimum* (Table 6.3). Strain *P. fluorescens* CHA0 is an antibiotic-producing biocontrol PGPB that limits the severity of the damage to cucumber plants

Table 6.3 Reduction in the severity of damage (damping-off) of cucumber plants caused by *Pythium ultimum*

Treatment	Seed germination rate (%)	Shoot fresh weight (g)	Root fresh weight (g)
None	100	3.48	2.85
P. ultimum	31	0.69	0.21
P. ultimum + CHA0	79	2.11	0.95
P. ultimum + CHA0/ ACC-D	87	2.27	1.27
P. ultimum + UW4	62	1.53	0.62

CHA0, the biocontrol PGPB *Pseudomonas fluorescens* CHA0; *CHA0/ACC-D*, the biocontrol PGPB *P. fluorescens* CHA0 transformed with a broad-host-range plasmid that contains an ACC deaminase gene from *P. putida* UW4; *UW4*, the PGPB strain *P. putida* UW4

Table 6.4 Effect of *Pseudomonas fluorescens* CHA0 and *P. fluorescens* CHA0/ACC-D on the soft rot of potatoes by *Erwinia carotovora*

Treatment	Weight of rotted potatoes, g/slice
None	15.6
P. fluorescens CHA0	14.5
P. fluorescens CHA0/ACC-D	7.5

caused by the fungal pathogen *Pythium ultimum*; it does so by means of antibiotic synthesis. When strain CHA0 is genetically transformed so that it expresses ACC deaminase activity, the damage to cucumber plants is reduced even further (i.e., the two activities act synergistically). Moreover, strain *P.* sp. UW4, which does not synthesize any antibiotics and does not have any other biocontrol activity, protects cucumber plants to a significant extent against the inhibitory effects of the fungal pathogen. This result is consistent with the lowering of stress ethylene by the bacterial ACC deaminase.

Strains *P. fluorescens* CHA0 and *P. fluorescens* CHA0/ACC-D (a transformant of strain *P. fluorescens* CHA0 that expresses an exogenous ACC deaminase gene) were also used in an effort to protect potato slices against damage by the bacterial pathogen *Erwinia carotovora* (Table 6.4). In this case, the biocontrol PGPB strain CHA0 did not significantly affect the extent of damage to potatoes due to the bacterial pathogen. However, the CHA0 transformant that expresses ACC deaminase lowers the level of stress ethylene and also decreases the damage to the potatoes by approximately 50%. In this experiment, the antibiotic synthesized by strain CHA0 was ineffective against the bacterial pathogen. Despite the "failure" of one biocontrol mechanism (i.e., antibiotic production), lowering the level of stress ethylene significantly (but not completely) protected the potatoes against damage by the bacterial pathogen.

The bacterial genus *Agrobacterium* includes several related strains that are the causative agents of crown gall and hairy root disease that affects a wide range of dicotyledonous, as well as some monocotyledonous, plants. For example, strains of *Agrobacterium vitis* are the predominant cause of crown gall tumors of grape in various grape-growing regions worldwide. Moreover, the galls that are induced by

Fig. 6.17 Effect of ACC deaminase-producing bacteria on tumor (crown gall) size (**a**) and ethylene production (**b**) in tomato plants infected with the bacterial pathogen *Agrobacterium vitis*. Abbreviations: UW4, the rhizospheric PGPB *Pseudomonas putida* UW4; UW4 mutant, an ACC deaminase minus mutant of strain *P. putida* UW4; PsJN, the endophytic PGPB *Burkholderia phytofirmans* PsJN; PsJN mutant, an ACC deaminase minus mutant of strain *B. phytofirmans* PsJN

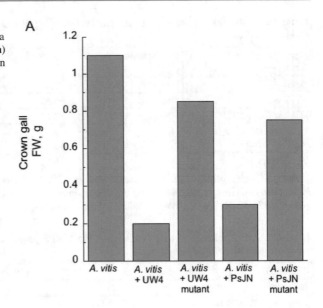

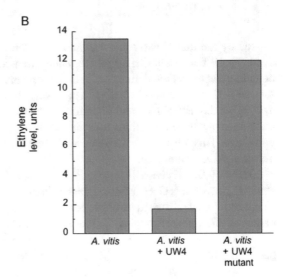

Agrobacterium infection produce very high levels of ethylene. This observation is consistent with the possibility that ethylene may facilitate or at least exacerbate crown gall development. Based on this logic, one may predict that by limiting ethylene production in *Agrobacterium*-infected plants it should be possible to significantly decrease the damage to those plants that occurs as a consequence of *Agrobacterium* infection. In fact, researchers in two separate laboratories have demonstrated that ACC deaminase-containing PGPB can significantly lower both plant ethylene levels and the size of crown gall tumors in *Agrobacterium*-infected plants (Fig. 6.17). In addition, PGPB in which the ACC deaminase gene has either

Table 6.5 Effect of an ACC deaminase-containing biocontrol PGPB strain on the development of southern blight disease in tomato plants

Measurement	+ S. rolfii	+ P. lentimorbus + S. rolfii
Disease index	87.2%	63.1%
Dead plants	55.4%	33.3%
Plant dry weight	1.1 g	2.0 g
Shoot length	22 cm	28 cm
Chlorophyll content	0.9 mg/g fresh weight	1.5 mg/g fresh weight
Proline content	105 mg/g fresh weight	88 mg/g fresh weight
Superoxide dismutase	3.7 units/mg protein	2.8 units/mg protein
Catalase	75 units/mg protein	44 units/mg protein
Ethylene content	3.2 μmoles/h/g fresh weight	2.3 μmoles /h/g fresh weight

All plants were infected when they were 30 days old and measurements were taken 7 days later

been interrupted or deleted are no longer able to lower ethylene levels or prevent the growth of crown gall tumors.

Tomato plants are highly susceptible to a number of different fungal-caused diseases including *Fusarium* wilt (caused by *Fusarium oxysporum*), *Verticillium* wilt (caused by *Verticillium dahlia*), early blight (caused by *Alternaria solani*), late blight (caused by *Phytophthora infestans*) and, most problematically, southern blight (caused by *Scelerotium rolfsii*). Unfortunately, it is extremely difficult to control these diseases using chemical fungicides because of the fact that the fungal spores are capable of surviving for many years in the soil notwithstanding the presence of a chemical fungicide. However, biocontrol PGPB that utilize several different mechanisms of biocontrol have the potential to dramatically limit the damage to plants from fungal pathogens. In particular, when fungal pathogens are not completely killed by other mechanisms and they are still able to infect susceptible plants, PGPB that contain ACC deaminase can significantly lower the damage to infected plants. As mentioned above, this is because a significant portion of the damage to the plant occurs as a consequence of the synthesis of stress ethylene by the plant. This was shown to be the case in one recent study that employed an ACC deaminase-producing strain of the PGPB *Paenibacillus lentimorbus* to protect tomato plants from southern blight disease (Table 6.5). In this case, when the biocontrol PGPB with ACC deaminase was present, both plant death and the plant disease index decreased significantly. In addition, the presence of the PGPB resulted in an increase in plant dry weight (and also fresh weight; not shown in this table), shoot length and chlorophyll content, all indications that treatment with this PGPB produced a healthier plant. Moreover, the presence of the ACC deaminase-containing PGPB resulted in a lowering of the proline, superoxide dismutase, catalase and ethylene contents, indicating that the plant was experiencing a lower level of stress. Here, it is important to bear in mind that many experiments of this sort assume, but do not prove, that the positive results that are observed using a biocontrol PGPB that contains ACC deaminase are a consequence of the presence of ACC deaminase. However, in the experiments summarized in Table 6.5, by measuring the changes in the plant ethylene level (including following different times of pathogen infection) the researchers have directly correlated the positive results that they have observed with the lowering of stress ethylene levels in the plant.

Xanthomonas euvesicatoria (as well as other *Xanthomonas* species) is the causative agent of bacterial spot disease which affects a number of different crop plants, including pepper and tomato plants, causing spots on the leaves, stem and fruit. To avoid damage to plants infected with this pathogen farmers have employed both chemical pesticides and resistant varieties of the selected crop plants. More recently, some researchers have shown that some biocontrol endophytic PGPB can significantly lower the damage to plants infected with this pathogen. In one set of experiments, four different endophytic bacterial strains, including *Ochrobactrum* sp., *Pantoea agglomerans*, *Bacillus thuringiensis*, and *Pseudomonas fluorescens*, were able to provide significant protection against bacterial spot disease of both pepper and tomato. When these fours strains were examined in detail, it became clear that the one antipathogenic mechanism that these four strains had in common was the possession of ACC deaminase activity. Therefore, it was not surprising that in addition to acting as biocontrol agents, these PGPB strains also promoted the growth of noninfected plants. Thus, the results of these experiments are consistent with the notion that ACC deaminase from endophytic (as well as rhizospheric) bacteria can decrease the damage to plants from both fungal and bacterial pathogens.

6.9   Systemic Resistance

In many plants, long-lasting and broad-spectrum systemic resistance to a variety of disease-causing agents including fungal and bacterial pathogens can be induced by treating the plant with either PGPB or with the pathogen itself (Fig. 6.18). Systemic resistance that is induced by the pathogen itself is called systemic acquired resistance (SAR) while the resistance induced by a PGPB is called induced systemic resistance (ISR). SAR is typically characterized by the coordinate activation of the expression of a set of "pathogenesis-related" (PR) genes that encode proteins with antimicrobial activity. In some reports ISR has been associated with enhancement of lignification and increases in peroxidase and superoxide dismutase activity within the treated plant. In addition, several bacterial molecules have been implicated as signals for the induction of systemic resistance including the *O*-antigenic side chain of the bacterial outer membrane component lipopolysaccharide, bacterial siderophores and, in a few instances, bacterial salicylic acid. Both ethylene and jasmonic acid have been implicated in the ISR response, however, the evidence for this involvement is a bit circumstantial since measured plant levels of jasmonate and ethylene do not change. However, plant mutants that are defective in aspects of jasmonate or ethylene signaling do not develop ISR. Thus, it has been suggested that ISR is based on increased sensitivity and not increased production of these hormones.

Most studies of systemic resistance have been carried out using fungal pathogens; however, this approach is also useful in the control of bacterial diseases. In one

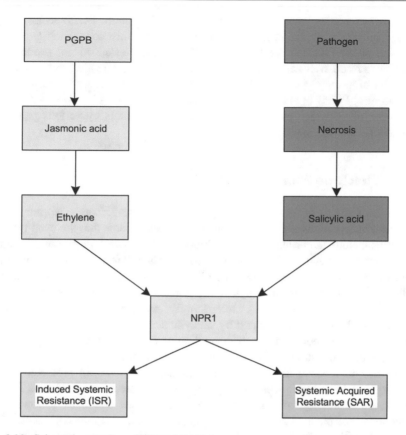

Fig. 6.18 Schematic overview of ISR and SAR

study, inoculation with two PGPB strains (*P. putida* and *Serratia marcescens*) protected cucumber plants against the pathogen *P. syringae* pv. lachrymans, the causal agent of bacterial angular leaf spot. Treatment of seeds or cotyledons was found to result in a significant decrease in both lesion number and size, and in the leaf population of the pathogen. Importantly, PGPB have been shown to induce systemic resistance in the field as well as in laboratory and greenhouse conditions.

At the present time a thorough and detailed knowledge of ISR and SAR is still very much incomplete so that it is not always possible, a priori, to predict the response of a plant to a particular PGPB strain. For example, one strain of *P. fluorescens* was effective in inducing systemic resistance in *Arabidopsis thaliana* but not in radish, protecting the plant against both a fungal (*Fusarium oxysporum*) and a bacterial (*P. syringae* pv. tomato) phytopathogen. Another *P. fluorescens* strain induced systemic resistance in radish but not in *Arabidopsis*. This suggests that different biocontrol bacteria are "differentially active in eliciting induced systemic resistance in related plant species."

Numerous studies have concluded that various biocontrol PGPB reduce fungal damage to plants by inducing systemic resistance. This conclusion is often a reflection of the fact that the biocontrol PGPB being tested do not produce any antibiotics or cell wall-degrading enzymes (or otherwise directly inhibit pathogen growth). Unfortunately, this conclusion generally fails to consider the possibility that the biocontrol PGPB in question may be able to produce IAA or cytokinin or ACC deaminase, all of which can decrease the damage to plants by various pathogens without affecting the pathogen itself.

6.9.1   Effects on Plant Gene Expression

A number of strains of filamentous fungi from the genus *Trichoderma* are used as biocontrol agents in agriculture. The biocontrol mechanisms that are used by these organisms include the production of antibiotics and/or hydrolytic enzymes, competition for nutrients and the fungal stimulation of plant defense responses including both ISR and SAR. Root colonization by *Trichoderma* triggers a systemic response mediated by jasmonic acid and ethylene in a manner similar to that described for ISR induced by PGPB (Fig. 6.18). Interestingly, systemic defenses induced by *Trichoderma* may also be mediated by salicylic acid (similar to SAR). In the latter case, the defense response may vary depending on the plant genetic background, the interaction time, the strain of *Trichoderma* and inoculum size.

To better understand how a particular strain of *Trichoderma* protected tomato plants against fungal pathogens, one group of researchers used real-time PCR to monitor the expression of plant genes involved in ISR and SAR following the interaction of the plant with *Trichoderma*. To do this, they incubated plant seedlings for 6 days with *Trichoderma* and compared the response of those plants with untreated control plants. Treated plants had significantly increased expression levels of ethylene signaling and lipoxygenase genes (part of the jasmonate and ethylene signaling pathways), while the salicylic acid responsive pathogenesis-related gene expression level was reduced, and salicylic acid biosynthesis gene expression was not affected. These results indicate that the *Trichoderma* inoculation stimulated an ISR but not an SAR response.

Using real-time PCR of numerous known genes to assess which genes are up regulated and which are downregulated presupposes that we know ahead of time which genes may alter their expression in response to either pathogens or growth-promoting bacteria or fungi. However, it is quite likely that more than a few plant genes (and possibly even hundreds to thousands) will exhibit altered expression levels as a consequence of interacting with either pathogens or growth-promoting bacteria or fungi. Thus, to better understand the detailed nature of how the plant is responding, it is necessary to monitor the expression of hundreds to thousands of plant proteins or mRNAs using the techniques of proteomics or transcriptomics, respectively. Thus, in one recent study of the biocontrol mechanisms used by a strain of *Bacillus velezensis* against *Fusarium* wilt of watermelon revealed nearly a thousand differentially expressed genes in the process of the biocontrol strain triggering ISR. For a start, the biocontrol PGPB strain was shown to inhibit the proliferation of several pathogenic fungi on agar plates including *Fusarium oxysporum* f. sp. *niveum*, *Fusarium oxysporum* f. sp. *cubense*, *Fusarium graminearum*, *Mycosphaerella melonis*, and *Botrytis cinerea*. In addition, the biocontrol PGPB strain could inhibit the germination of *F. oxysporum* spores and contained several hydrolase activities including protease, glucanase, and cellulase. However, highly effective biocontrol PGPB often prevent pathogen damage to plants using several different mechanisms. Thus, a transcriptomic analysis of watermelon plants treated with *Fusarium oxysporum* and the biocontrol PGPB showed that the biocontrol strain induced the expression of the enzymes: catalase, peroxidase, and superoxide dismutase which protect plants against activated oxygen species. Moreover, the biocontrol strain induced the expression of genes involved in the synthesis and functioning of several phytohormones including ethylene, jasmonate, salicylate, abscisic acid and brassinosteroids. The induction of so many mRNAs indicates that ISR (i.e., plant genes induced by the biocontrol PGPB) and SAR (i.e., plant genes induced by the presence of the pathogen) is a complex response resulting in a plant that is able to survive the pathogen attack.

6.10 Bacteriophages

In some parts of the world, sprays of either antibiotics (typically either tetracycline or streptomycin) or copper are used to minimize the damage to crops from bacterial pathogens. However, there are serious problems with both of these approaches. The continued use of antibiotics can result in the development of antibiotic-resistant pathogens and the subsequent spread of the resistance genes to other bacteria in the environment. The long-term use of copper sprays can lead to the accumulation of toxic levels of copper in the environment. Therefore, in some countries workers have sought to replace the use of antibiotics and copper sprays with bacteriophages, viruses that infect bacteria.

On a practical level, it is important to isolate diverse bacteriophages from a wide range of host bacterial strains that represent the diversity of the pathogens involved in the targeted disease. It is also suggested, when isolating new bacteriophages that, when they are tested under laboratory conditions, strains that produce clear plaques should always be selected, as this will decrease the possibility of selecting lysogenic bacteriophages that do not (completely) lyse their bacterial hosts but rather are somewhat prone to undergo a lysogenic cycle (Fig. 6.19). While the host range of most bacteriophages is usually quite narrow, with new isolates this needs to be tested directly to be sure that the selected bacteriophage lyses only the target pathogen and does not affect any bacterial strains that might facilitate plant growth. Ideally, bacteriophage should be delivered to the plant as a cocktail of several different bacteriophage strains all directed against the targeted bacterial pathogen. In one study, a cocktail of four different bacteriophages decreased the incidence of disease in tomato plants caused by the pathogen *Ralstonia solanacearum* by around 80% in both green house and field experiments. In this case, while a single bacteriophage

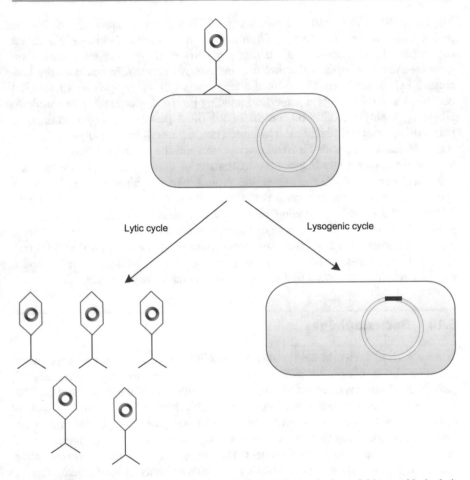

Fig. 6.19 Schematic representation of a bacteriophage infecting its bacterial host, with the lytic cycle resulting in the complete lysis of the bacterial host and simultaneous production of additional bacteriophage molecules, and the lysogenic cycle resulting in the bacteriophage genomic DNA being integrated into the chromosomal DNA of the host bacterium without the bacterium being lysed. The red circles inside of the bacteriophage heads represent bacteriophage genomic DNA

significantly reduced the incidence of bacterial disease of tomato, using a bacteriophage cocktail reduced the disease incidence to a much greater extent. The use of a cocktail of bacteriophages makes it less likely that bacteriophage-resistant mutants of the targeted pathogen will develop since each of the different bacteriophages will likely bind to a different site on the bacterial surface. It is also important, prior to their widespread use, for the DNA genome sequences of the selected bacteriophages to be completely determined (RNA bacteriophages are generally not used for this purpose) so that they can be accurately tracked in the environment. When testing biocontrol bacteriophages in the field, the presence of other microbes, including other pathogens, may influence the effectiveness of the added bacteriophages. In

Table 6.6 The number of enteric bacteria (per defined weight of strawberries) found on the surface of strawberries with and without bacteriophage treatment, 48 h after treatment

Sample	Without bacteriophage	With bacteriophage
Strawberries with organic amendments	94×10^7	29×10^3
Surface sterilized supermarket strawberries infected with one bacterial isolate	54×10^7	14×10^3
Surface sterilized supermarket strawberries infected with three antibiotic-resistant bacterial isolates	64×10^7	42×10^3

addition, prior to treating affected plants, it is essential to ascertain precisely which organism caused the disease symptoms since in some cases pathogens of a different genus or species can cause very similar disease symptoms but will not be killed by the bacteriophage cocktail that has been developed.

In one study, it was found that the surface of strawberries was contaminated with a number of different enteric bacteria that could make these fruits dangerous for people to consume. These bacteria included strains belonging to the genera *Escherichia*, *Enterobacter*, *Raoultella*, *Klebsiella*, *Pantoea*, *Shigella*, *Citrobacter*, and *Cronobacter*. In addition, the contaminating levels of these bacteria were particularly high when organic amendments such as composted material and animal manure were used in the cultivation of the strawberries. To remove the bacteria that adhered to their surface, fruits were treated with either antibiotics or with cocktails of several different *Enterobacteriaceae* bacteriophages. As expected, the antibiotics significantly decreased the number of bacteria on the surface of the strawberries; however, continued use of antibiotics could potentially select for bacteria that were antibiotic resistant. And, as some pesticide-degrading pathogenic bacteria might be present on chemically treated fruit, this approach might present a risk to consumers. On the other hand, the bacteriophage treatment dramatically reduced, but did not completely abolish, the number of viable bacteria on the surface of these fruits. Importantly, bacteriophage treatment dramatically lowered the number of antibiotic-resistant bacteria that had been deliberately added to previously disinfected store bought fruit (Table 6.6). In all cases, bacteriophage treatment reduced the number of contaminating bacteria down to acceptable levels. The experiments using bacteriophages described here are still preliminary and both the long-term efficacy of this approach and the persistence of bacteriophages on treated fruit remains to be determined before this approach may be considered for more widespread use.

In the field, as a consequence of the UV light sensitivity of most bacteriophages, the application of this type of biocontrol agent is often done at dusk when the UV light intensity is low. This results in an increase in bacteriophage environmental longevity and effectiveness. Even so, in the field some bacteriophages need to be applied weekly, or even daily, to be effective. In addition to there being some problems with bacteriophage stability in the field, it has sometimes been observed that following the rapid lysis of a very large portion of the population of a bacterial pathogen, a small number of bacteriophage-resistant cells sometimes remain. If these resistant cells, admittedly a very small number, are able to proliferate then resistant

Table 6.7 Some reported successful experiments using bacteriophage as bacterial pathogen biocontrol agents

Pathogen	Plant	Disease
Pectobacterium spp.	Potato	Soft rot
Dickeya solani	Potato	Soft rot/blackleg
Streptomyces scabies	Potato	Common scab
Ralstonia solanacearum	Tomato	Bacterial wilt
Xanthomonas campestris pv. *vesicatoria*	Tomato, peppers	Bacterial spot
Xylella fastidiosa	Grapevine	Pierce's disease
Xanthomonas axonopodis pv. *allii*	Onion	Xanthomonas leaf blight
Pectobacterium carotovorum ssp. *carotovorum*	Lettuce	Soft rot
Streptomyces scabies	Radish	Common scab
Xanthomonas axonopodis pv. *citri*	Grapefruit	Asiatic citrus canker
Xanthomonas axonopodis pv. *citrumelo*	Orange	Citrus bacterial spot
Pseudomonas syringae pv. *porri*	Leek	Bacterial blight
Pseudomonas syringae pv. *tomato*	Tomato	Bacterial speck
Pseudomonas tolaasil	Mushrooms	Brown blotch disease
Erwinia amylovora	Pear, apple trees	Fire blight

bacteria can predominate at the next generation, ultimately negating the effectiveness of this approach. Whether this scenario occurs in the field as well as in the lab is not known at the present time.

At the present time, a number of bacteriophage-based biocontrol agents have been successfully tested in the laboratory (Table 6.7) and a few have been licensed for use. For example, there are a range of licensed bacteriophage products for the biocontrol of the bacterial pathogen *Xanthomonas campestris* pv. vesicatoria which causes bacterial spot of tomatoes and peppers, and *Pseudomonas syringae* pv. tomato, which is the causative agent of bacterial speck on tomatoes.

6.11   Quorum Sensing and Quorum Quenching

In the environment, using the mechanism of quorum sensing, bacterial cells are able to sense the presence of other similar, as well as different, organisms. Quorum sensing is a form of intercellular communication that is used by many different species of bacteria in the environment. Once bacterial cell densities attain a certain critical level, quorum sensing enables bacteria to switch on different sets of genes thereby facilitating concerted interactions between the cells, enabling groups of bacteria to act in a coordinate manner.

Most quorum sensing systems behave similarly. Initially, low molecular weight molecules called autoinducers are synthesized inside of bacterial cells. Some of these molecules are eventually released outside of the cells. When the number of bacterial cells in a population increases, the extracellular concentration of these autoinducers also increases. When the extracellular concentration of autoinducers exceeds some

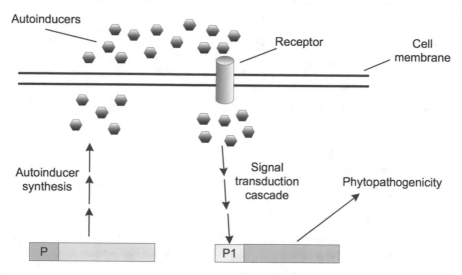

Fig. 6.20 A generalized overview of the process of quorum sensing. Autoinducer are molecules synthesized inside a bacterium (by a varying number of steps depending upon the particular molecule) before being transported (passively or actively) outside of the cell. When the external concentration of autoinducer molecules reaches a threshold value, the autoinducers are transported into cells where, as a consequence of a signal transduction cascade, they activate the transcription of one or more target genes. P and P1 indicate different bacterial promoters. When the autoinducer molecule has an acyl homoserine lactone structure (see Fig. 6.20), homologs of the long-chain fatty acid transporter system which facilitate uptake of long-chain fatty acids, may act as a receptor protein

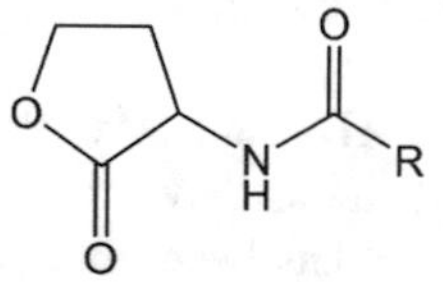

Fig. 6.21 The acyl homoserine lactone (AHL) structure. The acyl chain may range from C_4 to C_{18} in length with occasional side chain modifications at the C_3 position

threshold level, the autoinducers bind to a cellular receptor and trigger a signal transduction cascade that causes population-wide changes in bacterial gene expression thereby enabling a group of cells to act in unison, e.g., as a plant pathogen (Fig. 6.20). While there are a number of different classes of molecules that can act as autoinducers, acyl homoserine lactones (AHLs) are a major class of autoinducer signals used by Gram-negative proteobacteria for quorum sensing (Fig. 6.21). While much of the literature on quorum sensing is directed toward AHLs, other molecules that can act as quorum sensing signals include but are not limited to 3-hydroxypalmitic acid methyl ester, dodecacenoic acid, 3-hydroxybenzoic acid, and 12-methyl-tetradecanoic acid (Fig. 6.22).

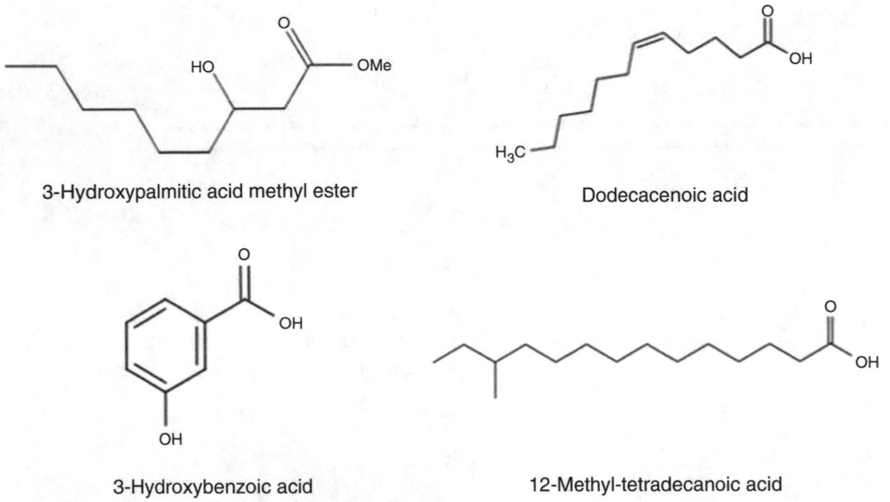

3-Hydroxypalmitic acid methyl ester

Dodecacenoic acid

3-Hydroxybenzoic acid

12-Methyl-tetradecanoic acid

Fig. 6.22 Structures of some quorum sensing molecules not related to homoserine lactones

Among other activities, quorum sensing facilitates the formation of bacterial biofilms, bacterial communities that develop within an extracellular matrix of polysaccharides produced by the bacteria. This facilitates bacterial survival in the somewhat hostile soil environment as biofilms protect bacteria against desiccation, UV radiation, predation, and antibiosis. Moreover, quorum sensing (leading to biofilm formation) is commonly employed by both phytopathogens and rhizobia. Thus, quorum sensing is often a key component of the interaction of phytopathogens and rhizobia with plant roots.

Notwithstanding the fact that AHLs are key components in bacterial signaling and sensing, a number of recent studies have found that many plants can perceive and respond to these signals. This has been shown to be the case for both AHLs produced by bacteria and for synthetic AHLs with the plant's response being dependent on the type of AHL and the particular plant involved; different acyl chain lengths appear to induce ISR in different plants (Fig. 6.23). Thus, AHL molecules can induce many plant defense responses. There are also some reports that AHLs can induce auxin biosynthesis thereby promoting plant growth, possibly as a response to pathogen inhibition of plant growth.

The ability to disrupt quorum sensing networks is termed quorum quenching and it is one way in which one bacterial species in the environment may gain an advantage over another. This may be particularly useful for facilitating crop growth if the quorum quenching system is part of a plant growth-promoting bacterium while the quorum sensing network is part of a bacterial phytopathogen. Quorum quenching can be achieved by several different approaches, including inhibition or inactivation of AHL production (Fig. 6.24). This occurs when a biocontrol bacterium that produces the enzyme lactonase (or a bacteriophage that has been engineered to produce lactonase) degrades some of the AHLs molecules. Some lactonases degrade

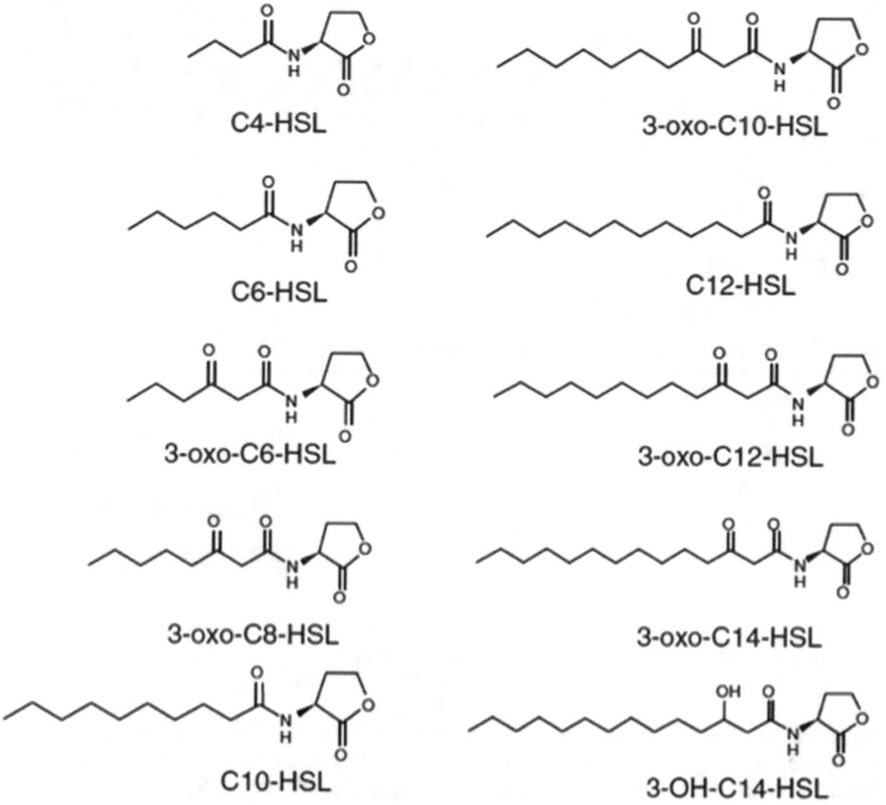

Fig. 6.23 Structures of acyl homoserine lactones (AHLs) that can induce systemic resistance in plants. Abbreviations: HSL is homoserine lactone

only one or two specific AHLs while others have a much broader specificity. Degradation of the AHLs disrupts quorum sensing, and ultimately decreases the phytopathogenicity of the phytopathogen. In addition, one group of scientists found that volatile organic compounds synthesized by two different strains of rhizospheric PGPB, *Pseudomonas fluorescens* and *Serratia plymuthica*, were responsible for decreasing the amount of acyl homoserine lactones produced by various phytopathogens thereby thwarting the ability of these strains to act as pathogens. Thus, quorum quenching may be viewed as yet another tool in the arsenal that may be employed in an effort to limit the damage to crop plants from various phytopathogens.

Finally, it is interesting to note that in a number of instances scientists have found that some quorum quenching biocontrol bacteria may also be active in quorum sensing. Importantly, the AHL degrading enzymes produced by these bacteria do not degrade the AHLs that they themselves produce. By preventing other rhizospheric bacteria from proliferating and at the same time facilitating their own

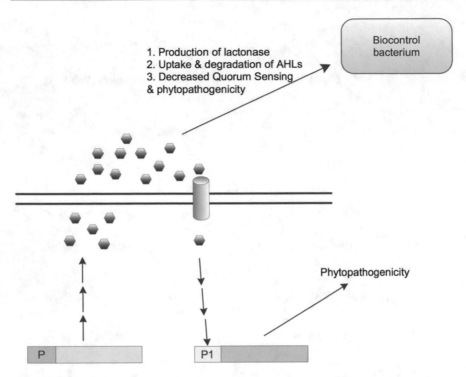

Fig. 6.24 A schematic overview of quorum quenching. A biocontrol bacterium that produces the enzyme lactonase takes up and degrades some of the acyl homoserine lactone molecules (AHLs) thereby decreasing the formation of quorum sensing, and ultimately phytopathogenicity, by the phytopathogen

proliferation, these bacteria can better establish their own niche on the plant surface and potentially be more effective biocontrol agents.

Questions
1. How can the bacterium *Agrobacterium radiobacter* be engineered to be a more effective biocontrol agent?
2. How can biocontrol PGPB be engineered to be more effective biocontrol agents?
3. What is a zone of clearance and how can it be used to indicate the effectiveness of a particular antibiotic against a specific pathogen?
4. How would you select for biocontrol strains of *P. fluorescens* that overproduce the antibiotic phenazine-1-carboxylate?
5. How can the modification of rare codons be used to increase the production of the antibiotic pyoluteorin?
6. What are siderophores and how do they prevent the proliferation of phytopathogens?

7. Which enzymes secreted by PGPB contribute to their ability to act as biocontrol agents? How do these enzymes contribute to biocontrol?
8. How would you isolate novel endophytic biocontrol PGPB from banana plants?
9. How does competition contribute to biocontrol activity?
10. Describe a scheme for identifying *Sphingomonas* sp. proteins that are involved in protecting plants against bacterial phyllosphere pathogens.
11. How can one pathogenic bacterium inhibit the functioning of a similar pathogenic bacterium?
12. What is *Muscodor albus* and how does it function in biocontrol?
13. What is induced systemic resistance? What is systemic acquired resistance?
14. Briefly describe the use of bacteriophages as biocontrol agents against bacterial phytopathogens.
15. What is quorum sensing and how can its inhibition protect plant against bacterial pathogens?

Bibliography

Akkörprü A, Çakar K, Husseini A (2018) Effects of endophytic bacteria on disease and growth in plants under biotic stress. YYU J Agric Sci 28:200–208

Bakker PAHM, Pieterse CMJ, van Loon LC (2007) Induced systemic resistance by fluorescent *Pseudomonas* spp. Phytopathology 97:239–243

Barahona E, Navazo A, Martínez-Granero F, Zea-Bonilla T, Perez-Jimenez RM, Martín M, Rivilla R (2011) *Pseudomonas fluorescens* F113 mutant with enhanced competitive colonization ability and improved biocontrol activity against fungal root pathogens. Appl Environ Microbiol 77:5412–5419

Barbey C, Crepin A, Bergeau D, Ouchiha A, Mijouin L, Taupin L, Orange N, Feuilloley M, Dufour A, Burini J-F, Latour X (2013) *In planta* biocontrol of *Pectobacterium atrosepticum* by *Rhodococcus erythropolis* involves silencing of pathogen communication by the rhodococcal gamma-lactone catabolic pathway. PLoS One 8(6):e66642

Biles CL, Abeles FB, Wilson CL (1990) The role of ethylene in anthracnose of cucumber, *Cucumis sativus*, caused by *Colletotrichum lagenarium*. Phytopathology 80:732–736

Bowman SM, Free SJ (2006) The structure and synthesis of the fungal cell wall. Bioessays 8:799–808

Bozsoki Z, Cheng J, Feng F, Gysel K, Vinther M, Andersen KR, Oldroyd G, Blaise M, Radutoiu S, Stougaard J (2017) Receptor-mediated chitin perception in legume roots is functionally separable from nod factor perception. Proc Natl Acad Sci USA 114:E8118–E8127

Brezezinska MS, Jankiewicz U, Burkowska A, Walczak M (2014) Chitinolytic microorganisms and their possible application in environmental protection. Curr Microbiol 68:71–81

Buttimer C, McAuliffe O, Ross RP, O'Mahony J, Coffey C (2017) Bacteriophages and bacterial plant diseases. Front Microbiol 8:34

Chan K-G, Atkinson S, Mathee K, Sam C-K, Chhabra SR, Camara M, Koh C-L, Williams P (2011) Characterization of *N*-acylhomoserine lactone-degrading bacteria associated with the *Zingiber officinale* (ginger) rhizosphere: co-existence of quorum quenching and quorum sensing in *Acinetobacter* and *Burkholderia*. BMC Microbiol 11:51

Chernin L, Ismailov Z, Haran S, Chet I (1995) Chitinolytic *Enterobacter agglomerans* antagonistic to fungal plant pathogens. Appl Environ Microbiol 61:1720–1726

Chernin LS, de la Fuente L, Sobolev V, Haran S, Vorgias CE, Oppenheim AB, Chet I (1997) Molecular cloning, structural analysis, and expression in *Escherichia coli* of a chitinase gene from *Enterobacter agglomerans*. Appl Environ Microbiol 63:834–839

Chernin L, Toklikishvili N, Ovadis M, Kim S, Ben-Ari J, Khmel I, Vainstein A (2011) Quorum-sensing quenching by rhizobacterial volatiles. Environ Microbiol Rep 3:698–704

Chet I, Inbar J (1994) Biological control of fungal pathogens. Appl Biochem Biotechnol 48:37–43

Choudhary DK, Varma A (eds) (2016) Microbial-mediated induced systemic resistance in plants. Springer, Singapore, 226 pp

Chowdhury SP, Hartmann GXW, Borriss R (2015) Biocontrol mechanism by root-associated *Bacillus amyloliquefaciens* FZB-42. Front Microbiol 6:780

Compant S, Duffy B, Nowak J, Clement C, Barka EA (2005) Use of plant growth-promoting bacteria for biocontrol of plant diseases: principles, mechanisms of action, and future prospects. Appl Environ Microbiol 71:4951–4959

Cook RJ (1993) Making greater use of introduced microorganisms for biological control of plant pathogens. Annu Rev Phytopathol 31:53–80

Couillerot O, Prigent-Combaret C, Caballero-Mellado J, Moenne-Loccoz Y (2009) *Pseudomonas fluorescens* and closely-related fluorescent pseudomonads as biocontrol agents of soil-borne pathogens. Lett Appl Microbiol 48:505–512

Cui W, He P, Munir S, He P, He Y, Li X, Yang L, Wang B, Wu Y, He P (2019) Biocontrol of soft rot of Chinese cabbage using an endophytic bacterial strain. Front Microbiol 10:1471

Dixit R, Agrawal L, Gupta S, Kumar M, Yadav S, Chauhan PS, Nautiyal CS (2016) Southern blight disease of tomato control by 1-aminocyclopropane-1-carboxylate (ACC) deaminase producing *Paenibacillus lentimorbus* B-30488. Plant Signal Behav 11:e1113363

Dowling DN, O'Gara F (1994) Metabolites of *Pseudomonas* involved in the biocontrol of plant disease. Trends Biotechnol 12:133–141

Dowling DN, Sexton R, Fenton A, Delany I, Fedi S, McHugh B, Callanan M, Moënne-Loccoz Y, O'Gara F (1996) Iron regulation in plant-associated Pseudomonas fluorescens M114: implications for biological control. In: Nakazawa T, Furukawa K, Haas D, Silver S (eds) Molecular biology of pseudomonads. American Society for Microbiology Press, Washington, DC, pp 502–511

Elad Y (1990) Production of ethylene in tissues of tomato, pepper, French-bean and cucumber in response to infection by *Botrytis cinerea*. Physiol Mol Plant Pathol 36:277–287

Ezra D, Hess WM, Strobel GA (2004) New endophytic isolates of *Muscodor albus*, a volatile-antibiotic-producing fungus. Microbiology 150:4023–4031

Frampton RA, Pitman AR, Fineran PC (2012) Advances in bacteriophage-mediated control of plant pathogens. Int J Microbiol. https://doi.org/10.1155/2012/326452

Fravel D, Olivain C, Alabouvette C (2003) *Fusarium oxysporum* and its biocontrol. New Phytol 157:493–502

Fujiwara A, Fujisawa M, Hamasaki R, Kawasaki T, Fujie M, Yamada T (2011) Control of *Ralstonia solanacearum* by treatment with lytic bacteriophages. Appl Environ Microbiol 77:4155–4162

Gandhi PM, Narayanan KB, Naik PR, Shakthivel N (2009) Characterization of *Chryseobacterium aquaticum* strain PUPC1 producing a novel antifungal protease from rice rhizosphere soil. J Microbiol Biotechnol 19:99–107

Geoghegan I, Steinberg G, Gurr S (2017) The role of the fungal cell wall in the infection of plants. Trends Microbiol 25:957–967

Glick BR, Bashan Y (1997) Genetic manipulation of plant growth-promoting bacteria to enhance biocontrol of phytopathogens. Biotechnol Adv 15:353–378

Großkinsky DK, Tafner R, Moreno MV, Stenglein SA, Garcia de Salamone IE, Nelson LM, Novák O, Strnad M, van der Graaff E, Roitsch T (2016) Cytokinin production by *Pseudomonas fluorescens* G20-18 determines biocontrol activity against *Pseudomonas syringae* in *Arabidopsis*. Sci Rep 6:23310

Gururani MA, Venkatesh J, Upadhyaya CP, Nookaraju A, Pandey SK, Park SW (2012) Plant disease resistance genes: current status and future directions. Physiol. Mol Plant Pathol 78:51–68

Haas D, Defago G (2005) Biological control of soil-borne pathogens by fluorescent pseudomonads. Nat Rev Microbiol 3:307–319

Haas D, Keel C (2003) Regulation of antibiotic production in root colonizing *Pseudomonas* spp. and relevance for biological control of plant disease. Annu Rev Phytopathol 41:117–153

Hamdan H, Weller DM, Thomashow LS (1991) Relative importance of fluorescent siderophores and other factors in biological control of *Graeumannomyces graminis* var. *tritici* by *Pseudomonas fluorescens* 2-79 and M4-80R. Appl Environ Microbiol 57:3270–3277

Hao Y, Charles TC, Glick BR (2007) ACC deaminase from plant growth promoting bacteria affects crown gall development. Can J Microbiol 53:1291–1299

Hao Y, Charles TC, Glick BR (2011) An ACC deaminase containing *A. tumefaciens* strain D3 shows biocontrol activity to crown gall disease. Can J Microbiol 57:278–286

Helman Y, Chernin L (2015) Silencing the mob: disrupting quorum sensing as a means to fight plant disease. Mol Plant Pathol 16:316–329

Hert AP, Marutani M, Momol MT, Roberts PD, Olson SM, Jones JB (2009) Suppression of the bacterial spot pathogen *Xanthomonas euvesicatoria* on tomato leaves by an attenuated mutant of *Xanthomonas perforans*. Appl Environ Microbiol 75:3323–3330

Hollensteiner J, Wemheuer F, Harting R, Kolarzyk AM, Diaz Valerio SM, Poehlein A, Brzuszkiewicz EB, Nesemann K, Braus-Stromeyer SA, Braus GH, Daniel R, Liesegang H (2017) *Bacillus thuringiensis* and *Bacillus weihenstephanesis* inhibit the growth of phytopathogenic *Verticillium* species. Front Microbiol 7:2171

Howie WJ, Suslow TV (1991) Role of antibiotic biosynthesis in the inhibition of *Pythium ultimum* in the cotton spermosphere and rhizosphere by *Pseudomonas fluorescens*. Mol Plant Microbe Interact 4:393–399

Hunziker L, Bönisch D, Groenhagen U, Bailly A, Schultz S, Weisskopf L (2015) *Pseudomonas* strains naturally associated with potato plants produce volatiles with high potential for inhibition of *Phytophthora infestans*. Appl Environ Microbiol 81:821–830

Innerebner G, Knief C, Vorholt JA (2011) Protection of *Arabidopsis thaliana* against leaf-pathogenic *Pseudomonas syringae* by *Sphingomonas* strains in a controlled model system. Appl Environ Microbiol 77:3202–3210

Islam S, Akanda AM, Prova A, Islam MT, Hossain MM (2016) Isolation and identification of plant growth promoting rhizobacteria from cucumber rhizosphere and their effect on plant growth promotion and disease suppression. Front Microbiol 6:1360

Janahiraman V, Anandham R, Kwon SW, Sundaram S, Pandi VK, Krishnamoorthy R, Kim K, Samaddar S, Sa T (2016) Control of wilt and rot pathogens of tomato by antagonistic pink pigmented facultative methylotrophic *Delftia lacustris* and *Bacillus* spp. Front Plant Sci 7:1626

Jiang C-H, Yao X-F, Mi D-D, Li Z-J, Yang B-Y, Zheng Y, Qi Y-J, Guo J-H (2019) Comparative transcriptome analysis reveals the biocontrol mechanism of *Bacillus velezensis* F21 against *Fusarium* wilt on watermelon. Front Microbiol 10:652

Jin X, Wang J, Li D, Wu F, Zhou X (2019) Rotations with Indian mustard and wild rocket suppressed cucumber Fusarium wilt disease and changed rhizosphere bacterial communities. Microorganisms 7:57

Keel C, Schnider U, Maurhofer M, Voisard C, Laville J, Burger U, Wirthner P, Haas D, Défago G (1992) Suppression of root diseases by *Pseudomonas fluorescens* CHAO: importance of the bacterial secondary metabolite 2,4-diacetylphloroglucinol. Mol Plant Microbe Interact 5:4–13

Khalaf EM, Raizada MN (2018) Bacterial seed endophytes of domesticated cucurbits antagonize fungal and oomycete pathogens including powdery mildew. Front Microbiol 9:42

Kiani T, Khan SA, Noureen N, Yasmin T, Zakria M, Ahmed H, Mehboob F, Farrakh S (2019) Isolation and characterization of culturable endophytic bacterial community of stripe rust-resistant and stripe rust-susceptible Pakistani wheat cultivars. Int Microbiol 22:191–201

Kloepper JW, Leong J, Teintze M, Schroth MN (1980) Enhanced plant growth by siderophores produced by plant growth-promoting rhizobacteria. Nature 286:885–886

Koby S, Schickler H, Chet I, Oppenheim AB (1994) The chitinase encoding Tn7-based *chi*A gene endows *Pseudomonas fluorescens* with the capacity to control plant pathogens in soil. Gene 147:81–83

Krol E, Becker A (2014) Rhizobial homologs of the fatty acid transporter FadL facilitate perception of long-chain acyl-homoserine lactone signals. Proc Natl Acad Sci USA 111:10702–10707

Kurtböke DI, Palk A, Marker A, Neuman C, Moss L, Streeter K, Katouli M (2016) Isolation and characterization of *Enterobacteriaceae* species infesting post-harvest strawberries and their biological control using bacteriophages. Appl Microbiol Biotechnol 100:8593–8606

Lapidot D, Dror R, Vered E, Mishli O, Levy D, Helman Y (2015) Disease protection and growth promotion of potatoes (*Solanum tuberosum* L.) by *Paenibacillus dendritiformis*. Plant Pathol 64:545–551

Liu L, Kloepper JW, Tuzun S (1995) Induction of systemic resistance in cucumber against bacterial angualar leaf spot by plant growth-promoting rhizobacteria. Phytopathology 85:843–847

Liu Z, Ma P, Holtsmark I, Skaugen M, Eijsink VGH, Brurberg MB (2013) New type of antimicrobial protein produced by the plant pathogen *Clavibacter michiganensis* subsp *michiganensis*. Appl Environ Microbiol 79:5721–5727

Macedo-Raygoza GM, Valdez-Salas B, Prado FM, Prieto KR, Yamaguchi LF, Kato MJ, Canto-Canché BB, Carrillo-Beltrán M, Di Masio P, White JF, Beltrán-Garcia MJ (2019) *Enterobacter cloacae*, an endophyte that establishes a nutrient-transfer symbiosis with banana plants and protects against the black Sigatoka pathogen. Front Microbiol 10:804

Mohamad OAA, Li L, Ma J-B, Hatab S, Xu L, Guo J-W, Rasulov BA, Liu Y-H, Hedlund BP, Li W-J (2018) Evaluation of the antimicrobial activity of endophytic bacterial populations from Chinese traditional medicinal plant licorice and characterization of the bioactive secondary metabolites produced by *Bacillus atrophaeus* against *Verticillium dahliae*. Front Microbiol 9:924

Molina L, Constantinescu F, Michel L, Reimmann C, Duffy B, Defago G (2003) Degradation of pathogen quorum-sensing molecules by soil bacteria: a preventative and curative biological control mechanism. FEMS Microbiol Ecol 45:71–81

Müller T, Behrendt U, Ruppel S, von der Waybrink G, Müller MEH (2016) Fluorescent pseudomonads in the phyllosphere of wheat: potential antagonists against fungal phytopathogens. Curr Microbiol 72:383–389

Nascimento FX, Vicente CSL, Barbosa P, Espada M, Glick BR, Oliveira S, Mota M (2013) The use of the ACC deaminase producing bacterium *Pseudomonas putida* UW4 as a biocontrol agent for pine wilt disease. Biocontrol 58:427–433

Neilands JB, Leong SA (1986) Siderophores in relation to plant growth and disease. Annu Rev Plant Physiol 37:187–208

Ng W-L, Bassler BL (2009) Bacterial quorum-sensing network architectures. Annu Rev Genet 43:197–222

O'Sullivan DJ, O'Gara F (1992) Traits of fluorescent *Pseudomonas* spp. involved in suppression of plant root pathogens. Microbiol Rev 56:662–676

Omoboye OO, Oni FE, Batool H, Yimer HZ, De Mot R, Höfte M (2019) *Pseudomonas* cyclic lipopeptides suppress the rice blast fungus *Magnaporthe oryzae* by induced resistance and direct antagonism. Front Plant Sci 10:901

Pei R, Lamas-Samanamud GR (2014) Inhibition of biofilm formation by T7 bacteriophages producing quorum-quenching enzymes. Appl Environ Microbiol 80:5340–5348

Pieterse CMJ, Leon-Reyes A, Van der Ent S, Van Wees SCM (2009) Networking by small-molecule hormones in plant immunity. Nat Chem Biol 5:308–316

Raaijmakers JM, Mazzola M (2012) Diversity and natural functions of antibiotics produced by beneficial and plant pathogenic bacteria. Annu Rev Phytopathol 50:403–424

Raaijmakers JM, de Bruijn I, de Kock MJD (2006) Cyclic lipopeptide production by plant-associated *Pseudomonas* species: diversity, activity, biosynthesis and regulation. Mol Plant Microb Interact 19:699–710

Raaijmakers JM, Paulitz TC, Steinberg C, Alabouvette C, Moe Moenne-Loccoz Y (2009) The rhizosphere: a playground and battlefield for soilborne pathogens and beneficial microorganisms. Plant Soil 321:341–361

Ramete A, Moenne-Loccoz Y, Defago G (2006) Genetic diversity and biocontrol potential of fluorescent pseudomonads producing phloroglucinols and hydrogen cyanide from Swiss soils naturally suppressive or conducive to *Thielaviopsis basicola*-mediated black root rot of tobacco. FEMS Microbiol Ecol 55:369–381

Raza W, Wang J, Wu Y, Ling N, Wei Z, Huang Q, Shen Q (2016) Effect of volatile organic compounds produced by *Bacillus amyloliquefaciens* on the growth and virulence traits of tomato bacterial wilt pathogen *Ralstonia solanacearum*. Appl Microbiol Biotechnol 100:7639–7650

Rubio MB, Quljada NM, Perez E, Dominguez S, Monte E, Hermosa R (2014) Identifying beneficial qualities of *Trichoderma parareesei* for plants. Appl Environ Microbiol 80:1864–1873

Rybakova D, Rack-Wetzlinger U, Cernava T, Schaefer A, Schmuck M, Berg G (2017) Aerial warfare: a volatile dialogue between the plant pathogen *Verticillium longisporum* and its antagonist *Paenibacillus polymyxa*. Front Plant Sci 8:1294

Saraf M, Pandya U, Thakkar A (2014) Role of allelochemicals in plant growth promoting rhizobacteria for biocontrol of phytopathogens. Microbiol Res 169:18–29

Sarniguet A, Kraus J, Henkels MD, Muehlchen AM, Loper JE (1995) The sigma factor σ^S affects antibiotic production and biological control activity of *Pseudomonas fluorescens* Pf-5. Proc Natl Acad Sci USA 92:12255–12259

Schnider U, Keel C, Blumer C, Troxler J, Défago G, Haas D (1995) Amplification of the housekeeping sigma factor in *Pseudomonas fluorescens* CHA0 enhances antibiotic production and improves biocontrol abilities. J Bacteriol 177:5387–5392

Shehata HR, Ettinger CL, Eisen JA, Raizada MN (2016) Genes required for the anti-fungal activity of a bacterial endophyte isolated from a corn landrace grown continuously by subsistence farmers sine 1000 BC. Front Microbiol 7:1548

Sivan A, Chet I (1992) Microbial control of plant diseases. In: Mitchell R (ed) Environmental microbiology. Wiley-Liss, New York, pp 335–354

Strobel G (2011) Muscodor species-endophytes with biological promise. Phytochem Rev 10:165–172

Timmusk S, Copolovici D, Copolovici L, Tedet T, Nevo E, Behers L (2019) *Paenibacillus polymyxa* biofilm polysaccharides antagonize *Fusarium graminearum*. Sci Rep 9:662

Toklikishvili N, Dandurishvili N, Tediashvili M, Giorgobiani N, Szegedi E, Glick BR, Vainstein A, Chernin L (2010) Inhibitory effect of ACC deaminase-producing bacteria on crown gall formation in tomato plants infected by *Agrobacterium tumefaciens* or *A. vitis*. Plant Pathol 59:1023–1030

Toral L, Rodríguez M, Béjar V, Sampedro I (2018) Antifungal activity of lipopeptides from *Bacillus* XT1 CECT 8661 against *Botrytis cinerea*. Front Microbiol 9:1315

Tuzun S, Kloepper J (1994) Induced systemic resistence by plant growth-promoting rhizobacteria. In: Ryder MH, Stephens PM, Bowen GD (eds) Improving plant productivity with rhizosphere bacteria. CSIRO, Adelaide, pp 104–109

Van der Ent S, Van Wees SCM, Pieterse CMJ (2009) Jasmonate signaling in plant interactions with resistance-inducing beneficial microbes. Phytochemistry 70:1581–1588

Van Loon LC (1984) Regulation of pathogenesis and symptom expression in diseased plants by ethylene. In: Fuchs Y, Chalutz E (eds) Ethylene: biochemical, physiological and applied aspects. Martinus Nijhoff/Dr W. Junk, The Hague, pp 171–180

Van Loon LC, Rep M, Pieterse C (2006) Significance of inducible-defense-related proteins in infected plants. Annu Rev Phytopathol 44:135–162

Vogel C, Innerebner G, Zingg J, Guder J, Vorholt JA (2012) Forward genetic in planta screen for identification of plant-protective traits of *Sphingomonas* sp. strain Fr1 against *Pseudomonas syringae* DC3000. Appl Environ Microbiol 78:5529–5535

Volpiano CG, Lisboa BB, São José JFB, Rotta de Oliveira AM, Beneduzi A, Passaglia LMP, Vargas LK (2018) *Rhizobium* strains in the biocontrol of the phytopathogenic fungi *Sclerotium* (*Athelia*) *rolfsii* on the common bean. Plant Soil 432:229–243

Walters DR, Ratsep J, Havis ND (2013) Controlling crop diseases using induced resistance: challenges for the future. J Exp Bot 64:1263–1280

Wang Y, Brown HN, Crowley DE, Szaniszlo PJ (1993) Evidence for direct utilization of a siderophore, ferrioxamine B, in axenically grown cucumber. Plant Cell Environ 16:579–585

Wang C, Knill E, Glick BR, Defago G (2000) Effect of transferring 1-aminocyclopropane-1-carboxylic acid (ACC) deaminase genes into *Pseudomonas fluorescens* strain CHA0 and its *gacA* derivative CHA96 on their growth-promoting and disease-suppressive capacities. Can J Microbiol 46:898–907

Wang X, Wei Z, Yang K, Wang J, Jousset A, Xu Y, Shen Q, Friman V-P (2019) Phage combination therapies for bacterial wilt disease in tomato. Nat Biotechnol 37:1513–1520

Wei G, Kloepper JW, Tuzun S (1996) Induced systemic resistance to cucumber diseases and increased plant growth by plant growth-promoting rhizobacteria under field conditions. Phytopathology 86:221–224

Yan Q, Philmus B, Hesse C, Kohen M, Chang JH, Loper JE (2016) The rare codon AGA is involved in regulation of pyoluteorin biosynthesis in *Pseudomonas protegens* Pf-5. Front Microbiol 7:497

Yi H-S, Ahn Y-R, Song GC, Ghim S-Y, Lee S, Lee G, Ryu C-M (2016) Impact of a bacterial volatile 2,3-butanediol on *Bacillus subtilis* rhizosphere robustness. Front Microbiol 7:993

Yu JM, Wang D, Pierson LS III, Pierson EA (2018) Effect of producing different phenazines on bacterial fitness and biological control in Pseudomonas chlororaphis 30-84. Plant Pathol J 34:44–58

Zhang L-N, Wang D-C, Hu Q, Dai X-Q, Xie Y-s, Li Q, Liu H-M, Guo J-H (2019) Consortium of plant growth-promoting rhizobacteria strains suppresses sweet pepper disease by altering the rhizosphere microbiota. Front Microbiol 10:1668

Biocontrol of Insects and Nematodes

7

Abstract

The biotic pathogens that inhibit plant growth are not limited to fungi and bacteria but also include many species of insects and nematodes. In this chapter, the biocontrol of insects and nematodes are discussed in detail. By far, the most well-known and well-studied biocontrol bacteria for the prevention of insect predation and damage includes the hundreds of subspecies of the bacterium *Bacillus thuringiensis*. Many of these highly specific and biodegradable bacterial strains target only a limited number of insects. This bacterium has been licensed and is extensively used commercially instead of chemical insecticides throughout the world. By contrast, the use of plant growth-promoting bacteria as biocontrol agents to limit nematode damage to plants is in its relative infancy. Nevertheless, many laboratory studies to control nematodes seem promising with future commercialization seemingly not far off.

7.1 Insects

Insects can have a negative effect on humans in several different ways. Most importantly, they are often the cause of a significant amount of crop damage, and they can act as carriers of both human and animal diseases. During the 1940s, several powerful broad-range chemical insecticides were developed in order to control the proliferation of some of the most problematic insects. One of the most effective chemical insecticides was the well-known colorless and odorless chlorinated hydrocarbon DDT (dichlorodiphenyltrichloroethane). It is estimated that at its peak approximately 40,000 tons per year of DDT were used worldwide in agriculture. Unfortunately, as a consequence of its lipophilic nature, DDT can bioaccumulate, especially in birds, is toxic to a wide range of living organisms including humans and it persists for many years in the environment. In 1972, based on mounting toxicity data, the use of DDT in agriculture was banned in the United States. Subsequently, the Stockholm Convention on Persistent Organic Pollution, signed in 2001, resulted in a worldwide ban on the agricultural use of DDT. DDT and other chlorinated hydrocarbons attack the nervous system and muscle tissues of insects. Following the US ban on DDT, other chlorinated hydrocarbons, such as dieldrin, aldrin, chlordane, lindane, and toxophene, were utilized on a very large scale worldwide against crop pests and insects carrying various infectious agents. Organophosphates, another class of chemical insecticides that includes malathion, parathion, and diazinon, were initially developed as chemical warfare agents. However, they were subsequently used to control insect populations by inhibiting the target insect's enzyme acetylcholinesterase. Treatment with this enzyme cleaves the neurotransmitter acetylcholine and as a result disrupts the functioning of motor neurons as well as neurons in the insect's brain. With essentially all of the current and past chemical insecticides there is an ever-present danger that as a consequence of their lack of specificity, nontarget beneficial insects such as honey bees, as well as target pest insects, will be affected and that toxicity to higher animals and humans may result.

Over time, many insect pest populations have become increasingly resistant to treatment with many of the abovementioned chemical insecticides so that it became necessary to employ higher concentrations of these insecticides to control the pests thereby exacerbating the problems inherent in the use of these insecticides. Given the many problems associated with the use of chemical insecticides, it became essential to develop alternative means of controlling harmful insects. In this regard, using insecticides that are produced naturally by either microorganisms or plants was an obvious choice to augment and/or replace chemical insecticides. Unlike chemical insecticides, biological insecticides are generally highly specific for a target insect species, biodegradable, and are typically slow to select for resistance. Unfortunately, mediating against the use of natural insecticides, their perceived low potency and often high cost of production (compared to chemical insecticides) are considered to be impediments to their greater use for some applications. On the other hand, the use of recombinant DNA technology may provide a means of overcoming some of the perceived negative attributes of natural insecticides. In fact, although it is not specifically discussed here, transgenic plants that express the insecticidal activities of the bacterium *Bacillus thuringiensis* have been developed into safe, specific, and highly effective insecticides. These transgenic plants are commercially available and are utilized widely throughout the world. Ironically, while *B. thuringiensis* has been utilized as an insecticide since 1938 in France and the 1950s in much of the rest of the world, as yet, no genetically engineered strains of *B. thuringiensis* have been approved for widespread environmental use.

The worldwide market for pesticides has been estimated to be worth more than $70 billion (US) per year in 2018 and continues to grow rapidly. Although biopesticides, mostly *B. thuringiensis* which currently makes up around 95% of the biopesticides market, make up only a little more than 1% of the worldwide pesticide market total, it is thought that a considerable amount of future growth in this field is likely to involve biological pesticides.

Table 7.1 Some common subspecies of *Bacillus thuringiensis*

Subspecies	Toxin size, kD	Target insects
Berliner	130–140	Lepidoptera
Kurstaki KTO	130–140	Lepidoptera
Entomocidus	130–140	Lepidoptera
Aizawai	130–140	Lepidoptera
Galleriae	130	Lepidoptera
Aizawai IC	135	Lepidoptera, Diptera
Kurstaki HD-1	71	Lepidoptera, Diptera
Tenebrionis (*san diego*)	66–73	Coleoptera
Morrisoni	125–145	Diptera
Israelensis	68	Diptera

7.1.1 *Bacillus thuringiensis*

The Gram-positive rod-shaped bacterium *B. thuringiensis* is by far the most studied, most effective, and most often utilized microbial insecticide. *B. thuringiensis* was first discovered in 1901 in Japan where it was shown to be the causative agent in the death of large numbers of silkworms. *B. thuringiensis* was rediscovered in 1911 as the causative agent of the death of Mediterranean flour moths. *B. thuringiensis* consists of a large number of strains and subspecies, each of which has the ability to synthesize a different toxin that is directed toward specific insects. There are estimated to be more than 700 known strains of *B. thuringiensis* (a few of the more common strains are shown in Table 7.1). The numbering/classification system for *B. thuringiensis* insecticidal (Cry) proteins is based on the degree of evolutionary divergence of their protein sequences.

 B. thuringiensis subsp. *kurstaki* kills lepidopteran larvae. *B. thuringiensis* subsp. *israelensis* kills diptera such as mosquitoes and blackflies. *B. thuringiensis* subsp. *tenebrionis* (also called *san diego*) is toxic to coleoptera (beetles). In addition, a few strains of *B. thuringiensis* have been shown to produce insecticidal toxins against hymenoptera (including sawflies, wasps, bees, and ants), orthoptera (grasshoppers, crickets, and locusts), and mallophaga (lice).

 The insecticidal activity of *B. thuringiensis* subsp. *kurstaki* (and many other of the strains that have been examined in detail) is contained within a very large structure called a parasporal crystal, that is synthesized during the process of bacterial sporulation, where sporulation produces dormant or storage forms of a bacterium (i.e., spores) that can become reactivated when environmental conditions change to enable vegetative growth to resume. While the parasporal crystal does not seem to play any critical role in the functioning of the bacterium, its synthesis enables the bacterium to persist in the environment because a dead insect can provide a sufficient amount of nutrients to allow germination and growth of a dormant bacterial spore. The parasporal crystal makes up a large part of a sporulated culture, i.e., approximately 20–30% of the dry weight. Moreover, a parasporal crystal typically consists of ~95% protein and ~5% carbohydrate. In fact, around 200 different parasporal crystal proteins (called Cry proteins and encoded by *cry* genes) are known. However, the parasporal crystal generally does not contain the active form of the insecticide. The parasporal crystal is typically an aggregate of protein that can be dissociated into subunits by treatment of the crystal with mild alkali which is similar to the pH of the insect gut (Fig. 7.1). Following mild alkali treatment, the crystal is solubilized into the protein subunits that are commonly a protoxin or inactive precursor of the active toxin. The subunits can be further dissociated in the laboratory by treatment with the chemical agent β-mercaptoethanol, which can reduce the disulfide linkages that hold the protein subunits together. Treatment of the dissociated protein with certain insect gut proteases converts the inactive protoxin to a lower molecular weight active insecticidal toxin.

For its insecticidal activity to be functional, a parasporal crystal must first be ingested by a target insect, then the protoxin is converted into an active toxin within the insect's gut by the combination of its gut alkaline pH and its specific digestive proteases. The active toxin protein is then inserted into the gut epithelial cell membranes of the insect which results in the creation a membrane ion channel, eventually leading to the loss of large amounts of cellular ATP (Fig. 7.2). About 15 min after the ion channel forms, insect cellular metabolism ceases; as a consequence, the insect stops feeding within a few hours, becomes dehydrated, and eventually dies within a few days. The requirement for both alkaline pH and the presence of specific proteases, to convert the protoxin into an active toxin, makes it is highly unlikely that nontarget species, such as humans and other animals (with a gut pH that is highly acidic), will be adversely affected should they accidently consume this biological insecticide.

In order to kill an insect pest, it is absolutely necessary for *B. thuringiensis* parasporal crystals to be ingested thereby limiting the susceptibility of nontarget insects. Thus, direct physical contact of either the bacterium or the insecticidal toxin with the surface of an insect does not inhibit or kill the insect. *B. thuringiensis* is generally applied by spraying and it is typically formulated as a mixture that includes insect attractants. This increases the probability that the target insect will ingest the toxin (which is present on a plant's leaf or stem surface). On the other hand, it is quite unlikely that insects that bore into plants or attack plant roots will ingest a *B. thuringiensis* toxin that has been sprayed onto a host plant. Thus, other, more complex, strategies have been devised to control such pests. Moreover, when *B. thuringiensis* parasporal crystals are sprayed onto plants, the exposure of the plant surface to sunlight can cause the degradation of over 60% of the tryptophan residues of the parasporal crystal within a single day, with the unfortunate result that the insecticidal toxin protein becomes inactive. Of course, in cloudy conditions or when the treated plant is in the shade, the level of sunlight that is present is decreased so that the parasporal crystals may persist under these conditions for as long as a month. The fact that the insecticidal protoxin does not persist for long periods of time in the natural environment means that (unlike chemical insecticides that may persist in the environment for years) the development of resistant insects is highly unlikely, although not impossible.

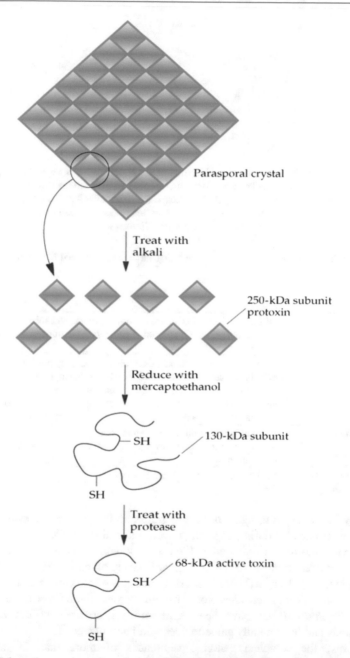

Parasporal crystal

Treat with alkali

250-kDa subunit protoxin

Reduce with mercaptoethanol

— SH

130-kDa subunit

SH

Treat with protease

68-kDa active toxin

— SH

SH

Fig. 7.1 Schematic representation of a *B. thuringiensis* parasporal crystal composed of Cry protoxin proteins. Each of the 250-kDa protein subunits of the parasporal crystal contains two 130-kDa polypeptides. Conversion of the 130-kDa protoxin into an active 68-kDa toxin requires the combination of a slightly alkaline pH (7.5–8) and the action of a specific protease(s), both of which are found in the insect gut. The active toxin binds to protein receptors on the surface of the insect gut epithelial cell membrane. © 2010 American Society for Microbiology. Used with permission. No further reproduction or distribution is permitted without the prior written permission of American Society for Microbiology

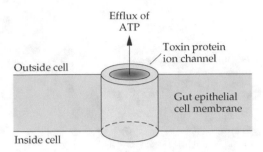

Fig. 7.2 Schematic representation of the insertion of the *B. thuringiensis* toxin into the membrane of an insect gut epithelial cell. The toxin forms an ion channel between the cell cytoplasm and the external environment thereby allowing ATP to escape from the cell. © 2010 American Society for Microbiology. Used with permission. No further reproduction or distribution is permitted without the prior written permission of American Society for Microbiology

Table 7.2 Some subspecies of *B. thuringiensis* that have been approved for use in the field and some of their targets

B. thuringiensis subspecies	Targets
Aizawai	Fruits and nuts, berries, peppers, tomato, root crops, tobacco, beans, corn cotton, cabbage, eggplant melons cucumber, cauliflower, broccoli, ornamentals
Kurstaki	Berries, fruits, nuts, melons, cucumber squash, eggplant, tomato, broccoli, cabbage, kale, mustard, parsley, spinach, turnip lettuce, stored grain, stored crop seed, ornamentals, cotton, celery, peanut, sugar beet, tobacco, avocado, onion, carrot forestry products, grape, canola, sorghum, wheat forage crops, corn, sunflower, root crops, cranberry
Israelensis	Mosquito breeding habitat including rice fields, ponds, pastures, ditches, salt marshes, tidal water, sewage lagoons; lakes; ornamental and nursery plants; mushrooms
Tenebrionis	Eggplant, tomato, potato, ornamentals

A key limitation in using *B. thuringiensis* as an insecticide is that it can kill a susceptible insect only during a specific developmental stage. Generally, this means that the toxin has to be applied when the insect population is at the larval stage. In addition to the requirement that the *B. thuringiensis* toxin be sprayed while the target insect in in its larval stage, *B. thuringiensis* treatments often costs from 1.5 to 3 times as much as chemical insecticides. Notwithstanding these limitations, several subspecies of *B. thuringiensis* have been approved for commercial use in several jurisdictions and have rapidly gained widespread acceptance (Table 7.2).

Recently, the complete genomic sequences of more than 30 strains of *B. thuringiensis* have been reported. While the genome size difference of these strains, ranging from 5.31×10^6 bp to 6.77×10^6 bp, is not very great, the number of plasmids that these strains contain ranges from 1 to 14. Moreover, some Cry proteins that have nematicidal activity have been discovered. In addition, several strains of *B. thuringiensis* produce compounds that (either directly or indirectly) are

Latin name	Common name
Spodoptera exigua	Beet armyworm
Spodoptera frugiperda	Fall armyworm
Helicoverpa zea	Corn earworm
Chilo partellus	Spotted stalk borer
Phthorimaea opercullela	Potato tuber moth
Heliothis virescens	Tobacco budworm
Manduca sexta	Carolina sphinx moth
Ephestia kuehniella	Mediterranean flour moth
Spodoptera litura	Tobacco cutworm
Trichoplusia ni	Cabbage looper
Chrysodeixis chalcites	Tomato looper
Lobesia botrana	European grapevine moth
Mamestra brassica	Cabbage moth
Spodoptera littoralis	African cotton leafworm
Ectomyelois ceratoniae	Locust bean moth
Plutella xylostella	Diamondback moth

Table 7.3 Economically important Lepidoptera pests that are targeted by Vip3

known to enhance plant growth; these compounds include ACC deaminase, IAA, siderophores, bacteriocins, chitinase, and phosphate solubilization enzymes.

In addition to Cry proteins, many strains of *B. thuringiensis* can also synthesize some other insecticidal proteins, in this case during the vegetative phase of bacterial growth. These proteins are typically secreted into the growth medium and are called vegetative insecticidal proteins (Vip) and the secreted insecticidal protein (Sip). The Vip proteins may act synergistically with Cry proteins (with which it shares no sequence homology) to kill their target insects, providing a double-barreled approach to insect toxicity making it more difficult for susceptible insects to develop resistance. The binary toxin comprising Vip1 and Vip2 and the Sip toxin are not toxic to lepidoptera but exhibit insecticidal activity against some coleopterans and hemiptera ("true bugs" including insects such as aphids). On the other hand, Vip3 targets several major lepidopteran pests (Table 7.3). No target insects have as yet been identified for Vip4. Vip1 together with Vip2 constitute a binary toxin while Vip3 proteins are single chain and not binary.

7.1.1.1 Producing *B. thuringiensis* Cry Protein During Vegetative Growth

Generally, most *B. thuringiensis* protoxin proteins are synthesized only during the sporulation phase of growth. That is, only a portion of the growth cycle of the bacterium is devoted to sporulation and parasporal crystal synthesis. One possible means of increasing the parasporal crystal yield and simultaneously decreasing the amount of time that it takes to produce this protein would be to ensure that transcription and translation of the toxin gene occurred during the vegetative rather than the stationary phase of bacterial growth. In addition, synthesis of the insecticidal toxin protein during vegetative growth would enable scientists to utilize a continuous fermentation process, potentially significantly decreasing the cost of insecticide production. That is because continuous fermentations require much smaller sized, and therefore less expensive, bioreactors than conventional batch fermentations. Moreover, the downstream processing equipment that is used for continuous fermentations is also much smaller and less expensive than comparable equipment for conventional batch fermentations.

Interestingly and unlike most other *B. thuringiensis cry* genes, the expression of the *cry3A* gene is normally controlled by a vegetative promoter, rather than by a sporulation promoter. The *cry3A* gene codes for an insecticidal toxin that targets coleopteran larvae. When a naturally occurring mutant strain of *B. thuringiensis* that was unable to form spores was genetically transformed with a plasmid carrying a cloned *cry3A* gene and its normal transcriptional regulatory region, the insecticidal toxin encoded on the plasmid was both overproduced and stabilized in comparison to when this protein was produced in the wild-type strain. This result suggests the possibility that, in addition to *cry3A*, other *cry* genes that are regularly expressed during sporulation might be placed under the control of the *cry3A* promoter and thereby overproduced by introducing these constructs into a sporulation-defective *B. thuringiensis* mutant so that the Cry protein of interest would only be expressed during vegetative growth.

During the sporulation of *Bacillus* strains such as *B. thuringiensis,* a specific transcription initiation factor (a sigma factor) interacts with the transcriptional promoters of genes that are active only within the sporulation phase of the bacterial life cycle. This sigma factor initiates the transcription of the messenger RNAs that are unique to the process of sporulation. In fact, when a *cry* gene that contains a sporulation specific promoter is cloned and expressed in other strains of *Bacillus* including *Bacillus subtilis, Bacillus megaterium* and *B. thuringiensis*, mRNA transcription of the introduced *cry* gene occurs only during sporulation. Moreover, only by placing the *cry* gene under the transcriptional control of a promoter that is active during vegetative growth can the insecticidal protoxin be expressed during vegetative growth of the bacterium. This has been done successfully in laboratory experiments (Fig. 7.3). When a protoxin gene was cloned into a *Bacillus* plasmid under the control of a constitutively active promoter from a tetracycline resistance gene and then introduced into a strain of *B. thuringiensis* that did not synthesize the protoxin protein, the protoxin protein was produced continuously throughout the bacterial growth cycle, including during both the vegetative and sporulation phases. In addition, when the plasmid that was carrying the protoxin gene under the transcriptional control of the tetracycline promoter was used to transform a sporulation-defective mutant of *B. thuringiensis*, protoxin synthesis occurred in the absence of sporulation, throughout vegetative growth of the bacterium, and at a much higher level than in the wild-type cells. Unfortunately, such a genetically manipulated strain of *B. thuringiensis* requires the approval of regulatory agencies who are extremely reluctant to allow the release of genetically manipulated bacteria into the environment.

In a separate experiment, scientists constructed a chimeric *cry1C–cry1Ab* gene by genetic engineering, placed the chimeric gene under the transcriptional control of the

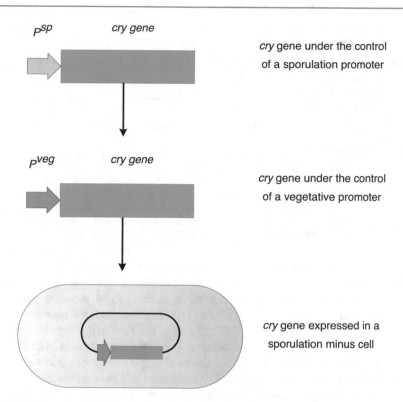

Fig. 7.3 Schematic representation of the construction of a strain of *B. thuringiensis* that overproduces an insecticidal protoxin. The wild-type strain of *B. thuringiensis* contains a *cry* gene under the transcriptional control of a sporulation-specific promoter (p^{sp}). The native promoter is then replaced with a vegetative promoter (p^{veg}) and the construct is placed onto a *Bacillus* plasmid that is subsequently introduced into a strain of *B. thuringiensis* that does not carry a *cry* gene. The final transformed *B. thuringiensis* strain that is unable to sporulate and therefore expresses the protoxin constitutively

vegetative *cry3A* promoter, and then integrated the entire construct into the chromosomal DNA of a nonsporulating derivative of a strain of *B. thuringiensis* subsp. *kurstaki*. Integration of the genetic construct into chromosomal DNA yields a modified *B. thuringiensis* strain that is more stable than if the chimeric gene had been introduced into the bacterium on a plasmid that might eventually be lost from the bacterium. The nonsporulating strain had a disrupted *sigK* gene, that encodes the transcription initiation factor σ28, a key determinant of sporulation-specific transcription. The chimeric *cry1C–cry1Ab* gene contained ~2.2 kb of *cry1C* gene DNA plus ~1.3 kb of *cry1Ab* gene DNA. In this case, the mature insecticidal toxin that is produced following enzymatic cleavage of the chimeric protoxin is identical to the toxin that is produced from the intact *cry1C* gene. Interestingly, the mature insecticidal toxin that was produced from the chimeric protoxin was much more active against three different target insects than the identical toxin produced from the cry1C gene (Table 7.4). Following testing with several different insects, the

Table 7.4 Activities of Cry1C insecticidal protoxin and a chimeric Cry1C–Cry1Ab protoxin against three different insect species

Insect species	LC$_{50}$ of Cry1C (ng)	LC$_{50}$ of Cry1C–Cry1Ab (ng)
Spodoptera littoralis (Egyptian cotton leafworm)	378	103
Plutella xylostella (Diamondback moth)	174	4.6
Ostrinia nubilalis (European corn borer)	3200	822

The LC$_{50}$ values reflect the amount of insecticidal toxin (in nanograms) that are required to kill half of the insect population being tested under a defined set of conditions. A smaller LC$_{50}$ value indicates a more potent insecticidal toxin. Here it is important to keep in mind that notwithstanding the differences between the protoxins, the active toxin is identical in both cases. © 2010 American Society for Microbiology. Used with permission. No further reproduction or distribution is permitted without the prior written permission of American Society for Microbiology

Cry1C–Cry1Ab protoxin was found to be 3–34 times more active than the Cry1C protoxin. This unexpected result may occur because of the increased stability to proteolytic digestion of the Cry1Ab portion of the hybrid protoxin protein. In addition, the chimeric protoxin Cry1C–Cry1Ab was encapsulated within the bacterial cells, rendering the protein significantly more resistant to the degradative effect of ultraviolet (UV) light in the environment. The UV light normally inactivates the protoxin that is secreted outside of the bacterial cell during sporulation. However, despite increased potency and greater UV resistance, the environmental persistence of the nonsporulating mutant was significantly decreased compared with that of the sporulating wild-type bacterial strain. From a practical perspective, this lack of environmental persistence may be advantageous since it is less likely that the nonsporulating and less environmentally persistent mutant will transfer any of its DNA to other organisms in the environment.

7.1.1.2 Transferring *Cry* Genes

Because many crop plants may be attacked by a variety of different insects, it would be beneficial if one could engineer bacterial insecticides that are effective against several different insect pests. One way to do this, that would not include genetic engineering, might include the transfer some of the naturally occurring *B. thuringiensis* plasmids that encode Cry proteins into different *B. thuringiensis* subspecies. Most country's regulatory agencies do not consider plasmid transfer to be genetic engineering so that this manipulation is subject to a lower level of regulatory scrutiny. This was done in the laboratory, transferring *cry* genes on their native plasmids from subspecies *tenebrionis* and *aizawai* into subspecies *kurstaki, israelensis* and *tenebrionis* (Fig. 7.4). These modified strains were subsequently tested for toxicity to three different insect species. The results of those experiments indicated that for all strains, the toxicity of the native host toxin protein(s) was maintained, and in most instances, the introduced toxin gene (on the plasmid) also expressed an active toxin with the same specificity as the toxin produced by the source bacterium (Table 7.5). Moreover, when the

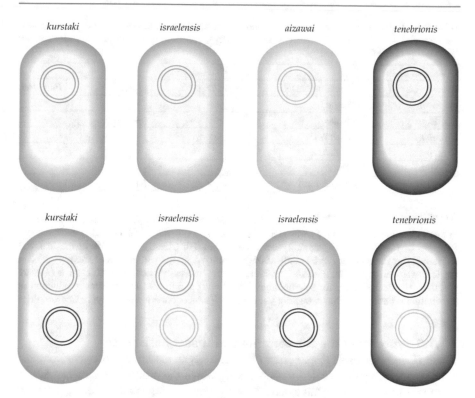

Fig. 7.4 Naturally occurring (top row) and transformed (bottom row) subspecies of *B. thuringiensis*. The oval shapes represent various strains of bacteria, while the circles represent insecticide-encoding plasmids. The plasmids are colored the same as the bacterial subspecies from which the toxin gene originated. © 2010 American Society for Microbiology. Used with permission. No further reproduction or distribution is permitted without the prior written permission of American Society for Microbiology

B. thuringiensis subsp. *tenebrionis* toxin gene was introduced into *B. thuringiensis* subsp. *israelensis*, the resultant transformant strain was somewhat toxic to *Pieris brassicae*, the cabbage white butterfly, against which neither of the gene products alone had previously displayed any insecticidal activity.

Insects such the sugarcane borer (*E. saccharina*) that attack the internal regions of plants such as sugar cane are impervious to *B. thuringiensis*-based insecticides that are sprayed onto the outer surface of the plant. This notwithstanding, some researchers have introduced *B. thuringiensis cry* genes strain into a soil bacterium that can colonize either plant roots or interior surfaces. The idea behind this is that the insecticidal toxin is delivered to the part(s) of the plant that are attacked by the target insect. However, for such a genetic manipulation to be acceptable to regulatory agencies for use in the environment, it would be advantageous for the insecticidal toxin gene to be introduced to its new host on its own naturally occurring plasmid.

Table 7.5 Toxicities of naturally occurring and transformed subspecies of *B. thuringiensis* against the insects *Pieris brassicae* (cabbage white butterfly), *Ades aegypti* (mosquito), and *Phaedon cochleariae* (beetle)

Source of toxin		Toxicity to:		
Host DNA	Introduced DNA	*Pieris*	*Aedes*	*Phaedon*
Aizawai	None	++	+	−
Israelensis	None	−	++	−
Israelensis	*Aizawai*	++	++	−
Israelensis	*Tenebrionis*	+	++	++
Kurstaki	None	++	+	−
Kurstaki	*Tenebrionis*	++	+	++
Tenebrionis	None	−	−	++
Tenebrionis	*Aizawai*	++	+	+

In these experiments, the toxicity was graded as follows: ++, 0–5% of the leaf was consumed (for *Phaedon* or *Pieris*) or 100% mortality occurred within 1 h (*Aedes*); +, 5–50% of the leaf was consumed (for *Phaedon* or *Pieris*) or 50–100% mortality occurred within 24 h (*Aedes*); −, 50% of the leaf was consumed (for *Phaedon* or *Pieris*) or no mortality occurred within 24 h (*Aedes*). © 2010 American Society for Microbiology. Used with permission. No further reproduction or distribution is permitted without the prior written permission of American Society for Microbiology

7.1.1.3 Preventing the Development of Insect Resistance

When *B. thuringiensis* subsp. *kurstaki* was used as an insecticide in an environment where sunlight was prevented from breaking down the protoxin, e.g., when stored grain was treated to protect it against insect predation, insects that were resistant to the lethal effects of Cry proteins developed within a few generations. This inherited resistance has been attributed to an alteration in one of the insect's midgut membrane proteins that normally acts as a receptor for the Cry protein. In these instances, resistant insects accumulate because the protoxin persists in this environment for a relatively long period of time thereby selecting for resistant insects. In this particular case, the simplest way to avoid selecting for resistant insects in the absence of sunlight is to limit the use of this bacterial insecticide to field applications where there is lots of sunlight. However, extensive regular use of any *B. thuringiensis* subspecies, even under field conditions, may result in a relatively high level of environmental persistence of the bacterium, a level that is high enough to allow for the selection of insect resistance to the Cry protein to occur. Certainly, as greater quantities of *B. thuringiensis* subspecies are used over a wider geographical area (including the extensive synthesis of protoxin proteins in transgenic plants), the probability that resistant strains of insects will be selected increases significantly. In an effort to avoid the development of resistant insects, scientists have examined several strategies. These approaches, which can be effective with either with *B. thuringiensis* bacteria that are sprayed or with transgenic plants expressing the insecticidal protoxin, include the following:

1. It is possible for scientists to construct bacterial strains that produce two or more *B. thuringiensis* insecticidal toxins that simultaneously target the same insect. The

key to success in this regard is that these toxins must bind to different insect midgut receptors. In this case, it becomes extremely unlikely that an insect will develop resistance to both toxins at the same time. This is because most of the insect resistance to particular *B. thuringiensis* insecticidal toxins results from mutations to specific receptor proteins that are part of the insect's midgut. When this strategy is used in transgenic plants, it is typically called "gene pyramiding." And, while this approach is employed less commonly when *B. thuringiensis* insecticidal toxins are sprayed, it is likely to be equally as effective as gene pyramiding.

2. Strains of *B. thuringiensis* that produce an insecticidal toxin may be delivered together with traditional chemical insecticides. The idea behind this strategy is that almost no insect will survive these two very different treatments, and insect resistance will not develop to either compound. This approach is used relatively commonly, and typically includes a lower level of chemical insecticide than when the use of a chemical insecticide is the only treatment. Thus, for example, transgenic plants that produce a *B. thuringiensis* insecticidal toxin are often treated with chemical insecticides. However, the number of chemical insecticide treatments that are required is significantly reduced when the plants produce a *B. thuringiensis* insecticidal toxin compared to the number of treatments that are required with a nontransgenic plant. For example, in Florida, nontransgenic corn plants are often sprayed with chemical insecticides as many as ten times per growing season. On the other hand, in the same locale, transgenic plants that produce a *B. thuringiensis* insecticidal toxin are more likely to be sprayed with chemical insecticides only about three or four times a growing season.

3. *B. thuringiensis* insecticidal toxin may be applied at the same time as another biologically based insecticidal protein. Again, it is extremely unlikely that the target insects will survive both types of insecticides. And, since the two bioinsecticides will be chosen to have very different targets, it is highly unlikely that both targets will mutate simultaneously.

4. Two toxins with different modes of action may be employed. For example, simultaneous application of either chitinase or a chitinase producing (and secreting) bacterial strain can dramatically increase the effectiveness of a *B. thuringiensis* insecticidal toxin-producing strain. The chitinase enzyme is thought to cause perforations in the chitin-containing peritrophic membrane of the insect larvae, partially lysing the membrane and debilitating the larvae, and at the same time increasing the accessibility of the midgut membranes to the *B. thuringiensis* insecticidal toxin.

5. A strategy that is commonly used includes the use of refugia (small tracts of land where the crop is not treated with the microbial insecticide). In this case, ~20% of a crop is not sprayed with *B. thuringiensis* (or 20% is nontransgenic) with the remaining 80% of the plants being sprayed or transgenic and producing a *B. thuringiensis* insecticidal toxin. Of the untreated insects, the wild-type insects can proliferate in the absence of the *B. thuringiensis* insecticidal toxin, while only (a very small number of) mutant insects that are resistant to the high levels of *B. thuringiensis* insecticidal toxin survive in the presence of the toxin. Upon

mating, the very small number of *B. thuringiensis* resistant insects will all mate with the much larger number of *B. thuringiensis* sensitive insects, so that the next generation will contain mostly homozygous *B. thuringiensis* sensitive insects and a very small number of heterozygous *B. thuringiensis* sensitive insects. This strategy assumes that resistance to the *B. thuringiensis* insecticidal toxin is inherited as a recessive trait. In fact, this approach has been used in the field for a number of years, with all of the available evidence indicating that little to no resistance to any *B. thuringiensis* insecticidal toxins has developed.

Activate Cry toxins bind to specific proteins (cadherins) on the surfaces of the microvilli of the insect midgut epithelial cells. Binding of toxin molecules to cadherins, which are transmembrane glycoproteins containing 12 cadherin repeating domains and one membrane-proximal extracellular domain (Fig. 7.5), aids in the formation of a multimeric form of the toxin monomers and development of a pore in the membrane. Loss of cadherins or mutation of cadherin genes typically results in resistance to *B. thuringiensis*. In one set of experiments, a fragment of a cadherin protein with the 12 repeating units and the membrane-proximal extracellular domain was mixed with *B. thuringiensis* toxin Cry1A and fed to insect larvae. This experiment was originally performed with the expectation that the cadherin protein fragment would block the binding of the Cry1A protein to the midgut epithelial cells. Instead, the addition of the cadherin protein fragment dramatically enhanced the Cry1A-induced insect mortality. To understand this apparent anomaly, it was thought that the cadherin peptide fragment may first bind to insect midgut microvilli and then attract Cry1A molecules; this then increases the probability of the insecticidal toxin interacting with the bona fide receptor. It is therefore possible that the simultaneous application of Cry proteins and a peptide containing a portion of the receptor protein will overcome or significantly delay the development of insect resistance by increasing Cry protein insect toxicity. However, production of the peptide containing a portion of the receptor protein may add significant expense to the use of *B. thuringiensis* as an insecticidal spray.

It has been observed that *B. thuringiensis* subsp. *israelensis* actively and intrinsically thwarts insect resistance. In contrast to what has been observed with other strains of *B. thuringiensis*, no instances of field resistance of mosquitoes to *B. thuringiensis* subsp. *israelensis* have ever been reported, and only very low levels of insect resistance have been observed in laboratory studies. This somewhat unexpected lack of insect resistance may be a consequence of several different phenomena. For a start, various strains of *B. thuringiensis* subsp. *israelensis* synthesize at least three different Cry proteins, i.e., Cry4A, Cry4B, and Cry11A. *B. thuringiensis* subsp. *israelensis* also produces a protein called Cyt1A, a highly hydrophobic endotoxin that is not at all homologous to any of the known Cry proteins and appears to have a completely different mode of action. While Cry proteins bind to the protein component of the insect midgut epithelial membrane, Cyt1A binds to the lipid component of the insect midgut epithelial membrane, especially the unsaturated fatty acids. In terms of insect toxicity, Cyt1A acts synergistically with the Cry proteins. In experiments with purified insecticidal proteins, it

Fig. 7.5 Schematic representation of a cadherin protein molecule embedded within an insect midgut epithelial membrane. Each of the 12 cadherin domains are labeled with numbers. Domains 7, 11, and 12, all of which are highlighted, have been implicated as being binding sites for Cry proteins. The purple-colored region represents the membrane-proximal extracellular domain; the transmembrane domain is shown in light gray; and the cytoplasmic domain is shown in red. The arrow indicates the junction between the membrane-proximal extracellular domain and the transmembrane domain. The cadherin peptide analogue includes the 12 numbered cadherin domains as well as the membrane-proximal extracellular domain. © 2010 American Society for Microbiology. Used with permission. No further reproduction or distribution is permitted without the prior written permission of American Society for Microbiology

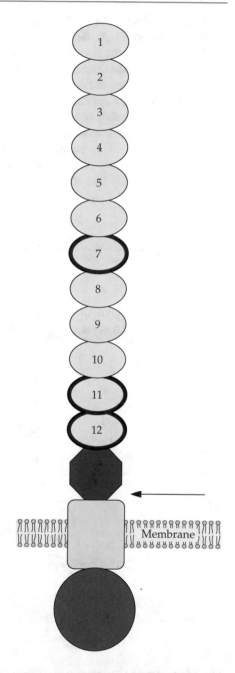

was shown that with the addition of the Cyt1A protein, insects that had previously become resistant to any or all of Cry4A, Cry4B, and Cry11A were still killed when they were treated with *B. thuringiensis* subsp. *israelensis*. Subsequently it was demonstrated that following the binding of Cyt1A to the lipid portion of the midgut epithelial membrane, the Cyt1A protein can act as a receptor for some of the Cry proteins encoded by *B. thuringiensis* subsp. *israelensis*. Thus, the reasons that *B. thuringiensis* subsp. *israelensis* is a highly effective insecticidal bacterium may be because the strain not only carries several different insecticidal proteins (with each normally binding to a different receptors) but also contains a protein that can act as a receptor for these insecticidal proteins. It is therefore extremely unlikely that any target insect will be able to develop resistance to *B. thuringiensis* subsp. *israelensis*. When regulatory agencies eventually allow the use of genetically engineered strains of *B. thuringiensis* in the environment, it will be important to determine whether other strains of *B. thuringiensis* can be constructed to synthesize Cyt1A as well as specific Cry proteins and whether these engineered strains can also prevent the development of insect resistance (Fig. 7.6).

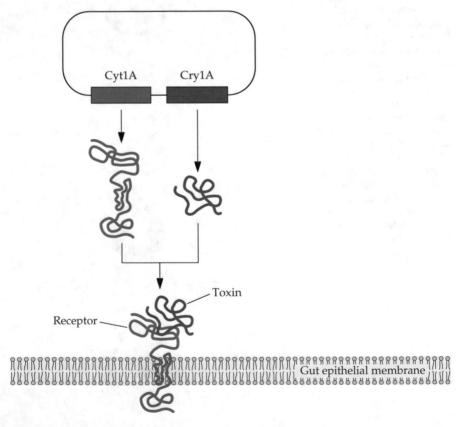

Fig. 7.6 Schematic representation of a strain of *B. thuringiensis* subsp. *israelensis* encoding both Cyt1A and Cry proteins. The Cyt1A protein inserts into the gut epithelial membrane and acts as a receptor for the Cry proteins. © 2010 American Society for Microbiology. Used with permission. No further reproduction or distribution is permitted without the prior written permission of American Society for Microbiology

7.1.2 Antifeeding Prophage

The New Zealand grass grub (*Costelytra giveni*) is a scarab beetle that is a common pasture pest whose larvae feeds on the roots of plants such as perennial ryegrass (*Lolium perenne*) and white clover (*Trifolium repens*). The adult beetles subsequently feed on tree leaves, shrubs and foliage of pasture crops. One commercially available biocontrol agent that results in killing of the insect and preventing crop damage is the bacterium *Serratia entomophila* which causes amber disease of the larvae. It should be emphasized here that this biocontrol agent is only effective against insect larvae and not against adult insects. In this case, following the ingestion of the biocontrol bacteria, the infected larvae stop eating within 2–5 days. The larvae subsequently acquire an amber coloration due to their stomach emptying. Unfortunately, the bacterially infected insects do not die for another 1–3 months. Recently, researchers identified a strain of *Serratia proteamaculans* (which they isolated from a diseased grass grub larva) that was specific for the New Zealand grass grub and the New Zealand manuka beetle (another coleopteran pest) where killing of these insect pests occurred within 5–12 days of the bacterium being ingested by the larvae (Fig. 7.7). These scientists subsequently showed that the new biocontrol bacterial strain that they had isolated carried an insect virus integrated into its chromosomal DNA and that this prophage (an inactive form of the virus) was responsible for the accelerated killing of the insect pest. In this case, once the bacterium has been ingested both the bacteria and the virus begin to replicate and within a few days the concentration of the virus was increased more than 100-fold. Following the replication of the virus inside the insect, the virus accumulates in the insect's hemocoel (the insect's body cavity containing its cellular fluid or blood). This leads to relatively rapid killing of the target insect. These results are certainly very encouraging and given the fact that a similar biocontrol bacterium has already been commercialized in New Zealand (i.e., *Serratia entomophila*), there is reason to expect that should field trials of this newly isolated bacterium be as successful as the initial experiments, approval of this highly specific bioinsecticide for environmental use should be forthcoming.

7.1.3 PGPB-Mediated Insect Killing

Somewhat surprisingly, some groups of *Pseudomonas* PGPB, in addition to suppressing fungal and bacterial diseases can also infect and kill insect larvae. In one study, wild-type and various mutants of *Pseudomonas protegens* CHA0, *Pseudomonas chlororaphis* PCL1391, and *Pseudomonas* sp. CMR12a were tested for their insecticidal activity. One insecticidal assay included injecting *Galleria mellonella* (greater wax moth) larvae with aliquots of washed bacteria and then monitoring the behavior of the larvae in Petri dishes. Once the larvae began to become melanized (i.e., they began to darken by infiltration of melanin) they were scored as either live or dead every hour. In another insecticidal assay, 1-week-old *Plutella xylostella* (diamondback moth) larvae were inoculated with washed

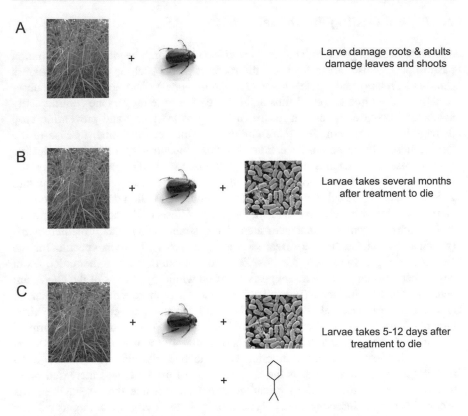

Fig. 7.7 Preventing the damage to pasture plants by New Zealand grass grubs (*Costelytra giveni*). (**a**) Perennial ryegrass and an adult grass grub. (**b**) Treatment of grass grub infested ryegrass with the commercial biocontrol agent *Serratia entomophila*. (**c**) Treatment of grass grub infested ryegrass with a newly isolated strain of *Serratia proteamaculans* which carries a viral prophage within its genome

bacterial cells and larvae were subsequently scored as alive or dead. For a start, the wild-types of all three biocontrol PGPB effectively infected and killed the insect larvae in both assays. In addition, it was found that both hydrogen cyanide production and the synthesis of different cyclic lipopeptides by these bacteria contributed to the insecticidal activity. Interestingly, mutants that lacked antimicrobial compounds such as 2,4-diacetylphloroglucinol, phenazine, pyrrolnitrin or pyoleteorin did not show any decreased insecticidal activity. Overall, the data suggest that the insecticidal activity of these strains is the result of several features of these bacterial strains, but it is not a consequence of the antimicrobial compounds that some of these bacteria produce and use to control bacteria and fungi.

In addition to the ability of some PGPB having the ability to directly kill agricultural insect pests, some PGPB can protect plants from insect pests by inducing systemic resistance within plants. In this way, PGPB induce the production of a variety of chemical compounds including camalexin (an indole alkaloid), glucosinolates (secondary metabolites produced by Brassicales, an order of flowering plants; also called mustard oil glycosides), flavonoids, and phenolic compounds (including lignin and gossypol).

7.1.4 Brevibacillus laterosporus

Brevibacillus laterosporus is a rod-shaped bacterium that has been reported to have insecticidal activity against a wide range of insects including coleoptera, lepidoptera, and diptera as well as some nematodes. In addition, some strains of *B. laterosporus* have been found to show antimicrobial activity against various bacteria and fungi. Similar to *B. thuringiensis*, *B. laterosporus* strains produce a parasporal crystal that contains the insecticidal activity. *B. laterosporus* insecticidal proteins are comparable in insect toxicity to similar proteins from *B. thuringiensis*. Recently, a strain of *B. laterosporus* was shown to be endophytic so that the majority of cabbage plants sprayed with this bacterial strain were subsequently found with this bacterium within cabbage tissues (roots, leaves, and stems). Since a strain of this bacterium has been registered as a bioinsecticide (and additional strains may be registered in the future), it is important to ascertain that these bacteria are safe for human consumption as they be ingested during consumption of treated cabbages (and other brassicas). Fortunately, to date, there is no indication of these bacteria being dangerous for humans.

7.2   Nematodes

Nematodes (or roundworms) are small (typically 0.2–2.5 mm in length) primitive animals that have existed for around 500 million to 1 billion years with the oldest known nematode fossils being around 120–135 million years old. Nematodes are highly abundant and extremely diverse with the majority feeding on bacteria, fungi, and protozoa while about 15% of known nematode species are plant parasites. The first described plant parasitic nematodes (in 1743) were in wheat seeds. At present, nematodes are recognized as a major agricultural pathogen causing approximately $125–150 billion (US) of crop damage worldwide, notwithstanding all of the methods that are currently used to control nematodes (mostly chemical nematicides such as fumigants, organophosphates, and carbamates). For many crops, it is estimated that nematodes may typically cause from 5 to 20% of crop loss with worldwide losses in agricultural productivity estimated to be around 12%. Generally, the most damaging nematodes in agriculture are root-knot nematodes. Following planting, nematode infestation occurs and nematodes feed on the root tissue. With time, as the damage to roots increases, plants wilt and eventually die.

Recent studies have suggested that nematode soil populations are likely to increase significantly as a consequence of global climate change. For example, in a rice paddy field treated with either elevated CO_2 concentrations (500 ppm), canopy warming of 2 °C or a combination of the two, nematode populations were observed to increase by around 30% (with no change in nematode diversity). Thus, should CO_2 levels and global temperatures continue to increase, there is a very real

Table 7.6 Some *B. thuringiensis* strains that contain nematicidal activity

B. thuringiensis strain	Protein	Nematicidal activity against
YBT-021	n.d.	*Ditylenchus destructor*
BMB171-15	Cry6Aa2	*Caenorhabdita elegans* and *Meloidogyne hapla*
BMB0224	Cry55Aa1	*Meloidogyne hapla*
DB27	Cry21Fa1	*Caenorhabdita elegans*
*Bt*7	n.d.	*Meloidogyne incognita*
Bt-64	n.d.	*Meloidogyne javanica*
YD5	n.d.	*Meloidogyne incognita*
*Bt*010	n.d.	*Meloidogyne hapla*

n.d. indicates not determined

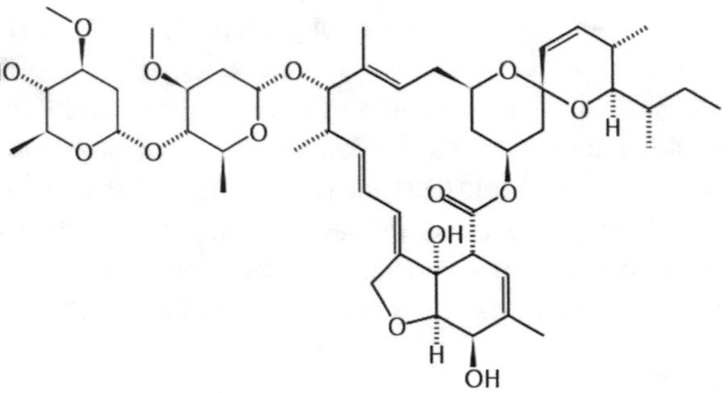

Fig. 7.8 The structure of avermectin

probability that crop yields will be negatively affected by a very large increase in soil nematode populations.

Interestingly, numerous *B. thuringiensis* strains, including those encoding Cry5, Cry6, Cry12, Cry13, Cry14, Cry21, and Cry55, have been shown to have some nematicidal activity (Table 7.6). In addition to Cry proteins, several other proteins that are synthesized by some strains of *B. thuringiensis* show nematicidal activity. These proteins, which include chitinases, collagenases, keratinolytic proteinases, serine proteases and neutral metalloproteinases, digest the nematode's tissues (especially the nematode's cuticle which completely surrounds this animal) thereby acting synergistically with the Cry proteins. The reported LC_{50} values for the Cry proteins shown in Table 7.6 are quite low (in the range of 10–100 µg/mL) suggesting that *B. thuringiensis* strains that produce these proteins might become effective nematocides. In addition, a number of other soil bacteria produce a range of small nonprotein molecules that are toxic to many nematodes. In particular, avermection (Fig. 7.8) and its derivatives, synthesized by *Streptomyces avermitilis*, which are 16-membered lactones, are used to treat both parasitic worms and insect pests.

7.2.1   PGPB as Nematode Biocontrol Agents

PGPB may inhibit the proliferation of plant pathogenic nematodes in several different ways. For example, antibiotics such as 2,4-diacetylphloroglucinol can significantly reduce the population of cyst nematodes, especially when the PGPB also synthesizes hydrogen cyanide. As mentioned above, PGPB sometimes produce cuticle lysing enzymes that, in addition to weakening adults, may also inhibit nematode egg hatching. PGPB may also induce systemic resistance initiating the production of a number of different plant defensive enzymes. Finally, PGPB may stimulate plant growth through a variety of mechanisms and thereby make it more difficult for the nematode to kill the plant.

7.2.1.1 Pine Wilt Disease

Pine wilt disease is one of the major threats affecting conifer forests worldwide. This disease is caused by the pinewood nematode, *Bursaphelenchus xylophilus*, including pinewood nematode associated bacteria and an insect vector, *Monochamus* spp., which is responsible for the pinewood nematode tree-to-tree dissemination. In Europe, pine wilt disease was first reported in 1999 in the Iberian Peninsula. Unfortunately, some scientists believe that this disease may become one of the most common diseases threatening European conifer forests in the near future. At the present time, there are no effective means of preventing this disease. Neverthe-less, based on the observation that ethylene production significantly increases in pine trees during pinewood nematode invasion, increased ethylene levels may be, at least partially, responsible for the development of pine wilt disease symptoms. One possible way of decreasing the damage caused by the pinewood nematode is to limit the increased levels of ethylene in nematode-infected trees through the inocu-lation of pine trees with ACC deaminase-containing PGPB. To test this possibility, 3-month-old pine seedlings were first inoculated with the PGPB *Pseudomonas* sp. UW4 (or an ACC deaminase minus mutant of this strain) applied to its roots, and then after 1 week, the seedlings were challenged with a nematode suspension of approximately 500 animals. One month after the nematode inoculation, the pine seedlings were harvested, and their root and shoot growth were assessed. In fact, the PGPB *P.* sp. UW4 not only induced pine tree seedling growth (both roots and shoots), in the presence of the pinewood nematode, it also significantly reduced disease symptoms caused by the nematode (Fig. 7.9). Moreover, an ACC deaminase minus mutant of *P.* sp. UW4 neither promoted plant growth nor prevented pine wilt disease symptoms, thereby confirming the key role played by ACC deaminase in the biocontrol of this phytopathogenic nematode. Unfortunately, a problem that exists with this approach is that despite the fact that lowering plant ethylene decreases the damage from the pinewood nematode, it does not prevent the nematode from infecting the plant. Thus, this treatment only partially addresses the problem. It is nevertheless very encouraging that a biocontrol strategy that prevents the anticipated widespread damage from the pinewood nematode can be developed. Moreover, it remains to be determined whether (i) an ACC deaminase-containing PGPB that is endophytic is superior to *P.* sp. UW4, a rhizospheric PGPB, and (ii) whether a bacterium that acts as a nematocide might work synergistically with an ACC deaminase-containing PGPB.

In a preliminary experiment, a group of researchers isolated 238 bacterial endophytes from four different pine tree species across 18 different sampling sites

Fig. 7.9 The effect of the ACC deaminase producing PGPB *Pseudomonas* sp. UW4 on pine tree seedlings infected with pinewood nematodes. The nematodes were applied to the seedlings through a piece of cotton that was taped to the seedling stem. Panel A, pine seedling shoots and roots in the absence of any added nematodes; panel B, pine seedling shoots and roots in the presence of added nematodes; panel C, pine seedling shoots and roots in the presence of added nematodes and an ACC deaminase minus mutant of *P.* sp. strain UW4; panel D, pine seedling shoots and roots in the presence of added nematodes and the wild-type strain *P.* sp. UW4

in Korea. Following analysis by 16S ribosomal DNA sequencing, metabolites were extracted from several isolates and tested for nematicidal activity against the pine wood nematode. Two bacterial isolates showed significant nematicidal activity, killing approximately 60% of the nematodes in the test samples. On the basis of these experiments, it was suggested that one or more of these bacterial strains might be used as biocontrol agents for the pine wood nematode. The perceived advantage of using these strains is that not only are they nematicidal, but in addition, as endophytes they are likely to persist in tree tissues for an extended period of time.

7.2.1.2 Endophytic Fungi and Bacteria

The nematode *Radopholus similis* (commonly known as the burrowing nematode) is a pest of a variety of agricultural crops, most notably banana and citrus. This nematode can severely damage the plant roots thereby reducing the ability of the tree to take up water and nutrients. As a consequence, *R. similis* is considered to be a major root pathogen of banana trees and has been reported to cause yield losses of from 30% to 60%. These nematodes are currently controlled by chemical fumigation of the soil and by ensuring that new sprouts are raised from clean tissue culture. In an effort to avoid the use of chemical fumigation, scientists have observed that some endophytic fungi and some endophytic bacteria are able to partially limit the damage to the plant that is caused by this nematode. The various endophytes were isolated from the roots of banana plants infected with nematodes but nevertheless displaying a decreased level of nematode-caused disease. Moreover, when several different endophytic fungi together with several different endophytic bacteria were tested (one bacterium and one fungus at a time) for biocontrol activity against *R. similis*, it was observed that, with some combinations of fungi and bacteria, up to a 90% reduction in damage to plant roots was observed (Fig. 7.10). In addition to providing

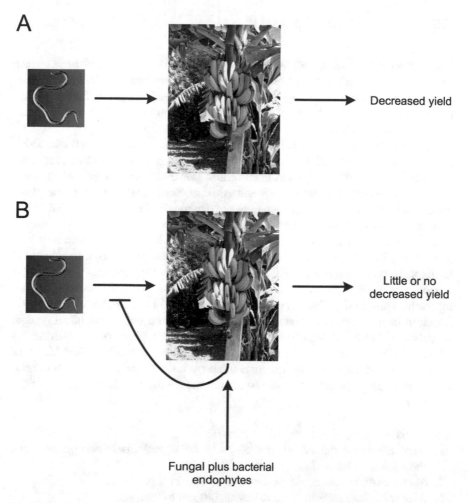

Fig. 7.10 An overview of protecting banana trees from damage caused by nematodes. (**a**) Nematodes infect and damage tree roots resulting in a decreased yield of both biomass and fruit. (**b**) In the presence of certain fungal endophytes together with bacterial endophytes, nematode damage to plants is dramatically decreased and plant and fruit yield is not appreciably affected

Table 7.7 The effects of different inoculants on the 1-month growth of tomato plants under greenhouse conditions in the presence of root-knot nematodes

Treatment	Length, cm		Biomass, dry weight in grams	
	Shoots	Roots	Shoots	Roots
Control	28.13	17.13	1.29	0.39
Mycorrhiza	36.73	22.87	1.58	0.45
PGPB	35.73	22.63	1.62	0.44
Mycorrhiza + PGPB	41.23	23.47	1.90	0.47
Nematodes	21.33	16.87	0.73	0.15
Nematodes + Mycorrhiza	24.60	18.97	1.17	0.36
Nematodes + PGPB	24.47	20.87	1.11	0.31
Nematodes + Mycorrhiza + PGPB	27.60	20.73	1.36	0.39

biocontrol, the successful fungal–bacterial combinations also promoted plant growth from 20% to 50%. While these observations are important and may provide the basis for the development of biological alternatives to chemical fumigation, they are preliminary as scientists have not as yet developed an understanding of the mechanistic basis for this biocontrol behavior.

In another recent study, nematode-infected wheat plants were treated with *Pseudomonas* spp. PGPB and arbuscular mycorrhizal fungi (AMF). In this case, both treatments significantly improved plant health; however, surprisingly, when wheat plants were treated with both PGPB and AMF no additional benefits to the plants were observed, i.e., these treatments were not synergistic. Here again, the mechanistic basis of this biocontrol behavior was not elaborated.

In yet another recent study, the growth of tomato plants infected with root-knot nematodes (*Meloidogyne incognita*) was compared to the growth of the same plants in the absence of nematodes with the addition of either mycorrhiza (*Rhizophagus irregularis*), PGPB (*Pseudomonas synxantha* plus *Pseudomonas jessenii*) or both mycorrhiza and PGPB. The experiment was conducted for 1 month after nematode infection in a greenhouse. Plants that were treated with either individual or dual inoculants showed both significantly increased plant growth and reduced nematodes infection (Table 7.7). In addition, the treated plants exhibited high levels of phenolics and defensive enzymes concomitant with nematode infection. Unfortunately, no attempt was made to understand the mechanistic basis for this biocontrol behavior.

Questions

1. What are some the advantages of using biological insecticides rather than chemical insecticides?
2. Briefly explain how *B. thuringiensis* functions as a biological insecticide.
3. Why is *B. thuringiensis* toxic to insects but not to humans?
4. How would you use genetic engineering to improve the usefulness/effectiveness of a particular *B. thuringiensis* protoxin?

5. What strategies have scientists developed to decrease or avoid the onset of insect resistance to a particular *B. thuringiensis* insecticidal toxin?
6. What features of the bacterium make it unlikely that mosquitoes will ever develop resistance to *B. thuringiensis* subsp. *israelensis* strains?
7. What are cadherins, and how might they be employed to enhance the toxicity of a particular Cry protein?
8. What are antifeeding prophage and how do they work?
9. How can PGPB that do not encode a specific insecticidal toxin act as an insecticide?
10. What are nematodes and why are they a problem for agriculture?
11. How can ACC deaminase from PGPB protect trees against pine wilt disease?
12. How can endophytes be used to protect against nematode damage to plants?
13. How can Cry proteins protect plants against damage by nematodes?

Bibliography

Aronson AI, Shai Y (2001) Why *Bacillus thuringiensis* insecticidal toxins are so effective: unique features of their mode of action. FEMS Microbiol Lett 195:1–8

Azizoglu U (2019) *Bacillus thuringiensis* as a biofertilizer and biostimulator: a mini-review of the little-known plant growth-promoting properties of Bt. Curr Microbiol 76:1379–1385

Ben Gharsa H, Bouri M, Gannar A, Mougou Hamdane A, Glick BR, Rhouma A (2018) Evaluation of the interspecific competition within *Agrobacterium* spp. in the soil and rhizosphere of tomato (*Solanum lycopersicum*) and maize (*Zea mays*). J Plant Pathol 100:505–511

Bravo A, Gill SS, Soberon M (2007) Mode of action of *Bacillus thuringiensis* Cry and Cyt toxins and their potential for insect control. Toxicon 49:423–435

Bravo A, Gómez I, Porta H, García-Gómez BI, Rodriguez-Almazan PL, Soberón M (2012) Evolution of *Bacillus thuringiensis* Cry toxins insecticidal activity. Microb Biotechnol 6:17–26

Chakroun M, Banyuls N, Bel Y, Escriche B, Ferre J (2016) Bacterial vegetative insecticidal proteins (Vip) from Entomopathogenic bacteria. Microbiol Mol Biol Rev 80:329–350

Chaves NP, Pocasangre LE, Elango F, Rosales FE, Sikora R (2009) Combining endophytic fungi and bacteria for the biocontrol of *Radopholus similis* (Cobb) Thorne and for effects on plant growth. Sci Hortic 122:472–478

Chen J, Jua G, Jurat-Fuentes JL, Abdullah MA, Adang MJ (2007) Synergism of *Bacillus thuringiensis* toxins by a fragment of a toxin-binding cadherin. Proc Natl Acad Sci U S A 104:13901–13906

De Maagd RA, Bravo A, Crickmore N (2001) How *Bacillus thuringiensis* has evolved specific toxins to colonize the insect. Trends Genet 17:193–199

Ferre J, Real MD, Van Rie J, Jansens S, Peferoen M (1991) Resistance to the *Bacillus thuringiensis* bioinsecticide in a field population of *Plutella xylostella* is due to a change in a midgut membrant receptor. Proc Natl Acad Sci U S A 88:5119–5123

Flury P, Vesga P, Péchy-Tarr M, Aellen N, Dennert F, Hofer N, Kupferschmied KP, Kupferschmied P, Metla Z, Ma Z, Siegfried S, de Weert S, Bloemberg G, Höfte M, Keel CJ, Maurhofer M (2017) Antimicrobial and insecticidal: cyclic lipopeptides and hydrogen cyanide produced by plant beneficial *Pseudomonas* strains CHA0, CMR12a, and PCL1391 contribute to insect killing. Front Microbiol 8:100

Harun-Or-Rashid M, Chung YR (2017) Induction of systemic resistance against insect herbivores in plants by beneficial soil microbes. Front Plant Sci 8:1816

Hurst MRH, Beattle A, Jones SA, Laugraud A, van Koten C, Harper L (2018) *Serratia proteamaculans* strain AGR96X encodes an antifeeding prophage (Tailocin) with activity against grass grub (*Costelytra giveni*) and manuka beetle (*Pyronota species*) larvae. Appl Environ Microbiol 84:e02739-17

Imperiali N, Chirboga X, Schlaeppi K, Fesselet M, Villacrés D, Jaffuel G, Bender SF, Dennert F, Blanco-Pérez R, van der Heijden MGA, Maurhofer M, Mascher F, Turlings TCJ, Keel CJ, Campos-Herrera R (2017) Combined field inoculations of Pseudomonas bacteria, arbuscular mycorrhizal fungi, and entomopathogenic nematodes and their effects on wheat performance. Front Plant Sci 8:1809

Jouzani GS, Valijanin E, Sharafi R (2017) *Bacillus thuringiensis*: a successful insecticide with new environmental features and tidings. Appl Microbiol Biotechnol 101:2691–2711

Lereclus D, Delecuse A, Lecadet MM (1993) Diversity of *Bacillus thuringiensis* toxins and genes. In: Entwistle PF, Cory JS, Bailey MJ, Higgs S (eds) *Bacillus thuringiensis*, an environmental biopesticide: theory and practice. Wiley, Chichester, UK, pp 37–69

Liu Y, Ponpandian LN, Kim H, Jeon J, Hwang BS, Lee SK, Park S-C, Bae H (2019) Distribution and diversity of bacterial endophytes from four *Pinus* species and their efficacy as biocontrol agents for devastating pine wood nematodes. Sci Rep 9:12461

Mettus AM, Macaluso A (1990) Expression of *Bacillus thuringiensis* δ-endotoxin genes during vegetative growth. Appl Environ Microbiol 56:1128–1134

Nascimento FX, Vicente CSL, Barbosa P, Espada M, Glick BR, Oliveira S, Mota M (2013) The use of the ACC deaminase producing bacterium *Pseudomonas putida* UW4 as a biocontrol agent for pine wilt disease. BioControl 58:427–433

Ormskirk MM, Narciso J, Hampton JG, Glare TR (2019) Endophytic ability of the insecticidal bacterium *Brevibacillus laterosporus* in Brassica. PLoS One 14(5):e0216341

Palma L, Muñoz D, Berry C, Murillo J, Caballero P (2014) *Bacillus thuringiensis* toxins: an overview of their biocidal activity. Toxins 6:3296–3325

Proença DN, Schwab S, Vidal MS, Baldani JI, Xavier GR, Morais PV (2019) The nematicide *Serratia plymuthica* M24T3 colonizes *Arabidopsis thaliana*, stimulates plant growth, and presents plant beneficial potential. Braz J Microbiol 50(3):777–789. https://doi.org/10.1007/s42770-019-00098-y

Raymond B, Federici BA (2017) In defence of *Bacillus thuringiensis*, the safest and most successful microbial insecticide available to humanity—a response to EFSA. FEMS Microbiol Ecol 93: fix084

Schnepf HE, Whiteley HR (1981) Cloning and expression of the *Bacillus thuringiensis* crystal protein gene in *Escherichia coli*. Proc Natl Acad Sci U S A 78:2893–2897

Schnepf E, Crickmore N, Van Rie J, Lereclus D, Baum J, Feitelson J, Zeiger DR, Dean DH (1998) *Bacillus thuringiensis* and its pesticidal crystal proteins. Microbiol Mol Biol Rev 62:775–806

Schünemann R, Knaak N, Fiuza LM (2014) Mode of action and specificity of *Bacillus thuringiensis* toxins in control of caterpillars and stink bugs in soybean culture. ISRN Microbiol 2014:135675

Sharma IP, Sharma AK (2017) Physiological and biochemical changes in tomato cultivar PT-3 with dual inoculation of mycorrhiza and PGPR against root-knot nematode. Symbiosis 71:175–183

Tabashnik BE, Brévault T, Carrière Y (2013) Insect resistance to *Bt* crops: lessons from the first billion acres. Nat Biotechnol 31:510–521

Wang J, Li M, Zhang X, Liu X, Li L, Shi X, Hu H-W, Pan G (2019a) Changes in soil nematode abundance and composition under elevated [CO_2] and canopy warming in a rice paddy field. Plant Soil 445:425–437

Wang Y, Wang J, Fu X, Nageotte JR, Silverman J, Bretsnyder EC, Chen D, Rydel TJ, Bean GJ, Li KS, Kraft E, Gowda A, Nance A, Moore RG, Pleau MJ, Milligan JS, Anderson HM, Asiimwe P, Evans A, Moar WJ, Martinelli S, Head GP, Haas JA, Baum JA, Yang F, Kerns DL, Jerga A (2019b) *Bacillus thuringiensis* Cry1Da_7 and Cry1B.868 protein interactions with novel receptors allow control of resistant fall armyworms, *Spodoptera frugiperda* (J.E. Smith). Appl Environ Microbiol 85:e00579-19

Whalon ME, Wingerd BA (2003) Bt: mode of action and use. Arch Insect Biochem Physiol 54:200–211

Environmental Interactions

8

Abstract

The study of plant growth-promoting bacteria and the mechanisms that they employ is not just an interesting laboratory exercise. Therefore, this chapter emphasizes how an understanding of these mechanisms can facilitate the use of plant growth-promoting bacteria to address real world problems in the natural environment. In this chapter, the following issues are addressed: how plant growth-promoting bacteria with various activities can facilitate plant growth in conditions of high salt concentrations or extensive drought or flooding or extreme cold; how plants growing in nature can select for bacteria with specific useful traits; how diazotrophic endophytic bacteria may fertilize some crop plants; and how the addition of specific plant growth-promoting bacteria can alter plant gene expression.

8.1 Overview of Stress

For the most part, plants that are grown in the field under natural conditions are subject to a more or less constant onslaught of one stress after another, all of which can potentially inhibit plant growth and development (Fig. 8.1). These stresses may be caused by biotic factors such as viruses, nematodes, insects, bacteria, or fungi (see Chaps. 6 and 7). Alternatively, abiotic factors such as extremes of temperature, high light, flooding, drought, salt, the presence of toxic metals and environmental organic contaminants, radiation, and wounding can also be inhibitory to plant growth and development. While various plants may respond differently to different stresses, e.g., some plants are much more sensitive to increasing levels of salt, nearly all plants respond to stress by producing ethylene (Fig. 5.18). As discussed in Chap. 6 when considering the synthesis of stress ethylene in response to fungal pathogen infection, abiotic stresses also induce the synthesis of two peaks of ethylene (Fig. 6.15). Moreover, as documented for biotic stresses, lowering the second (larger) ethylene

© Springer Nature Switzerland AG 2020
B. R. Glick, *Beneficial Plant-Bacterial Interactions*,
https://doi.org/10.1007/978-3-030-44368-9_8

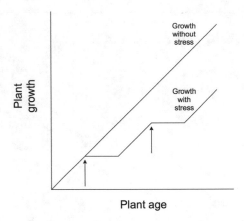

Fig. 8.1 Schematic representation of plant development in the presence and absence of environmental stress (biotic and/or abiotic). The arrows indicate the onset of a stress. The period of no growth following each stress represents the time that it takes the plant to recover from the stress. Following a stress, plants may grow more slowly than non-stressed plants, they may not grow at all (as depicted) or they may lose biomass

peak can significantly decrease the severity of the negative consequences of the stress. Ideally, it would seem to be advantageous for a plant to synthesize the first (smaller, defensive) ethylene peak but not the second (senescence-causing) ethylene peak. In fact, this is readily achieved when plants are treated with PGPB that produce the enzyme ACC deaminase. This occurs because these bacteria contain a low, constitutive, level of ACC deaminase that does not significantly alter the magnitude of the first ethylene peak. However, between the first and second ethylene peaks, as additional ACC is synthesized, the increasing level of ACC induces the synthesis of more ACC deaminase which results in a lowering of the level of ACC and hence a lowering of the second peak of ethylene (Figs. 5.18 and 6.15).

It is worth pointing out that while ethylene signaling is required for induced systemic resistance in plants elicited by PGPB (Fig. 6.17), high levels of ethylene are not required for this to occur. Thus, lowering of plant ethylene levels by bacterial ACC deaminase is not incompatible with induced systemic resistance. In fact, many PGPB strains that possess ACC deaminase also induce systemic resistance.

8.2   Salt and Drought

Soil salinity is an enormous worldwide problem for agriculture, especially for crops that are grown under irrigation. This is because salt is inhibitory to the growth of a very large number of different plants. At the present time, the amount of salt-inhibited land worldwide is estimated to be >900 million hectares which is approximately 6% of the total global land mass, or about 20% of the world's cultivated area. Disturbingly, around half of the world's land that is devoted to the growth of irrigated crops is somewhat adversely affected by salt. Soil salinity is caused, in part, by inefficient irrigation and drainage systems. As a consequence of the lack of available water, some areas are under-irrigated causing salts to accumulate in the soil; the salt remains behind in the soil when water is either taken up by plants or lost to evaporation.

Corn (also known as maize) is one of the most important crops in the world with its total production surpassing rice or wheat. Ironically, the bulk of the corn that is produced worldwide is not used as a food for humans but rather it is used as either animal feed or as a feedstock for corn ethanol, corn starch, or high fructose corn syrup. More than one-third of worldwide corn production is in the United States. Until very recently, the corn yield per acre of land has continued to increase. This is generally attributed to the fact that an increasing number of corn plants are planted per acre of land. However, an unfortunate consequence of this strategy is that a decreasing amount of soil water is available to each plant. In practice, this means that corn plants are becoming increasingly sensitive to drought.

The initial responses of plants to drought and salinity are very similar; both responses are attributed to water stress. When plants are exposed to high salinity, a decrease in the growth rate followed by a gradual recovery to a new reduced growth rate is the plant's first response to the decrease in water potential caused by salt, rather than to any salt-specific toxicity. Once salt has been taken up by plants, sodium ions may be translocated to shoots in the transpiration stream in the xylem. Although sodium ions have the ability to return to roots via the phloem, in most plants this flow is generally minimal so that leaves and shoots typically accumulate higher concentrations of sodium ions than roots. In addition, most crops translocate only a very limited amount of sodium ions to reproductive or storage structures such as fruits and seeds as these organs are fed mainly through the phloem. However, plant vegetative tissues are supplied mainly by the flow through the xylem and, as a result, have higher sodium ion levels. As a consequence of this accumulation, the metabolic toxicity of sodium ions is primarily attributed to the sodium ions competing with potassium ions for binding sites that are essential for cellular functioning.

Plants utilize a variety of different biochemical strategies to cope with salt stress including (i) selective accumulation or exclusion of sodium ions, (ii) control of sodium ion uptake by roots and transport into leaves, (iii) compartmentalization of sodium ions at the cellular and whole-plant levels, (iv) synthesis of compatible solutes such as trehalose or proline, (v) alteration of some membrane structures, (vi) induction of antioxidative enzymes such as superoxide dismutase and peroxidase, and (vii) modulation of plant hormone levels including auxin, cytokinin, and ethylene. However, it is important to bear in mind that the salt tolerance of any particular plant species can vary with the intrinsic salt tolerance of a particular cultivar, the growth phase of the plant, the ionic composition of the soil, the health of the plant, and the presence of rhizospheric or endophytic PGPB.

When considering the various mechanisms that PGPB utilize it is important to keep in mind that many of these mechanisms are interconnected and affect one another. In addition to providing direct protection against the inhibitory effects of abiotic stress (e.g., by stimulating plant growth with bacterial IAA or lowering plant ethylene levels as a consequence of the functioning of ACC deaminase), PGPB may also alter plant gene expression so that the plant is less likely to succumb to the abiotic stress. For example, interaction with various PGPB has been shown to cause an increase in the plant production of the "water-structuring" metabolites betaine, proline and trehalose as well as the synthesis of enzymes such as superoxide dismutase and catalase that can detoxify reactive oxygen species.

In one report, researchers isolated and characterized a strain of *Chryseobacterium gleum* and showed that it had a wide range of plant growth-promoting activities including the production of ACC deaminase, synthesis of IAA, production of siderophores, ammonia, hydrogen cyanide, and fungal cell wall degrading enzymes including cellulase and protease. Notably, this bacterium also produced the enzyme keratinase that allowed the bacterium to digest poultry feathers, a potential source of protein, nitrogen, phosphorous and several trace metals. This bacterium promoted wheat plant growth in both the absence and presence of high levels of salt (200 mM); however, it was not possible to attribute the ability to overcome salt stress to a single bacterial trait.

Many more studies have been focused on the ability of PGPB to ameliorate the effects of salt stress rather than drought stress. Nevertheless, many workers have also reported considerable success when using PGPB to overcome drought stress (Table 8.1).

In the examples given below, conferring salt tolerance upon plants is typically attributed to one or more specific PGPB mechanisms. However, the literature also contains a number of reports where the mechanism(s) employed by a PGPB strain in conferring salt tolerance to a plant is completely unknown. While the latter bacterial strains may be useful in certain applications, since their mechanisms are unknown, they do not provide researchers with any guidelines for selecting additional (possibly similar) PGPB that are able to confer salt tolerance to treated plants.

8.2.1   IAA

In one study it was shown that incubation of two different cultivars of corn with the PGPB *Azospirillum brasilense* or *A. lipoferum* (both known to secrete relatively high levels of IAA) enabled the treated plants to tolerate high levels of salt in the soil (Table 8.2). In this case, the *Azospirillum* with its activity of both nitrate reductase and nitrogenase in addition to its production of IAA, partially reversed the inhibitory effects of the added salt. While the *Azospirillum*-treated plants grew to a significantly greater extent at all of the salt concentrations that were tested and in all of the parameters measured, this study did not definitively identify the mechanism(s) of plant growth promotion that were employed by these PGPB under high salt conditions. In addition to the data shown in Table 8.2, in the absence of any added bacteria the shoot and root content of proline increased significantly along with the increasing level of salt as the plant attempted to compensate for the added osmotic stress. However, in the presence of *Azospirillum*, the rise in proline content was less

Table 8.1 Some of the bacterial strains that have been shown to help to overcome plant drought stress

Plant tested	PGPB strain
Arabidopsis	*Paenibacillus polymyxa*
Arabidopsis	*Phyllobacterium brassicacearum*
Arabidopsis	*Azospirillum brasilense*
Arabidopsis	*Pseudomonas chlororaphis*
Pepper	*Bacillus licheniformis*
Cucumber	*Bacillus cereus, Bacillus subtilis, Serratia* sp.
Sunflower	*Achromobacter xylosoxidans*
Sunflower	*Pseudomonas putida*
Common beans	*Rhizobium tropici* and *Paenibacillus polymyxa*
Hyoscyamus niger (medicinal plant)	*Pseudomonas putida, Pseudomonas fluorescens*
Pea	*Pseudomonas* spp.
Tomatoes, peppers	*Achromobacter piechaudii*
Peppers	*Bacillus licheniformis*
Potato	*Bacillus pumilus, Bacillus firmus*
Sorghum	*Bacillus* spp.
Wheat	*Azospirillum lipoferum*
Wheat	*Bacillus amyloliquefaciens, Azospirillum brasilense*
Wheat	*Bacillus thuringiensis, Paenibacillus polymyxa*
Green gram	*Pseudomonas fluorescens, Bacillus subtilis*
Mung bean	*Pseudomonas aeruginosa*
Maize (corn)	*Pseudomonas entomophila, Pseudomonas stutzeri, Pseudomonas syringae, Pseudomonas monteilli*
Maize	*Azospirillum lipoferum*
Maize	*Burkholderia phytofirmans, Enterobacter* sp.
Maize	*Bacillus amyloliquefaciens, Bacillus licheniformis, Bacillus thuringiensis, Paenibacillus favisporus*
Maize	*Burkholderia* sp.

dramatic, presumably reflecting the lower level of osmotic stress in PGPB-treated plants.

In an effort to create a PGPB strain that would produce more IAA and (hopefully) be more effective at protecting plants against inhibition by high salt concentrations, one group of scientists engineered a strain of *Sinorhizobium meliloti* with an additional pathway for IAA biosynthesis. This was done by expressing the *iaaM* gene from the phytopathogen *Pseudomonas savastanoi* on a plasmid in a strain of *S. meliloti* (Fig. 8.2). In this study, when the transformed *S. meliloti* strain was used to nodulate *Medicago truncatula* (barrel medic) plants, the resultant plants were more resistant to the inhibitory effects of high salt as well as to several other stresses such as UV and high temperature. However, since the increased IAA concentration induced plants to synthesize a higher level of trehalose (which is also known to

Table 8.2 Effect of various amounts of salt and *Azospirillum* inoculation on the growth of corn plants grown for 45 days

Treatment	NaCl, MPa	Plant dry weight, g		Leaf area, cm³	Water content, %	
		Shoot	Root		Shoot	Root
Control	0.0	5.8	1.9	272	42.1	7.0
Control	0.2	3.8	1.7	168	36.2	4.8
Control	0.6	2.9	0.65	142	18.5	0.75
Control	1.0	2.0	0.5	91	14.5	0.7
Control	1.6	1.9	0.4	57	13.5	0.6
Azospirillum	0.0	7.1	2.5	407	50.3	9.4
Azospirillum	0.2	6.2	2.0	277	49.2	6.1
Azospirillum	0.6	4.6	1.2	256	41.9	5.2
Azospirillum	1.0	3.4	1.1	220	27.0	2.6
Azospirillum	1.6	3.1	0.8	120	20.7	1.8

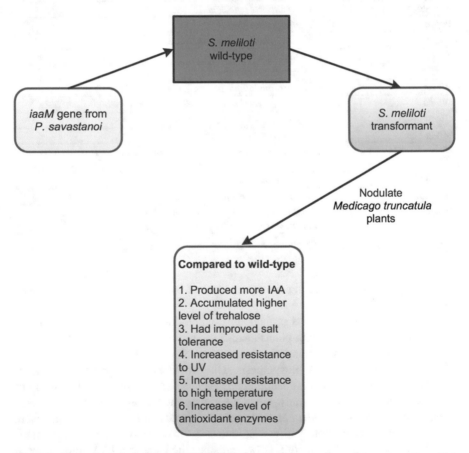

Fig. 8.2 Flow chart depicting how *S. meliloti* that has been engineered to overproduce IAA protects nodulated plants against a range of stresses

Fig. 8.3 The effect of adding the bacterium *P. extremorientalis* to *Silybum marianum* plants in the presence of increasing amounts of salt. Key: yellow bars, root length without bacteria; green bars, root length with bacteria; blue bars, shoot length without bacteria; red bars, shoot length with bacteria

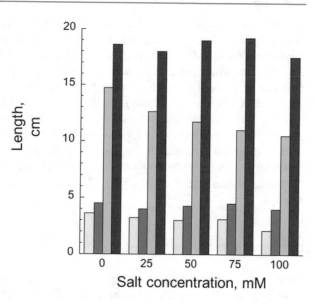

protect plants against osmotic stress), it is not clear whether the increase in plant stress resistance is a direct or indirect effect of the increased level of IAA.

In another set of experiments, the medicinal plant *Silybum marianum* (milk thistle) was inoculated with the root colonizing bacterium *Pseudomonas extremorientalis* and then grown in the presence of varying concentrations of salt. Increasing the salt concentration from 0 to 100 mM significantly decreased the plant's root and shoot length and fresh weight in the absence of the bacterium; however, root and shoot length (Fig. 8.3), as well as fresh weight, of salt-treated plants, were increased in the presence of the added bacterium. In addition, in this experiment, concentrations of IAA between 0.1 and 0.001 μM stimulated plant growth, in both the presence and absence of salt, to approximately the same extent as adding the bacterium. Since this bacterium does not produce ACC deaminase (see below), it was assumed that the observed tolerance to salt stress was a consequence of bacterially synthesized IAA (as no other mechanisms were considered).

In a separate but similar study with a different plant and different PGPB strains, it was shown that while 100 mM salt inhibited cotton seed germination as well as the growth of cotton seedling roots and shoots, the exogenous application of two different IAA-producing PGPB strains (*Pseudomonas putida* and *Pseudomonas chlororaphis*) alleviated a significant portion of the salt inhibition of plant growth. Moreover, the addition of varying amounts of purified IAA to the salt inhibited plants stimulated plant growth to an extent similar to the effect observed with the PGPB strains. This result is consistent with, although it does not prove, bacterial IAA being responsible for alleviating the salt inhibition of cotton seedlings.

In addition to protecting plants against salt stress inhibition, bacterial IAA also protects plants against drought stress inhibition. Alfalfa plants (*Medicago sativa*) subjected to drought stress (either from cessation of watering or the addition of polyethylene glycol) were treated with either the wild-type PGPB *Ensifer meliloti* (formerly *Sinorhizobium meliloti*) or an IAA-overproducing transformant of the wild-type (Table 8.3). Treatment of alfalfa plants with the IAA-overproducing transformant strain resulted in increased plant growth both in the absence and presence of drought stress. Moreover, both nitrogen assimilation (as indicated by the level of nitrogenase activity) and carbon assimilation (indicated by ribulose bisphosphate carboxylase activity) were increased by the presence of the transformed strain. Finally, the level of sucrose found in plant roots was significantly increased by the presence of the transformant, thus providing another means of protecting roots from the imposed drought. Overall, these experiments provide a very clear demonstration of the ability of IAA to protect alfalfa plants from drought stress.

8.2.2 Ethylene

Based on the model depicted in Fig. 5.18, it was predicted that any PGPB that produced the enzyme ACC deaminase would effectively lower the level of stress ethylene experienced by a plant and therefore decrease the damage to the plant as a consequence of the applied stress, including the presence of high salt. Furthermore, it was reasoned that PGPB isolated from hot, dry, and salty environments would themselves be salt and desiccation tolerant and be ideal candidates for protecting plants from salt and desiccation stress. Therefore, a salt-tolerant PGPB (isolated from the rhizosphere of plants growing in the Arava region of the Negev desert in Israel) that expressed ACC deaminase was tested for its ability to promote the growth of tomato plants at increasing levels of salt (Fig. 8.4). Not only did the bacterium decrease the production of ethylene by salt-treated tomato plants (to about 25% of the level in the absence of the bacterium), it also significantly increased tomato plant biomass production. Subsequently, a large number of other laboratories in several different countries around the world have performed successful variations of this experiment. Thus, many researchers have now isolated different ACC deaminase-containing PGPB strains and demonstrated their effectiveness in decreasing growth inhibition by various concentrations of salt with a variety of different plants. This body of evidence clearly demonstrates that using ACC deaminase-containing PGPB to lower plant ethylene levels is an effective strategy to protect plants against the salt caused inhibition of plant growth and development. Moreover, in addition to promoting salt tolerance under laboratory conditions, PGPB that produce ACC deaminase are also effective at promoting plant growth in the presence of salt in the field (Table 8.4).

ACC deaminase-producing bacteria are not limited to lowering plant growth inhibitory levels of ethylene. They have also been reported to increase the extent of seed germination, especially in the presence of high levels of salt. In one set of experiments, a number of endophytic PGPB strains that were previously isolated from within the tissues of the halophytic plant *Limonium sinense* (used in traditional Chinese medicine for the treatment of fever, hemorrhage, and other disorders) were

Table 8.3 The effect of treating droughted alfalfa plants with wild-type or an IAA-overproducing transformant of the PGPB *Ensifer meliloti*

Trait	Wild-type + no drought	Wild-type + drought	Transformant + no drought	Transformant + drought
Shoot fresh weight, mg	148	100	195	130
Nitrogenase activity, nmol C_2H_4 plant^{-1} min^{-1}	3.9	2.1	6.2	4.8
Ribulose bisphosphate carboxylase, pmol	0.82	0.79	1.20	0.96
Root sucrose content, mg/g dry weight	46.3	35.2	72.8	56.5

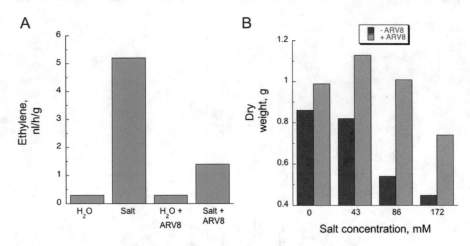

Fig. 8.4 The effect of salt and the PGPB *Achromobacter piechaudii* ARV8 on tomato plant ethylene production (**a**). The effect of salt and the PGPB *A. piechaudii* ARV8 on tomato plant biomass (**b**). The dry weights shown are per plant, comparing plants of the same age

Table 8.4 The effect of an ACC deaminase-containing strain of *Pseudomonas fluorescens* on the growth of groundnut (*Arachis hypogea*) plants in the field

Treatment	Plant height, cm	Number of pods/plant	Pod filling, %	Weight of 100 seeds, g	Total yield, kg/ha
No added bacteria	24.7	16.5	84.1	102.0	2352
Added bacteria	30.3	19.9	96.6	120.5	3002

Every one of these measurements yielded a statistically significant increase compared to the control without any added bacteria

tested for their effect on seed germination as well as plant growth. Of the 126 strains that were isolated, the behavior of four strains was examined in detail. The fours strains that were tested all produced ACC deaminase, synthesized IAA and siderophores, and were able to solubilize phosphate. While it was never demonstrated directly in this series of experiments, it was tacitly assumed that the key element in promoting germination and plant growth in the presence of high levels of salt was the presence of ACC deaminase. Regardless of how much of the stimulation of germination in the presence of high salt effect is due to the presence of ACC deaminase, all four of the endophytic PGPB that were tested significantly improved seed germination at the majority of the salt concentrations tested (Table 8.5).

In an effort to improve the growth of mungbean (*Vigna radiata*) plants under conditions of environmental stress (including high salt, drought, and high temperature), plants were co-inoculated with *Bradyrhizobium* sp. and three different PGPB (one at a time). The PGPB were all selected because of the high level of ACC deaminase that they produced as well as their resistance to high temperature,

Table 8.5 The effect of several different endophytic PGPB on the germination of *Limonium sinense* seeds in the presence of various amounts of salt

Bacterium added	0 mM NaCl	100 mM NaCl	200 mM NaCl	250 mM NaCl	300 mM NaCl	500 mM NaCl
None	83	30	33	28	15	12
Bacillus flexus	97	75	60	35	35	40
Isoptericola dokdonensis	85	53	43	28	37	28
Streptomyces pactum	78	58	55	42	38	27
Arthrobacter soli	77	51	42	45	35	25

The data presented are the percentage of the tested seeds that germinated

drought, and high salt concentrations. The impact of the three different PGPB on plant growth and development, under various stress conditions, was assessed. Following the stress, plant shoot, and root dry weight, nodule dry weight and nodule number were measured, and for all three stresses and with all three PGPB, the co-inoculation alleviated the bulk of the inhibition of plant growth that resulted from all of the stresses.

One group of researchers grew tomato plants for 3 weeks in the greenhouse, then stopped watering the plants for 7–12 days before watering of plants was resumed. Some plants were pretreated (prior to the imposition of drought conditions) with either *Pseudomonas putida* GR12-2, a PGPB previously isolated in the Canadian Arctic, *Achromobacter piechaudii* ARV8, a PGPB that had been isolated from the Negev desert in southern Israel, or with buffer alone. Both of these PGPB strains actively produced both IAA and ACC deaminase. As expected, tomato plants treated with strain GR12-2 grew to a much greater size than the control plants. Moreover, plants treated with strain ARV8 grew to an even larger size that was approximately five times the size of the control plants (Fig. 8.5). In this experiment, it was also shown that plants exposed to drought in the absence of added bacteria synthesized a higher amount of stress ethylene. From these experiments, it was concluded that bacterial ACC deaminase, possibly in concert with bacterial IAA, provided tomato plants with a significant level of protection against drought stress.

In one particular study, researchers showed that an endophytic strain of *Streptomyces* sp. (isolated from *Clerodendrum serratum*, an important medicinal plant) that contained ACC deaminase, could solubilize phosphate and produce siderophores (but not IAA) could help Thai jasmine rice plants overcome growth inhibition caused by 150 mM salt. To start, an ACC deaminase minus strain of the selected PGPB was constructed by insertional inactivation of the *acdS* gene. The wild-type PGPB (but not the ACC deaminase minus mutant) significantly increased plant growth, chlorophyll, proline, potassium, calcium, and water contents, and decreased plant ethylene levels, reactive oxygen species, and sodium levels both in the absence and presence of added salt. Gene expression patterns of plant genes by real-time PCR indicated that the expression of the plant enzyme ACC synthase was increased about fivefold in all salt-stressed plants compared to the non-salt-stressed controls. In

Fig. 8.5 The effect on tomato plant dry weight of inoculation either with no added bacterium, *Pseudomonas putida* GR12-2 (isolated from the rhizosphere of plants growing in the Canadian Artic) or *Achromobacter piechaudii* ARV8 (isolated from the rhizosphere of plants growing in the Negev desert). Both bacterial strains actively produce the enzyme ACC deaminase

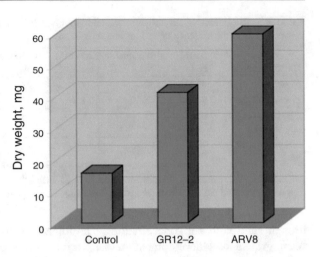

addition, the expression of the plant enzyme ACC oxidase was significantly lowered in the presence of salt compared to plants treated with the wild-type PGPB strain and increased 3–4 times in the presence of salt in plants treated with the ACC deaminase minus mutant strain. This work clearly demonstrates that, as expected, the presence of ACC deaminase alters plant gene expression resulting in a significant decrease in ethylene production and a consequent increase in plant growth.

It has been argued that ACC deaminase might be somewhat more effective if it were a secreted or a bacterial surface enzyme rather than a cytoplasmic enzyme. As a cytoplasmic enzyme one of the presumed limitations to the activity of this enzyme is the fact that the substrate ACC must be secreted or exuded outside of plant tissues and then taken up and transported into the ACC deaminase-containing bacterium. To overcome this perceived limitation, one research group genetically engineered ACC deaminase so that it was fused to another protein that was normally found to be anchored within the bacterial outer membrane. When this construct was expressed, the ACC deaminase portion of the molecule was now localized to the outer surface of two different endophytic PGPB strains (Fig. 8.6). Both engineered strains produced a significantly increased level of ACC deaminase, and in turn, inoculation of rice sprouts with the engineered strains promoted the growth of shoot and roots to a significantly greater extent than the wild-type both in the absence and presence of otherwise growth inhibitory levels of salt. Following these successful laboratory experiments, it will be interesting to see whether the engineered PGPB strains are as effective in the field as they are under controlled conditions.

A number of manuscripts have been published which clearly indicate that bacteria that synthesize IAA and/or produce the enzyme ACC deaminase can provide plants with a significant level of protection against both drought and salt stress. Based on this work, many researchers now take it for granted that for bacteria that synthesize IAA and/or produce the enzyme ACC deaminase, the mechanism of drought or salt protection has already been established and it is no longer necessary for them to generate IAA or ACC deaminase minus mutants in order to prove the involvement of

Fig. 8.6 Schematic representation of ACC deaminase, as part of a fusion protein, displayed on the surface of an endophytic PGPB thereby obviating the need for ACC to be transported across the bacterial cell membrane. The proteins are not drawn to size

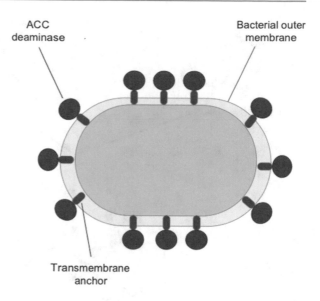

these mechanisms. Notwithstanding the fact that several other mechanisms may be involved in protecting plants against drought or salt inhibition, many workers are currently emphasizing the role of IAA and ethylene in drought and salt tolerance. In fact, there are now a large number of literature reports of plants being protected from drought stress by ACC deaminase/IAA including tomato, canola, pea, cotton, rice, chickpea, *Arthrocnemum macrostachyum* (a halophytic shrub), *Salicornia europaea* (common glasswort), peanut, bell peppers, cucumber, mung bean, ginseng, gray mangrove, soybean, barley, sunflower, sorghum, clover, pearl millet, groundnut, wheat, and maize. Moreover, depending on the plant and its intrinsic salt or drought tolerance, bacteria that produce IAA/ACC deaminase have been reported to promote plant growth at levels of salt as high as 200 mM NaCl.

If some ACC deaminase is beneficial to plant growth and survival during periods of stress, is the overproduction of this enzyme even more effective at protecting plants from stress? In one series of experiments, an ACC deaminase gene and its regulatory region from the PGPB *Streptomyces venezuelae* were cloned onto a plasmid and used to construct a strain of the wild-type bacterium that overproduced ACC deaminase. Both the wild-type and the overproducing strain were (separately) used to inoculate Thai jasmine rice plants that were grown in the presence of salt. The wild-type strain was found to reduce plant ethylene contents, decrease the uptake of sodium ions, and lower reactive oxygen species while simultaneously increasing the plant's potassium, chlorophyll, and biomass content. The ACC deaminase overproducing strain did not alter the chlorophyll content, but it did further increase the plant shoot length and biomass and as well as lowering the ethylene and sodium content of the plant compared to the wild-type strain. Based on these observations, an increased level of ACC deaminase activity provides an increased benefit to plants subjected to salt stress.

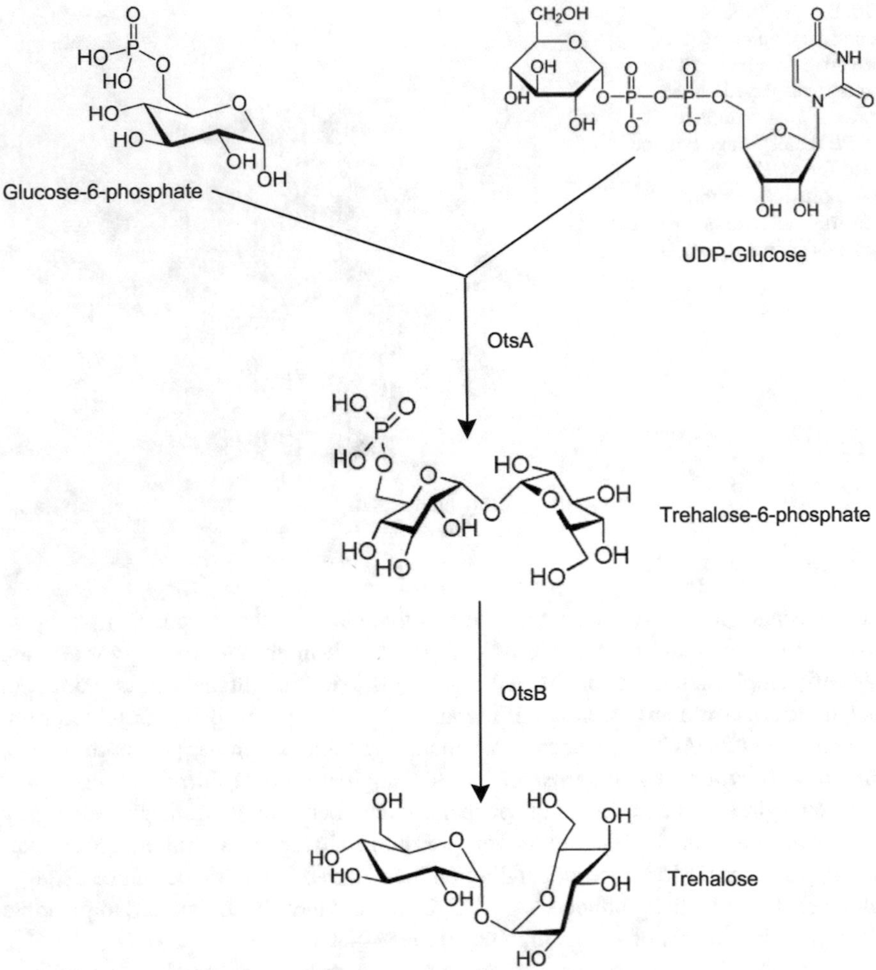

Fig. 8.7 Overview of one of the main pathways for the biosynthesis of trehalose. Here the synthesis of trehalose-6-phosphate from glucose-6-phosphate and UDP-glucose is catalyzed by the enzyme trehalose-6-phosphate synthase (OtsA) before trehalose-6-phosphate is converted to trehalose by the enzyme trehalose-6-phosphate phosphatase (OtsB)

8.2.3   Trehalose

Another way in which PGPB can protect plants from osmotic stress such as high salt and drought is through the synthesis of osmoprotectants such as proline and trehalose. Trehalose is an α,α-1,1-glucoside consisting of two molecules of α-glucose (Fig. 8.7). It is thought to form a gel phase (replacing water) as cells dehydrate thereby decreasing the damage to cells from drought and salt. Trehalose is a highly stable molecule that is resistant to both acid and high temperature. Salt or drought stress dramatically increases the expression of trehalose biosynthetic and biodegradative enzymes as well as the level of trehalose itself.

The most common biosynthetic pathway for trehalose is a two-step process in which glucose is first transferred from UDP-glucose to glucose-6-phosphate to form trehalose-6-phosphate and UDP in a reaction catalyzed by the enzyme trehalose-6-phosphate synthase encoded by the *otsA* gene (Fig. 8.7). Then, the trehalose-6-phosphate is dephosphorylated by the enzyme trehalose-6-phosphate phosphatase (encoded by the *otsB* gene). The first step in this synthesis reaction is rate-limiting, thus, trehalose may be overproduced by first isolating the *otsA* gene, and then the gene may be introduced into a trehalose-producing bacterium on a multi-copy number plasmid. Conversely, the synthesis of trehalose may be inhibited by disruption, removal, or other inactivation of the *otsA* gene. When several mutants of a strain of *Rhizobium etli* (a bacterium that nodulates bean plants) that both overproduce and underproduce trehalose were constructed, it was observed that an increase in trehalose caused an increase in the number of nitrogen-fixing nodules per plant while a decrease in trehalose resulted in a decrease in the nodule number (Fig. 8.8a). In addition, the trehalose overproducing strain had a higher level of nitrogenase activity than the wild-type while the underproducing strain exhibited a decrease in nitrogenase activity (Fig. 8.8b). Moreover, higher levels of trehalose had a dramatic effect on the survival (Fig. 8.8c) and very large positive effect on the yield (Fig. 8.8d) of bean plants nodulated by these rhizobial strains (compared to either the wild-type or the underproducing mutant strain), especially following periods of imposed drought.

In another series of experiments (Fig. 8.9), a strain of *A. brasilense* was transformed with a plasmid carrying a trehalose biosynthesis gene-fusion (i.e., *otsA* and *otsB*) from the yeast *S. cerevisiae*. The transformed strain grew well at 0.5 M NaCl (an extremely high level of salt) and accumulated trehalose, whereas wild-type *A. brasilense* did not tolerate this level of osmotic stress or accumulate trehalose. The reason for making a fusion gene is to ensure the optimal synthesis of trehalose without having to be concerned about producing too much or too little of one or the other of the trehalose biosynthesis proteins. When maize (corn) plants were inoculated with the strain of *A. brasilense* carrying this fusion gene, 85% of them survived drought stress compared to 55% of plants that survived when they were inoculated with the wild-type strain. Moreover, a 73% increase was observed in the biomass of maize plants inoculated with transformed *A. brasilense* compared with inoculation with the wild-type strain and subjected to drought conditions. There was also a significant increase of both leaf and root length in maize plants inoculated with the transformed *A. brasilense*. Therefore, from this experiment it is possible to conclude that inoculation of maize plants with *A. brasilense* containing higher levels of trehalose confers a significantly increased level of drought tolerance compared to the wild-type bacterium, and a significant increase in leaf and root biomass.

Since there are several mechanisms that PGPB can use to improve plant growth and metabolism in the presence of environmental stress, this raises the question as to whether any of these mechanisms can work synergistically. In one recent experiment, a group of researchers asked the question whether bacteria that produce both trehalose and ACC deaminase might synergistically protect tomato plants against salt stress. Starting with the PGPB strain *Pseudomonas* sp. UW4 which synthesizes

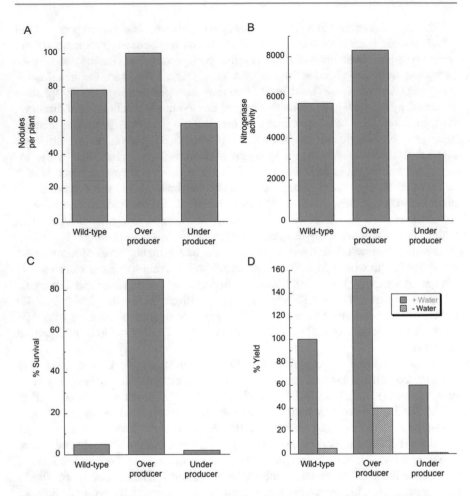

Fig. 8.8 The effect of wild-type, overproduced and underproduced levels of *Rhizobium etli* encoded trehalose on bean plant performance including the number of nodules formed per plant (**a**), the level of bacterial nitrogenase activity (**b**), plant survival following drought stress (**c**) and plant yield (normalized to the watered plants treated with the wild-type bacterium) following either watering or drought (**d**)

both trehalose and ACC deaminase, researchers constructed mutants that were devoid of either one or the other or both of these activities. In addition to testing these mutants and the wild-type strain, these researchers also constructed a strain that overproduced the *treS* gene thereby overproducing trehalose. In testing these strains, it was first shown that the survival of bacteria in the presence of high salt that had lost either of these genes was decreased ~50–1500-fold compared to the wild-type in the presence of salt (Table 8.6). Moreover, the survival of the double mutant decreased ~7500-fold in the presence of high salt while the survival of the trehalose overproducing strain was increased ~7.5-fold. This data is consistent with ACC

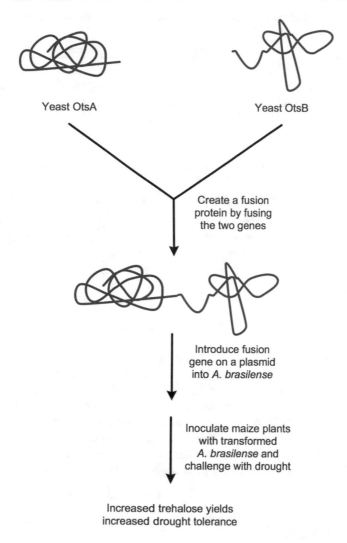

Fig. 8.9 Creating and testing a fusion protein that catalyzes trehalose synthesis

deaminase, and to a somewhat greater extent trehalose, facilitating the survival of this bacterium in the presence of high salt. More importantly, while the wild-type bacterium promoted a twofold increase in tomato plant growth in the presence of 200 mM NaCl compared to the control where no bacterium was added, the two single mutant strains each promoted only a small increase in plant growth (compared to no bacterial addition) while the double mutant did not promote any increase in plant growth (Table 8.7). Interestingly, the trehalose overproducing strain promoted a two and a half-fold increase in plant growth in the presence of 200 mM salt. These results clearly demonstrate that bacterial ACC deaminase and trehalose work

Table 8.6 Survival of the PGPB strain *Pseudomonas* sp. UW4 and several of its mutants in the presence of a high salt concentration

Strain	Survival without added salt, cells	Survival with 0.8 M salt, cells
Wild-type	2.2×10^8	2.8×10^6
acdS⁻	1.1×10^8	4.5×10^4
treS⁻	0.9×10^7	1.8×10^8
acdS⁻ and *treS⁻*	0.2×10^7	3.8×10^2
treS overexpression	3.2×10^8	2.1×10^7

Table 8.7 Growth of tomato plants treated with PGPB strain *Pseudomonas* sp. UW4 and several of its mutants in the presence of a 0.2 M salt

Strain	Plant dry weight with no added salt, g	Plant dry weight with 0.8 M salt, g
No bacterium	0.62	0.40
Wild-type	0.81	0.81
acdS⁻	0.62	0.50
treS⁻	0.65	0.55
acdS⁻ and *treS⁻*	0.62	0.40
treS overexpression	0.83	0.99

synergistically to facilitate plant growth in the presence of salt stress. Moreover, in this case it is not necessary to genetically engineer this PGPB strain as the wild-type bacterium already possesses both of these activities. These results suggest that PGPB that are found to confer high levels of salt tolerance to plants might be routinely screened for both ACC deaminase activity and trehalose synthesis to ensure the maximum benefit from PGPB addition to plants experiencing drought or high levels of salt. Finally, it remains to be demonstrated whether these two traits act synergistically under field conditions to help protect plants against salt and/or drought stress.

8.2.4   Cytokinin

Cytokinins are phytohormones that control many of a plant's physiological and developmental processes. In addition, it has been known for many years that treatment of plants with exogenous cytokinin can dramatically postpone their senescence. More recently, when scientists transformed plants with the *Agrobacterium tumefaciens ipt* gene (encoding the enzyme isopentenyltransferase) fused to various plant promoters, they observed that this was an efficient strategy in developing transgenic plants for increased yield by delaying the senescence of the whole plant and for protecting plants against drought and salt stress. Detailed investigation of transgenic *Arabidopsis* plants showed that elevated cytokinin levels leads to stress tolerance by maintaining the expression of genes that are normally associated with

Fig. 8.10 The effect on plant shoot biomass of inoculating alfalfa plants with *Sinorhizobium meliloti* non-transformed (wild-type), transformed with an *ipt* gene under the control of the *E. coli lac* promoter or with an *ipt* gene under the control of the *E. coli trp* promoter. Plants were weighed prior to the imposition of drought conditions, during drought and subsequent to drought after the plants had been rewatered

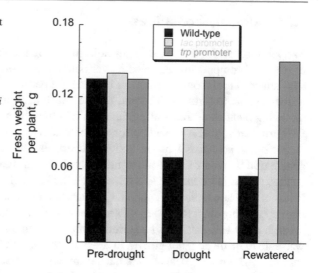

plant growth and metabolism and whose expression would otherwise decrease under stressful conditions.

Based on the successful experiments with transgenic plants, scientists engineered a strain of *Sinorhizobium meliloti* to overproduce cytokinin and then tested the engineered bacteria for the ability to protect alfalfa plants against the senescence that results from drought stress. The engineered strains, which expressed the *Agrobacterium ipt* gene under the control of either the *E. coli lac* or *trp* promoters, synthesized more zeatin (cytokinin) than the control strain under free-living conditions. The transformant that expressed the *ipt* gene under the control of the *lac* promoter produced approximately three times the level of cytokinin synthesized by the non-transformed bacterial strain while the transformant that included the stronger *trp* promoter produced nearly five times the level of cytokinin of the non-transformed strain. Following a 4-week inoculation period, the effects of the engineered strain that included the *trp* promoter on alfalfa growth were similar to those of the control (non-transformed) strain under non-drought conditions (Fig. 8.10). After being subjected to severe drought stress, the alfalfa plants inoculated with the engineered strains were significantly larger than the plants inoculated with the non-transformed strain with the strain with the *trp* promoter providing the greatest benefit to plants. When the plants subjected to drought were rewatered, under the conditions tested, plants inoculated with the strain with the *trp* promoter controlling the *ipt* gene grew to a level similar to plants that had not been drought stressed at all. This experiment is a clear demonstration of the ability of engineered rhizobial strains synthesizing higher levels of cytokinin to improve the tolerance of alfalfa (and possibly other crops) to severe drought stress. Given the reluctance of many regulators to approve the use of genetically engineered bacterial strains in the field, it will be interesting to determine whether rhizobial strains producing high cytokinin levels can either be directly selected from environmental samples or altered by conventional mutagenesis or CRISPR to perform this function.

8.2.5 Exopolysaccharides

Exopolysaccharides (EPS) are high-molecular weight compounds, of a wide range of chemical composition, secreted by various microorganisms into the environment. EPS attached to the outer surface of bacteria are partially responsible for the ability of the bacteria to form biofilms and for the attachment of bacterial cells to surfaces including plant roots and soil particles. In addition, by surrounding a bacterial cell EPS provides physical protection for the bacterium against high levels of salt. In one recent study, scientists isolated and characterized 110 new PGPB from the rhizosphere of a number of local salt contaminated soils in India (Fig. 8.11). The PGPB strains, all pseudomonads, were selected on the basis of the ability to produce high levels of EPS and to be salt tolerant. One particular strain, characterized by sequencing its 16S ribosomal RNA gene, *Pseudomonas aeruginosa*, was found to be able to grow in up to 12% salt. This bacterium, which also produced IAA and siderophores, was able to facilitate plant growth in the presence of (otherwise inhibitory levels of) salt. Chemical mutagenesis of the bacterium enabled researchers to select a mutant that produced a significantly decreased level of EPS (while the amount of IAA and siderophores was unchanged). Concomitant with the decrease in EPS, the mutant lost its ability to grow and promote plant growth in the presence of high levels of salt. This result is interpreted in terms of the EPS limiting the amount of salt that can enter the bacterium. With the loss of the EPS barrier, salt readily enters the bacterium that is now no longer able to promote plant growth in the presence of salt. These researchers believe that the stimulation of plant growth that is observed comes from IAA synthesis. However, they failed to assess the level of ACC deaminase activity or the possible presence of bacterially produced cytokinin. Given the fact that *P. aeruginosa* is a disease-causing bacterium, while this work helps to develop an understanding of the role of EPS, this bacterium cannot be used to commercially promote the growth of crop plants. Finally, several other workers have argued that it is precisely the production of EPS together with ACC deaminase activity that is responsible for conferring salt tolerance.

In another study, researchers isolated a halophilic bacterium, *Halomonas* sp., from the rhizosphere of mangrove plants and showed that it could readily grow in medium that contained 8% salt. Moreover, this growth was attributed to the copious amounts of EPS that this bacterium produced in the presence of salt; the bacterium produced ~1 g EPS/kg dry weight bacteria in the absence of salt and ~35 g EPS/kg dry weight bacteria in the presence of 8% salt. Clearly, not every PGPB can produce such high levels of EPS; however, under certain environmental conditions (such as in rice paddies) these organisms may be in a unique position to confer salt tolerance unto treated plants.

8.2.6 Gibberellins

One group of scientists endeavored to assess the possible role of gibberellins on the ability of certain PGPB to overcome plant growth inhibition caused by salt and

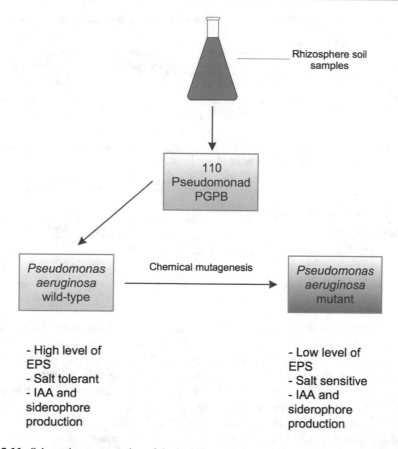

Fig. 8.11 Schematic representation of the isolation and characterization of an exopolysaccharide (EPS) producing strain of *Pseudomonas aeruginosa* and the relationship between EPS production and bacterial salt tolerance

drought stress. For a start, these researchers isolated several potential PGPB strains from a variety of agricultural fields in Korea. They then tested the isolated strains for the ability to promote the growth of a strain of gibberellin biosynthesis deficient mutant rice. Once the rice seedlings had attained the two leaf stage (~10 days after seed germination) a small aliquot of the culture filtrate from the isolated bacteria was applied to the apex of the plant seedlings. One bacterial strain that stimulated the growth of rice seedlings following 1 week of incubation (Table 8.8) was selected for additional study. This strain was determined to be *P. putida* and to produce the following gibberellins: GA1, GA4, GA9, and GA20 (where GA1 and GA4 are bioactive). In a key experiment, salt stress was induced by the addition of 120 mM NaCl to soybean plants while drought stress was induced by the addition of 15% polyethylene glycol. Under both sets of conditions, the addition of the newly isolated bacterium to the stressed soybean plants resulted in a significant promotion of plant growth (Table 8.9). While it is clear that this PGPB strain both promoted plant

Table 8.8 The effect of adding a newly isolated PGPB to gibberellin deficient mutant rice seedlings

Treatment	Shoot length, cm	Shoot fresh weight, g	Shoot dry weight, g	Chlorophyll content
Negative control	7.03	0.72	0.11	28.3
PGPB	7.71	0.79	0.13	29.7

Table 8.9 Effect of adding a gibberellin producing PGPB on the growth of soybean plants not stressed or stressed by either drought or salt

Treatment		Shoot length, cm	Plant fresh weight, g	Chlorophyll content
Control	No bacteria	30.7	16.3	33.1
	+ PGPB	36.0	17.6	36.9
Drought	No bacteria	26.6	14.4	32.0
	+ PGPB	30.2	16.2	32.6
Salt	No bacteria	25.2	14.3	29.6
	+ PGPB	28.0	16.3	30.8

growth and produced gibberellin, these researchers did not assay this bacterium to determine whether or not it contained other mechanisms of plant growth promotion. Therefore, although it is likely that the gibberellins produced by this bacterium played a role in promoting plant growth in the presence of salt and drought stress, it is unclear to what extent plant growth under these conditions may be attributed to bacterial gibberellin synthesis. Moreover, it would be of interest to determine whether gibberellin biosynthesis and other plant growth-promoting mechanisms could act synergistically in overcoming drought and salt stress.

8.2.7   Volatile Organic Compounds

One group of researchers demonstrated that the endophytic PGPB strain *Paraburkholderia phytofirmans* PsJN was able to stimulate the growth of *Arabidopsis thaliana* plants without there being any direct physical contact between the plant and the bacterium. That is, in this experiment the plant and the PGPB were grown in separate halves of a divided Petri plate with a physical separation between the two halves; the bacterial inoculum was between 1×10^4 and 1×10^6 CFU. Interestingly, plants that were treated with 1×10^6 CFU in a divided Petri plate and then were transplanted to soil grew at a similar rate to plants that were directly inoculated with strain PsJN. In addition, when plants that were treated with strain PsJN in a divided Petri plate and were transplanted to soil with 200 mM NaCl and 20 mM $CaCl_2$ they demonstrated the same tolerance to the salt as plants that had been constantly in contact with the bacterium. The volatile compounds responsible for this behavior were determined to be 2-undecanone, 1-heptanol, 3-methyl-butanol, and dimethyl disulfide. When germinating seeds were placed in divided Petri plates along with varying amounts of these organic compounds a comparable level of plant growth promotion and salt tolerance to what was observed with strain PsJN was found. Although a number of PGPB have been reported to synthesize volatile organic compounds that can promote plant growth, including, in this case, in the presence of high levels of salt, the mechanism of this growth promotion is unclear at the present time.

8.2.8   Trace Element Deficiency

A number of abiotic stresses including drought, heat stress, and a lack of trace element availability are major causes of oxidative damage in plants. High moisture conditions typically lead to higher levels of bioavailable metals in soil, potentially leading to growth inhibitory levels of metals in plants. On the other hand, low moisture conditions often result in the lack of plant uptake of some minerals which can reduce certain plant metabolic functions and thereby inhibit plant growth. Plants often become particularly susceptible to oxidative stress as the plant's antioxidative defense enzymes (e.g., catalases, peroxidases, and superoxide dismutases) require metal co-factors to function. In this regard, one of the ways that PGPB enhance plant growth is by increasing nutrient availability. In addition, many PGPB have their own antioxidant enzymes that can help to reduce a plant's oxidative stress and some of these enzymes respond differently than comparable plant enzymes to trace element deficiency. Thus, for example, some bacterial catalases lack a heme prosthetic group and instead contain a dimanganese cluster. As a result, conditions that inhibit a plant's antioxidative defense enzymes may not inhibit bacterial enzymes with similar activity.

8.3   Flooding and Anoxia

Deprivation of oxygen to the roots of plants is the main negative consequence of flooding. Under normal conditions, water dissolves about 230 mmol m^{-3} oxygen, and hypoxia occurs when the oxygen level falls below 50 mmol m^{-3}. Typically, shortly after the onset of flooding, soil microorganisms consume all of the available oxygen (anoxia). Plants subsequently respond to flooding with a wide range of physiological changes including adventitious root formation along the submerged stem portion, reduction in stem growth, wilting, chlorosis (loss of green coloration because of loss of chlorophyll), epinasty, and leaf abscission. Stressful, low oxygen levels stimulate ACC synthesis catalyzed by the enzyme ACC synthase and simultaneously arrest ACC oxidation to ethylene (by the enzyme ACC oxidase) in waterlogged plant roots as this second step requires oxygen. Stress-induced ACC is synthesized in anoxic plant roots and is then transported in the xylem to the shoots and leaves where it is rapidly oxidized to ethylene and subsequently induces epinastic leaf curvature, loss of biomass, and a decrease in the level of leaf

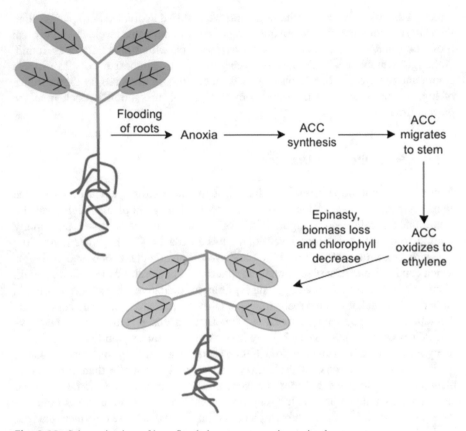

Fig. 8.12 Schematic view of how flooded roots cause epinasty in plants

chlorophyll (Fig. 8.12). ACC induces the expression of ACC oxidase genes in plant shoots. Despite the increase in ACC from flooded roots, ethylene synthesis in shoots of flooded plants is unaffected within the first photoperiod following the flooding. The time lag, from the appearance of ACC at the stem base until the ethylene burst in the upper leaves (typically around 12 h), is a characteristic of flooding stress. This delay might be partly explained by the formation of N-malonyl ACC since the N-malonyl ACC level increases in both roots and shoots of flooded plants as does the conjugation of ACC to glutamic acid.

The inhibition of plant growth that occurs as a consequence of flooding and anoxia may be partially alleviated by first treating the plants with ACC deaminase-containing PGPB. For example, in one series of experiments, tomato plants were grown for 2 months in a greenhouse with either no added bacteria, one of three different ACC deaminase-containing PGPB strains or a mutant of one of the PGPB strains that had been engineered to not manufacture ACC deaminase. At the end of this growth period, the dry weights of the plants that were treated with an ACC deaminase-containing PGPB were 10–20% greater than plants that did not come into

Fig. 8.13 The effect of ACC deaminase-containing PGPB on 2-month-old tomato plant height and shoot weight in (**a**) the absence of flooding and (**b**) the presence of flooding

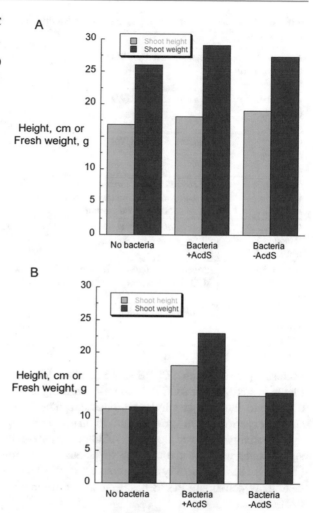

contact with this enzyme (Fig. 8.13a). This small, but significant, effect reflects the fact that under good conditions most PGPB provide only a very small benefit to plants with which they are associated. On the other hand, if tomato plants that were 2 months old were flooded for 9 days, they produced increased levels of ethylene and progressively lost biomass (Fig. 8.13b). On the other hand, plants that were treated with an ACC deaminase-containing PGPB strain lost only a small amount of biomass. In addition, following flooding, the plants that were treated with an ACC deaminase-containing PGPB strain had nearly double the chlorophyll content of plants that were either not treated with bacteria or treated with a PGPB strain that was unable to produce ACC deaminase. Moreover, flooding increased the ethylene production by tomato plant petioles (the stalk that attaches the leaf blade to the stem) by approximately five- to sixfold when the plants were either not treated with

Table 8.10 ACC deaminase-producing bacteria protecting plants from flooding stress

Plant	Bacteria	Conditions
Tomato (*Lycospersicon esculentum*)	*Pseudomonas* sp. UW4	Waterlogging stress applied on 55-day old tomato plants
Cucumber (*Cucumis sativus*)	*Pseudomonas* sp. UW4	Tested 72 h of hypoxic vs. non-hypoxic conditions
Canola (*Brassica napus*)	*Pseudomonas* sp. UW4	Field study with transgenic plants and high levels of Ni
Holy Basil (*Ocimum santum*)	*Achromobacter xylosoxidans, Serratia ureilytica, Herbaspirillum serpedicae, Ochromobactrum rhizosphaerae*	Waterlogging of 2 cm above the soil surface for 15 days
Chickpea (*Cicer arietinum*)	*Mesorhizobium ciceri* LMS-1	Waterlogging of 1 cm above the soil surface for 7 days
Marsh dock (*Rumex palustris*)	*Pseudomonas putida* wild-type and ACC deaminase minus mutant	Complete submergence for 3 days for short term stress and 17 days for long-term stress
Sesame (*Sesamum indicum*)	*Pseudomonas veronii* KJ	Waterlogging up to the soil surface for 10 days

bacteria or treated with a PGPB strain that was unable to produce ACC deaminase. Plants treated with an ACC deaminase-containing strain exhibited a two- to two and a half-fold increase in ethylene production. The bottom line of this and several similar experiments on various plants is that by lowering the production of ethylene that invariably followed flooding, ACC deaminase-containing PGPB were able to significantly decrease the damage to plants that would have otherwise occurred (Table 8.10). Importantly, the protection of plants against flooding by ACC deaminase-containing PGPB has been shown to be effective both in the lab and in the field.

8.4   Natural Selection

It has been widely demonstrated in laboratory experiments that many different plants exhibit a much higher level of resistance to a large range of environmental stresses when those plants have first been inoculated with ACC deaminase-containing PGPB. Based on these many observations, it is reasonable to ask whether these ACC deaminase-containing PGPB provide any advantage to uncultivated plants growing in the natural environment. In one key experiment, researchers isolated and characterized spore-forming PGPB from the rhizosphere of wild barley (*Hordeum spontaneum*) plants growing in a region of northern Israel termed "Evolution Canyon." The two slopes of Evolution Canyon, which is located at Mount

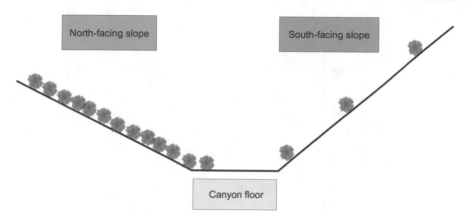

Fig. 8.14 Schematic representation of a cross section of Evolution Canyon

Table 8.11 Endospore forming bacterial distribution in the rhizosphere of wild barley from contrasting microclimates

	Number of isolates	Number using ACC	Biofilm formation	Phosphate solubilization	Halophilic growth
South-facing slope	35	35	30	16	28
North-facing slope	32	0	0	0	0

The isolated bacteria included *Paenibacillus polymyxa, Bacillus megaterium, Bacillus cerus* and *Bacillus pumilus*

Carmel near Haifa, are around 250 m apart at their bases. The south-facing slope is quite arid, has only very sparse plant growth and receives eight times the amount of sunlight as the north-facing slope that is more like a European forest and includes relatively lush plant growth (Fig. 8.14). While both the south- and north-facing slope contained similar bacterial genera and species in the wild barley rhizospheres, the bacteria from the much more highly stressed south-facing slope harbored numerous traits to make them more stress resistant (Table 8.11). These traits included the ability to utilize ACC as a nitrogen source (i.e., ACC deaminase activity), the ability to form biofilms, halophilic growth, and phosphate solubilization activity. This data has been interpreted as suggesting that these rhizosphere bacteria, together with the plant roots at the south-facing slope, appear to function as complex communities that afford the plant the ability to withstand the harsh conditions encountered. In other words, the drought conditions on the south-facing slope selects for bacteria that contain traits that allow both the PGPB and their plants hosts to better survive these harsh conditions. In addition, under the more moderate conditions found on the north-facing slope, while the same bacteria are present, they have mostly lost those traits that facilitate bacterial and plant survival under harsh conditions.

Several other recent studies have examined how, in the environment, overall plant fitness was largely a function of the soil microbial community. In this regard, it was

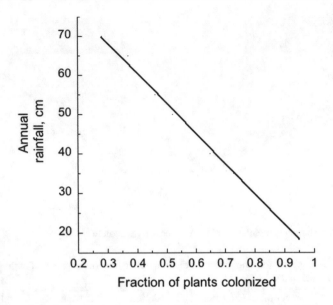

Fig. 8.15 A graph depicting the inverse relationship between the annual rainfall at various sites throughout Washington and Oregon and the fraction of wheat plants that were colonized by *Pseudomonas* spp. which encode phenazine-1-carboxylic acid

observed that plants were most fit when their current environmental conditions were the same as the historical environmental conditions of the associated microbial community. Thus for example, under drought conditions, plants were most fit when the soil microbial community was from soil where drought conditions were common rather than from well-watered soil. In one study, researchers assessed the geographic distribution of *Pseudomonas* spp. that produced the antibiotic phenazine-1-carboxylic acid in the rhizosphere of wheat plants grown in portions of the US states of Washington and Oregon. This study included an assessment of bacterial strains from 61 different commercial wheat fields within an area of approximately 22,000 km^2. In the first instance, every one of the fields that were sampled contained *Pseudomonas* spp. Secondly, a large proportion of the sampled fields contained *Pseudomonas* spp. that produced phenazine-1-carboxylic acid. Moreover, researchers noted that there was an inverse relationship between the annual rainfall at any particular site and the proportion of plants at that site that were colonized by phenazine-1-carboxylic acid-producing *Pseudomonas* spp. (Fig. 8.15). That is, the soil from areas where drought was common contained many more phenazine-1-carboxylic acid-producing *Pseudomonas* spp. than soils that were well watered. This result is consistent with the notion that environmental stress conditions (in particular drought) somehow selects for the presence of PGPB that decrease plant stress including the pressure from plant pathogenic fungi.

The observed behavior of plants and microbes in a natural environment has some important implications for plant survival and for agricultural productivity in an era of rapid and potentially extensive climate change. That is, it has been suggested, in attempting to respond to climate change, plants are not necessarily limited to strategies in which, to survive, they must either adapt (to the changed conditions) or else disperse (to locations where conditions are more favorable for their optimal growth). Rather, plant survival may be intricately dependent upon plant interaction with specific microbial communities.

8.4.1   Monitoring ACC Deaminase in Soil

To better understand the role of PGPB that contain ACC deaminase in natural environments it would be useful to have a simple and direct means of assessing their presence in various soils. To do this, one group of researchers developed an *acdS*-based PCR method to monitor *acdS* alleles in natural soils and plant systems. As a key step in the development of appropriate PCR primers, it was necessary to discard DNA sequences coding for D-cysteine desulfhydrases which have very similar DNA sequences to ACC deaminases but notably vary in five key amino acids that are known to be important for ACC deaminase activity. Initial results indicated that, as expected, PGPB that contain ACC deaminase are significantly enriched in the rhizosphere as compared to the bulk soil. In addition, when examining different species of *Poaceae* (a large family of monocotyledonous plants commonly known as grass), *acdS* transcript levels in the plants' rhizospheres varied from one species to another. This data provided initial evidence for a link between past *Poaceae* evolution and the functioning of root-associated PGPB. While this work is preliminary, it reinforces the notion of a strong linkage between the growth of various plants and the presence of PGPB in natural environments.

8.5   Cold Temperature

PGPB that can stimulate plant growth in the lab will probably not have any significant impact on plants in the field unless the bacteria can grow and persist in the natural environment. In countries with cold climates like Canada, these bacteria must be able to survive long cold winters and then grow at cool temperatures in the spring—in many cold countries, crops are planted in the spring when the soil temperature is 5–10 °C. In cold habitats, the main obstacle for soil bacteria to overcome under freezing conditions is typically a decrease in water availability (i.e., desiccation). Water dehydration can adversely affect the structure, dynamics, stability, and function of biological molecules and subsequently, cellular processes required for survival. In addition, formation of ice crystals, especially if the ice crystals form within a bacterial cell and lyse the cell membrane, can lead to cell death from the physical rupture of the cell. Moreover, the changes in osmotic pressures upon freezing can rupture cells.

Freezing and recrystallization rates affect the amount of damage incurred by a bacterial cell. Typically, a slower freezing rate is more detrimental to bacterial cells than flash freezing. Also, a high recrystallization rate is damaging. Smaller ice crystals with large surface areas are thermodynamically unstable, and thus they tend to aggregate and recrystallize thereby forming physically more damaging larger ice crystals.

8.5.1 Antifreeze Proteins

The main line of defense against freezing stress for bacteria is the production of cryoprotective molecules including low molecular weight cryoprotectants, antifreeze proteins, and (debatably) ice nucleation proteins. Common low molecular weight cryoprotectants include a number of alcohol sugars such as trehalose, glycerol, and sorbitol; sugars, such as lactose and sucrose; peptides; and amines such as glycine, betaine, arginine, and proline. Secreted cryoprotectant molecules can colligatively depress the freezing point of ice formation inducing a state of supercooled water without ice formation. Some of these molecules (such as trehalose and proline) also act as osmoprotectants facilitating an osmotic balance between the cell and the environment. Alternatively, in combination with cryoprotectants, some soil bacteria can protect themselves from freezing damage via the synthesis and secretion of antifreeze proteins that regulate the formation of ice crystals outside of the bacterium.

Antifreeze proteins adsorb onto the prism face of an ice crystal and limit ice growth along the alpha axes (Fig. 8.16). In concentrated solution, the preferred direction of ice crystal growth is along the c-axis forming a hexagonal bipyramid. Importantly, bacterial antifreeze proteins structure ice crystals so that they remain small and do not pierce the bacterial cell membrane.

To survive over-wintering in very cold environments, many bacteria secrete both ice nucleation and antifreeze proteins. In this way, ice crystals are formed in a controlled manner, outside of the bacterial cells. Moreover, in some instances, both antifreeze and ice nucleation activity are found on the same protein.

Only a very limited number of bacterial antifreeze proteins have been isolated and characterized. Moreover, even fewer bacterial antifreeze protein genes have been isolated and characterized. To isolate and characterize a bacterial (*Pseudomonas*) antifreeze protein gene produced by a PGPB strain the following procedure was utilized (Fig. 8.17). First, the bacterium was grown overnight at 30 °C (to produce a moderate amount of microbial biomass) and then for an additional 5 days at 4 °C (to induce antifreeze protein synthesis). The antifreeze protein, which was secreted to the growth medium, was purified using standard column chromatography and then partially digested with trypsin. It is important to note here that Gram-negative bacteria such as *Pseudomonas* spp. only secrete a relatively small number of proteins to the growth medium so that it is a relatively simple matter to purify antifreeze protein from the growth medium. The partial amino acid sequences of some of the peptide fragments (resulting from the trypsin digestion) of the antifreeze protein were determined. PCR primers based on the determined amino acid sequences were designed and then were used to amplify a portion of the antifreeze protein gene. The DNA sequence of this gene fragment was determined. The bacterial chromosomal

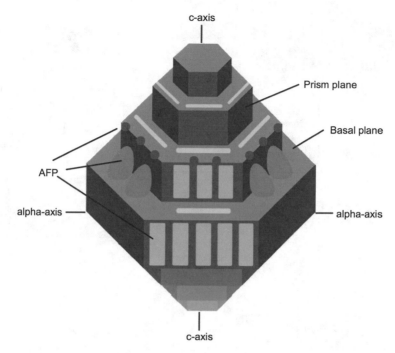

Fig. 8.16 Schematic view of ice crystal formation and the binding of antifreeze protein (AFP) to the crystal surfaces

DNA was partially restriction enzyme digested, diluted, and then self-ligated to form circles of DNA. At this point, a second set of PCR primers (i.e., inverse PCR primers) was then used to amplify the remainder of the antifreeze protein gene.

At the present time, fewer than ten bacterial antifreeze genes have been isolated and characterized. Unfortunately, the DNA sequences of these genes are quite different from one another. Thus, it is not yet possible to examine the database of psychrotrophic and/or psychrophilic bacteria DNA genomic sequences and ascertain with any certainty which open reading frames encode antifreeze proteins.

8.5.2   Chilling Stress

Plants may be stressed, not only when the temperature is increased but also when the temperature is decreased. This might occur when a cold nighttime temperature follows a warm daytime temperature, especially during the spring or fall season of the year. The resultant chilling stress can cause electrolytes to leak from plant roots, photosynthesis to decrease and an increase in the amount of the amino acid proline that the plant produces. The presence of some PGPB strains has been observed to limit the extent of damage to plants from chilling stress despite the fact that the

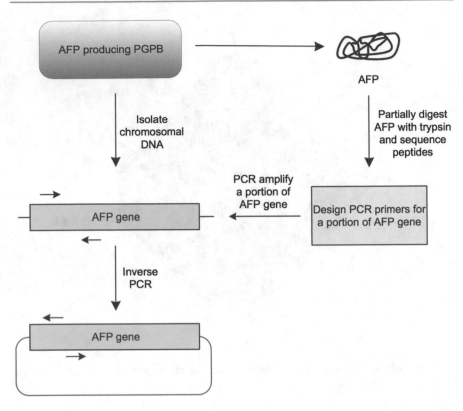

Fig. 8.17 Overview of a scheme used to isolate a bacterial antifreeze protein gene

mechanism of this protection is unclear although it could be related to the ability of the added bacterium to limit stress ethylene synthesis. For example, one group of researchers used the endophytic PGPB *Paraburkholderia phytofirmans* PsJN to protect grapevines against chilling stress and found that the bacterium significantly decreased the damage to plants when the temperature was lowered from 26 °C to 4 °C (Fig. 8.18). Thus, during chilling stress (i.e., at 4 °C), the presence of the PGPB strain resulted in grapevine plants that leaked significantly fewer electrolytes and fixed more carbon than untreated plants. Unfortunately, these workers did not ascertain the mechanistic basis behind the ability of the PGPB strain to protect plants against chilling stress. However, one might speculate that the ability of the endophytic strain PsJN to lower ethylene levels via ACC deaminase and/or stimulate plant growth through IAA production could be responsible for this chilling stress protection. To prove that ACC deaminase and/or IAA rather than a trait(s) that is unique to strain PsJN is responsible for this behavior, it would be necessary to repeat these experiments comparing the behavior of the wild-type strain and ACC deaminase and/or IAA minus mutants of this bacterium.

In a more definitive experiment regarding the role of ACC deaminase in overcoming chilling stress, other researchers introduced a foreign ACC deaminase

Fig. 8.18 Effect of chilling stress (4 °C; cool) on % electrolyte leakage or % of photosynthesis level of grapevine plantlets treated with the endophytic PGPB, *Paraburkholderia phytofirmans* PsJN. For both parameters, the data was normalized with the values at 26 °C (warm) being set to 100% where this normalized value is equivalent to 13% electrolyte leakage or 0.47 μmoles CO_2 cm^{-2} s^{-1}

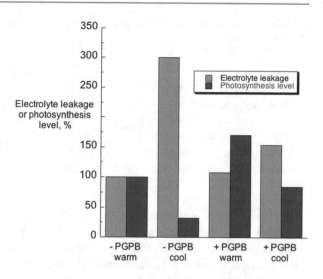

gene into a bacterium that lacked this activity and then inoculated tomato plants with either the wild-type or the transformed bacterium. The tomato plants were subsequently subjected to chilling stress (growth for 1 week at 10–12 °C). The foreign ACC deaminase gene was introduced into a psychrotolerant (able to grow at temperatures between 20 °C and 0 °C) strain of *Pseudomonas frederiksbergensis* along with its native regulatory region on a low copy number broad-host-range plasmid. When plants were inoculated with the transformant with an ACC deaminase gene, several indicators of chilling stress were reduced compared to treatment of plants with the wild-type bacterium (which did not contain ACC deaminase). These indicators included plant dry weight, plant ethylene levels, ACC oxidase activity, ACC accumulated in plants, and the expression of two cold-induced genes (Table 8.12). In summary, all of this data is consistent with the presence of ACC deaminase in the transformed bacteria lowering the chilling stress in the treated tomato plants.

8.6 Bacterial Cooperation

As discussed in Chap. 2, soil bacteria typically act as consortia so that to understand how a particular bacterium affects plants in the environment it is often necessary to understand how those bacteria interact with other soil bacteria, with mycorrhizae (see Chap. 9), and with the soil itself. One group of researchers has suggested that it may be possible to engineer consortia of cyanobacteria, microalgae (unicellular microscopic algae that include an estimated ~200,000–800,000 species, mostly not described), and bacteria to be functional in desert environments thereby improving soil fertility, water preservation, primary production, and soil stability. Deserts are characterized as having annual rainfall of less than 254 mm (10 inches); of these hyperarid and arid deserts (17.2% of global land area) are biologically unproductive

Table 8.12 Tomato plant traits altered by treatment with either a wild-type or transformed strain of *P. frederiksbergensis*

Trait	Control	Wild-type	Transformant
Plant dry weight, g	3.3	3.6	4.5
Ethylene, nmol mg $DW^{-1} h^{-1}$	27	29	15
ACO activity, pmol ethylene g $FW^{-1} h^{-1}$	13	12	9
ACC accumulation, nmol ACC g FW^{-1}	320	200	70
LeCBF1, normalized expression level	0.60	0.55	1.40
LeCBF3, normalized expression level	0.60	0.55	1.45

DW dry weight, *FW* fresh weight, *ACO* ACC oxidase, *LeCBF1*, and *LeCBF3* are cold-induced genes

while semiarid and dry subhumid deserts (23.9% of global land area) are well vegetated. Notwithstanding the lack of water, in some deserts, a consortium of microorganisms including cyanobacteria, microalgae, fungi, and bacteria colonize a few millimeters of top soil forming biological soil crusts. Some of the organisms in this consortium fix nitrogen, some are photosynthetic, and some produce metabolites that are excreted and used by other members of the consortium. These biological soil crusts help to stabilize desert soils from wind and water erosion. Preliminary experiments in artificially engineering these consortia have included co-immobilizing *Chlorella vulgaris* (a single-celled green algae) and *Azospirillum brasilense* (a diazotrophic PGPB) in alginate microbeads. Subsequently, the co-immobilized alginate beads (with a different species of *Chlorella*) were observed to increase soil quality in desert conditions as well as root and shoot development compared to inoculation with either of these immobilized organisms by itself (Table 8.13).

Notwithstanding the ecological advantages for bacteria to be able to move on solid surfaces, many bacteria are unable to do this. However, some soil bacteria (e.g., *Xanthomonas* spp.) can get around this limitation by taking advantage of motile bacteria in their vicinity. In one study, *Xanthomonas perforans* that was not motile was able to take advantage of the motile bacterium *Paenibacillus vortex*. In fact, the nonmotile bacteria attract the motile bacteria and then "rides them for dispersal." Following the demonstration that this behavior could occur both on agar plates and tomato leaves, this phenomenon was shown to occur between several different xanthomonads and motile bacterial species. This demonstration suggests that this so-called "hitchhiking" behavior might be widespread in nature and might be an ecologically important means of ensuring that nonmotile bacteria are dispersed within the soil.

8.7 Diazotrophic Endophytes

While plants require more nitrogen than any other nutrient, only a small fraction (typically around 2%) of the nitrogen that is present in most soils is available to plants. The remainder of the nitrogen in soil is present in organic forms. However,

Table 8.13 The effect of alginate bead immobilized *Azospirillum brasilense* + *Chlorella sorokiniana*, *Azospirillum brasilense* or *Chlorella sorokiniana* on the growth of sorghum plant roots and shoots in eroded soil

	Azospirillum + *Chlorella*	*Azospirillum*	*Chlorella*	Alginate beads	Negative control
Root, mg dw	155	110	105	55	48
Shoot, mg dw	50	40	37	32	30

dw is the plant tissue dry weight

some soil microorganisms can breakdown various organic forms of nitrogen and convert them to mineral forms such as nitrate and ammonia, thereby making more nitrogen available to plants. Unfortunately, in coarse textured soils much of the mineralized nitrogen may be leached away from (not very deep) plant roots and often ends up polluting waterways. Thus, it would appear that in addition to various strains of rhizobia that are efficient nitrogen fixers but are limited to a small fraction of agricultural plants, there is a potentially important ecological role to be played by diazotrophic bacteria. Previously, it was thought that some free-living diazotrophs such as *Azospirillum* spp. might act as an efficient means of providing nitrogen to plants. However, these bacteria provided plants with only a small fraction of their required fixed nitrogen. More recently, scientists have turned their attention to the possible use of diazotrophic endophytes. In this regard, it has been shown for example that the PGPB *Gluconacetobacter diazotrophicus* can provide a significant amount of the nitrogen required for sugar cane growth. This bacterium can fix nitrogen in this otherwise inhibitory aerobic environment (within the plant) as a consequence of the very high rate of respiration of this bacterium (which is fueled by its metabolism of sucrose from the sugar cane). This phenomenon is sometimes called respiratory protection and is not unique to *G. diazotrophicus*. That is, rhizospheric strains of diazotrophic *Azotobacter* spp. (a common Gram-negative soil bacterium) provides only limited amounts of fixed nitrogen to plants while endophytic strains of diazotrophic *Azotobacter* spp. can provide plants (e.g., rice) with significant amounts of fixed nitrogen.

The common trees, poplar and willow, are fast growing and economically important. In addition, these tree species have the ability to colonize nutrient poor environments. However, to grow rapidly on nutrient poor soils these trees require a source of fixed nitrogen. Thus, it was surmised that the internal tissues of these trees were probably colonized by a variety of diazotrophic endophytes. In fact, following the isolation of bacteria from the internal tissues of these trees (the cuttings were surface sterilized with 10% bleach) yielded a number of different diazotrophic endophytes including *Burkholderia* spp., *Rahnella* spp., *Sphingomonas* spp., and *Acinetobacter* spp. The potential of these isolates to fix nitrogen was determined by PCR amplification of the *nifH* gene from these isolates and subsequently ascertained using the acetylene reduction assay for nitrogenase activity. In another study with sugarcane plants, examination of eight different varieties of this plant yielded a total of 7198 operational taxonomic units (putative individual bacterial endophytes) of which around 24% were estimated to be diazotrophic. This data is consistent with the notion that in nature a significant amount of the fixed nitrogen is provided to plants by diazotrophic endophytes.

Not surprisingly, when diazotrophic endophytes are characterized in detail it is often observed that these bacteria contribute much more to plant growth and development than the mere provision of fixed nitrogen. Thus, following the isolation of 30 unique diazotrophic and endophytic pseudomonads from sugarcane (where all strains were initially selected by PCR of their *nifH* gene before they were shown to actively fix nitrogen), 30 synthesized IAA, 24 strains were found to solubilize inorganic phosphorus, 20 produced siderophores, 13 synthesized HCN, 18 possessed ACC deaminase activity, and 14 exhibited antifungal activity against the pathogen *Ustilago scitaminea*. The determination regarding which of the newly isolated strains is most effective under field conditions remains an open question and may require testing of these bacteria under a variety of field conditions.

8.8   Methane and Global Warming

It has been estimated that methane gas is currently responsible for ~25% of global warming. After carbon dioxide, methane is considered to be the second most important (i.e., problematic) greenhouse gas. There are several significant sources of methane in the environment including livestock, fossil fuel production, and wetland emissions. On the other hand, it is also possible to reduce environmental methane levels, either by atmospheric chemical oxidation or through oxidation by methanotrophs (bacteria that can metabolize methane as their only source of carbon and energy) in aerobic soils. Since changes to methane production and oxidation have a significant potential to alter global warming and climate change, scientists have increasingly sought to understand the mechanistic details of how methanotrophs oxidize methane. During the course of these studies, it was observed that ethylene is able to inhibit methane oxidation and that this inhibition occurs at ethylene concentration levels that are common in many soils. It has been suggested that this inhibition of methane oxidation is a consequence of competition between ethylene and methane for the active site of the enzyme methane monooxygenase, which carries out the conversion of methane to methanol. As discussed earlier in this chapter, the majority of plant stress responses initiate signaling pathways that trigger the upregulation of plant ethylene synthesis. This presents us with the possibility of creating a potentially dangerous autocatalytic mechanism of global warming. That is, as populations increase worldwide, climate change increases and more marginal land is used in an attempt to increase agricultural productivity, there is likely to be an increasing amount of plant stress on a worldwide scale (including increases in drought, flooding, the presence of salt, and phytopathogen damage). An increase in plant stress will almost certainly result in an increase in the concentration of ethylene that is released by plants into the soil. This increase in ethylene levels in the soil can potentially lead to a reduced soil capacity to oxidize methane thereby reducing the ability of many soils to act as a biological sink for methane. One group of researchers has suggested that it may be possible to limit the extent of the soil ethylene inhibition of soil bacterial methane monooxygenase through the increased usage of ACC deaminase-containing PGPB (Fig. 8.19). At the present time, this entire scenario is highly speculative so that it remains to be tested whether this idea works in practice.

8.9 Changing Plant Gene Expression

Using the techniques of transcriptomics (described in Chap. 3) it is possible to elaborate the changes in gene expression that occur within a plant as a consequence of that plant's interaction with PGPB, phytopathogens or environmental factors (such as drought, salt, metals, etc.). In one set of experiments, researchers used a transcriptomics approach to compare the effect of adding a wild-type PGPB strain (*Pseudomonas* sp. UW4) to canola seedlings compared to the effect of adding an ACC deaminase minus mutant of that PGPB strain. To examine the expression of as many canola genes as possible, mRNA isolated from inoculated canola plants (with either wild-type or mutant PGPB) was hybridized to an (commercially available) *Arabidopsis thaliana* oligonucleotide microarrary. The use of a heterologous *A. thaliana* microarray was necessitated because of the fact that a commercial canola oligonucleotide microarray was not available. The use of the heterologous microarray was justified by the fact that in exon sequences the identities between *A. thaliana* and canola DNA sequences ranged from 85% to 95%. This gave

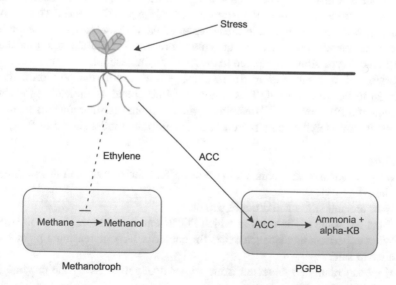

Fig. 8.19 Inhibition of methane oxidation by soil bacteria by ethylene may be partially overcome by the action of ACC deaminase-containing bacteria that effectively lower ACC and hence ethylene levels

researchers a high degree of confidence that most of the canola mRNAs that hybridized to the microarray represented the same gene as in *A. thaliana*. To ascertain that the gene assignments based on the *A. thaliana* genome sequence were largely correct, primers were designed (for nine different canola genes) to correspond to the portion of each mRNA that would bind the *A. thaliana* oligonucleotide sequences and the region was amplified by real-time PCR before its DNA sequence was determined. All nine of the partial canola genes that were sequenced were found to be nearly identical to the corresponding *A. thaliana* genes. In addition, expression values of the canola mRNAs that were measured with real-time (and reverse transcription) PCR correlated well with the values that were measured using the *A. thaliana* oligonucleotide microarray. This analysis revealed that bacterial inoculation of canola plants caused significant changes in plant gene expression, where the presence of strain UW4 (either wild-type or ACC deaminase minus mutant) caused 965 statistically significant expression changes in the shoots and 336 in the roots. Moreover, there were 559 expression changes in the shoots and 95 in the roots when comparing the effect of the wild-type versus the mutant strain, indicating that inoculation of plants with bacteria expressing or not expressing ACC deaminase caused different plant responses at the transcriptional level.

A detailed analysis of the changes in plant gene expression that occurred when the plant was treated with the wild-type versus the mutant strain revealed that the observed changes fell into a number of patterns. For example, upon incubation with the ACC deaminase minus mutant strain (compared to the wild-type strain), the expression of a number of plant stress-responsive genes were up regulated. These included various heat shock protein genes, several cytochrome P450 genes, some glutathione S-transferase genes, and some genes that had previously been identified as either senescence-associated or stress-inducible protein genes. This result is consistent with the notion that in the absence of ACC deaminase plants experience higher stress levels than when this enzyme is present. In addition, upon incubation with the wild-type strain (compared to the ACC deaminase minus mutant strain), the expression of a number of auxin responsive and auxin transport proteins was observed to be up regulated. This result was interpreted as indicating that in the presence of a functional ACC deaminase, auxin signal transduction was facilitated; this result ultimately helped researchers to develop the model shown in Fig. 5.18.

Questions
1. How do various environmental stresses affect the growth and development of plants?
2. How are salt stress and drought similar?
3. What are some of the ways in which PGPB can help to remediate salt stress?
4. What experimental evidence exists for bacterial IAA increasing a plant's tolerance to salt?
5. How can plants be protected from salt or drought stress by the enzyme ACC deaminase?
6. How might ACC deaminase-containing bacteria be distributed in soils in the natural environment?

7. How can ACC deaminase be produced on the bacterial outer surface? What is the advantage of doing this?
8. What is trehalose and how does it protect plant against growth inhibition caused by either drought or high salt?
9. How can *Azospirillum brasilense* be engineered to synthesize trehalose and protect plants against damage from drought?
10. How would you test whether trehalose and ACC deaminase work synergistically to protect plants from inhibition by salt stress?
11. How does cytokinin production affect plant growth under salt and drought stress?
12. What are exopolysaccharides and how might they protect plants from salt stress?
13. How would you test whether gibberellins can protect plants from growth inhibition by salt stress?
14. How would you test whether volatile organic compounds can protect plants from growth inhibition by salt stress?
15. How does flooding in roots affect plant shoots and leaves?
16. How can flooding stress be partially overcome using PGPB?
17. In a natural setting, outside of agriculture, under what conditions is the presence of PGPB advantageous to plants?
18. In the natural environment, how might the presence or absence of antibiotic-producing PGPB in plant rhizospheres be affected by drought conditions?
19. What are bacterial antifreeze and ice nucleation proteins?
20. What is chilling stress and how might PGPB protect plants from its effects?
21. What are biological soil crusts? Why are they a good thing?
22. How can nonmotile bacteria move in the environment?
23. How would you prove that diazotrophic endophytes are more effective than rhizospheric diazotrophs at providing fixed nitrogen to plants?
24. How might ACC deaminase-containing bacteria act to slow or decrease global warming?
25. How can microarray analysis be used to better understand the changes in plant gene expression that occur as a consequence of interaction of the plant with a specific PGPB?

Bibliography

Ahmad M, Zahair ZA, Nazli F, Akram F, Arshad M, Khalid M (2013) Effectiveness of halo-tolerant, auxin producing *Pseudomonas* and *Rhizobium* strains to improve osmotic stress tolerance in mung bean (*Vigna radiata* L). Braz J Microbiol 44:1341–1348

Ali S, Glick BR (2019) Plant-bacterial interactions that overcome abiotic stress. In: Singh JS (ed) New and future developments in microbial biotechnology and bioengineering. Elsevier, Amsterdam, pp 21–45

Ali S, Kim W-C (2017) Plant growth promotion under water: decrease of waterlogging-induced ACC and ethylene levels by ACC deaminase-producing bacteria. Front Microbiol 9:1096

Ali S, Charles TC, Glick BR (2014) Amelioration of damages caused by high salinity stress by plant growth-promoting bacterial endophytes. Plant Physiol Biochem 80:160–167

Barka EA, Nowak J, Clement C (2006) Enhancement of chilling resistance of inoculated grapevine plantlets with a plant growth-promoting rhizobacterium *Burkholderia phytofirmas* PsJN. Appl Environ Microbiol 72:7246–7252

Belimov AA, Zinovkina NY, Safronova VI, Litinsky VA, Nosikov VV, Zavalin AA, Tikhonovitch IA (2019) Rhizobial ACC deaminase contributes to efficient symbiosis with pea (*Pisum sativum* L.) under single and combined cadmium and water deficit stress. Environ Exp Bot 167:103859

Bhise KK, Bhagwat PK, Dandge PB (2017) Synergistic effect of *Chryseobacterium gleum* sp. SUK with ACC deaminase activity in alleviation of salt stress and plant growth promotion in *Triticum aestivum* L. 3. Biotech 7:105–101

Bianco C, Defez R (2009) *Medicago truncatula* improves salt tolerance when nodulated by an indole-3-acetic acid-overproducing *Sinorhizobium meliloti* strain. J Exp Bot 60:3097–3107

Bouffaud M-L, Renoud S, Dubost A, Moënne-Loccoz Y, Muller D (2018) 1-Aminocyclopropane-1-carboxylate deaminase producers associated to maize and other *Poaceae* species. Microbiome 6:114

Brígido C, Nascimento F, Duan J, Glick BR, Oliveira S (2013) Expression of an exogenous 1-aminocyclopropane-1-carboxylate deaminase gene in *Mesorhizobium* spp. reduces the negative effects of salt stress in chickpea. FEMS Microbiol Lett 349:46–53

Chandra D, Glick BR, Sharma AK (2018) Drought tolerant *Pseudomonas* spp. improves the growth performance of finger millet (*Eleusine coracana* (L.) Gaertn.) under non-stressed and drought-stressed conditions. Pedosphere 28:227–240

Cheng Z, Park E, Glick BR (2007) 1-Aminocyclopropane-1-carboxylate (ACC) deaminase from *Pseudomonas putida* UW4 facilitates the growth of canola in the presence of salt. Can J Microbiol 53:912–918

Cheng Z, Woody OZ, McConkey BJ, Glick BR (2012) Combined effects of the plant growth-promoting bacterium *Pseudomonas putida* UW4 and salinity stress on the *Brassica napus* proteome. Appl Soil Ecol 61:255–263

Defez R, Andreozzi A, Dickinson M, Charlton A, Tadini L, Pesaresi P, Bianco C (2017) Improved drought stress response in alfalfa plants nodulated by an IAA over-producing *Rhizobium* strain. Front Microbiol 8:2466

Dong M, Yang Z, Cheng G, Peng L, Xu Q, Xu J (2018) Diversity of the bacterial microbiome in the roots of four *Saccharum* species: *S. spontaneum*, *S. robustum*, *S. barberi*, and *S. officinarum*. Front Microbiol 9:267

Doty SJ, Oakley B, Xin G, Kang JW, Singleton G, Khan Z, Vajzovic A, Staley JT (2009) Diazotrophic endophytes of native black cottonwood and willow. Symbiosis 47:23–33

Egamberdieva D, Jabborova D, Mamadalieva N (2013) Salt-tolerant *Pseudomonas extremorintalis* able to stimulate growth of *Silybum marianum* under salt stress. Med Aromat Plant Sci Biotechnol 7:7–10

Egamberdieva D, Jabborova D, Hashem A (2015) *Pseudomonas* induces salinity tolerance in cotton (*Gossypium hirsutum*) and resistance to *Fusarium* root rot through the modulation of indole-3-acetic acid. Saudi J Biol Sci 22:773–779

Egamberdieva D, Wirth S, Bellingrath-Kimura SD, Mishra J, Arora NK (2019) Salt-tolerant plant growthpromoting rhizobacteria for enhancing crop productivity of saline soils. Front Microbiol 10:2791

Enez A, Hudek L, Brau L (2018) Reduction in trace metal mediated oxidative stress towards cropped plants via beneficial microbes in irrigated cropping systems: a review. Appl Sci 8:1953

Farwell AJ, Vesely S, Nero V, McCormack K, Rodriguez H, Shah S, Dixon DG, Glick BR (2007) Tolerance of transgenic canola (*Brassica napus*) amended with ACC deaminase-containing plant growth-promoting bacteria to flooding stress at a metal-contaminated field site. Environ Pollut 147:540–545

Forni C, Duca D, Glick BR (2017) Mechanisms of plant response to salt and drought stress and their alteration by rhizobacteria. Plant Soil 410:335–356

Gamalero E, Glick BR (2019) Plant growth-promoting bacteria in agriculture and stressed environments. In: Van Elsas D, Trevors JT (eds) Modern soil microbiology, 3rd edn. Kluwer, South Holland, pp 361–380

Gamalero E, Berta G, Glick BR (2009) The use of microorganisms to facilitate the growth of plants in saline soils. In: Khan MS, Zaidi A, Musarrat J (eds) Microbial strategies for crop improvement. Springer, Berlin, pp 1–22

Gepstein S, Glick BR (2013) Strategies to ameliorate abiotic stress-induced plant senescence. Plant Mol Biol 82:623–633

Golan Y, Shirron N, Avni A, Shmoish M, Gepstein S (2016) Cytokinins induce transcriptional reprograming and improve *Arabidopsis* plant performance under drought and salt stress conditions. Front Environ Sci 4:63

Gonzalez-Bashan LE, Lebsjy VK, Hernandez JP, Bustillos JJ, Bashan Y (2000) Changes in the metabolism of the microalga *Chlorella vulgaris* when coimmobilized in alginate with the nitrogen-fixing *Phyllobacterium myrsinacearum*. Can J Microbiol 46:653–659

Grichko VP, Glick BR (2001) Amelioration of flooding stress by ACC deaminase-containing plant growth-promoting bacteria. Plant Physiol Biochem 39:11–17

Hagai E, Dvora R, Havkin-Blank T, Zelinger E, Porat Z, Schultz S, Helman Y (2014) Surface motility induction, attraction and hitchhiking between bacterial species promote dispersal on solid surfaces. ISME J 8:1147–1151

Hamdia MAE, Shaddad MAK, Doaa MM (2004) Mechanisms of salt tolerance and interactive effects of *Azospirillum brasilense* inoculation on maize cultivars grown under salt stress conditions. Plant Growth Regul 44:165–174

Hegedus DD, Heydarian Z, Glick BR, Gruber M (2018) Gene expression patterns in roots of *Camelina sativa* with enhanced salinity tolerance arising from inoculation of soil with plant growth promoting bacteria producing 1-aminocyclopropane-1-carboxylate deaminase or expression of the corresponding *acdS* gene. Front Microbiol 9:1297

Heydarian Z, Yu M, Gruber M, Glick BR, Zhou R, Hegedus DD (2016) Inoculation of soil with plant growth promoting bacteria producing 1-aminocyclopropane-1-carboxylate deaminase or expression of the corresponding *acdS* gene in transgenic plants increases salinity tolerance in *Camelina sativa*. Front Plant Sci 7:1966

Husain KA, Kadhum NH, Sharqi MR (2019) Usage of some Halo bacteria species to alleviate sodium chloride toxicity in *Vigna radiate* L. cuttings in terms of rooting response. IOP Conf Ser Earth Environ Sci 388:012051

Ilangumaran G, Smith DL (2017) Plant growth promoting rhizobacteria in amelioration of salinity stress: a systems biology perspective. Front Plant Sci 8:1768

Iqbal N, Umar S, Khan NA, Khan MIR (2014) A new perspective of phytohormones in salinity tolerance: regulation of proline metabolism. Environ Exp Bot 100:34–42

Jaemsaeng R, Jantasuriyarat C, Thamchaipenet A (2018) Molecular interaction of 1-aminocyclopropane-1-carboxylate deaminase (ACCD)-producing endophytic *Streptomyces* sp. GMKU 336 towards salt-stress resistance of *Oryza sativa* L. cv. LDML105. Sci Rep 8:1950

Kang S-M, Radhakrishnan R, Khan AL, Kim M-J, Park J-M, Kim B-R, Shim D-H, Lee I-J (2014) Gibberellin secreting rhizobacterium, *Pseudomonas putida* H-2-3 modulates the hormonal and stress physiology of soybean to improve the plant growth under saline and drought conditions. Plant Physiol Biochem 84:115–124

Kim Y-C, Glick BR, Bashan Y, Ryu C-M (2013) Enhancement of plant drought tolerance by microbes. In: Aroca R (ed) Plant responses to drought stress: from morphological to molecular features. Springer, Heidelberg, pp 383–413

Kudoyarova G, Arkhipova T, Korshinova T, Bakaeva M, Loginov O, Dodd IC (2019) Phytohormone mediation of interactions between plants and non-symbiotic growth promoting bacteria under edaphic stress. Front Plant Sci 10:1368

Lau JA, Lennon JT (2012) Rapid responses of soil microorganisms improve plant fitness in novel environments. Proc Natl Acad Sci U S A 109:14058–14062

Ledger T, Rojas S, Timmermann T, Pinedo I, Poupin MJ, Garrido T, Richter P, Tamayo J, Donoso R (2016) Volatile-mediated effects predominate in *Paraburkholderia phytofirmans* growth promotion salt stress tolerance of *Arabidopsis thaliana*. Front Microbiol 7:1838

Li J, Sun J, Yang Y, Guo S, Glick BR (2012) Identification of hypoxic-responsive proteins in cucumber roots using a proteomic approach. Plant Physiol Biochem 51:74–80

Li J, McConkey BJ, Cheng Z, Guo S, Glick BR (2013) Identification of plant growth-promoting rhizobacteria-responsive proteins in cucumber roots under hypoxic stress using a proteomic approach. J Proteome 84:119–131

Li H-B, Singh RK, Singh P, Song Q-Q, Xing Y-X, Yang L-T, Li Y-R (2017) Genetic diversity of nitrogen-fixing and plant growth promoting *Pseudomonas* species isolated from sugarcane rhizosphere. Front Microbiol 8:1268

Liu Y, Cao L, Tan H, Zhang R (2017) Surface display of ACC deaminase on endophytic *Enterobacteriaceae* strains to increase saline resistance of host rice sprouts by regulating plant ethylene synthesis. Microb Cell Factories 16:214

Lorv JSH, Rose DR, Glick BR (2014) Bacterial ice crystal controlling proteins. Scientifica 2014:976895

Mavrodi DV, Mavrodi OV, Parejko JA, Bonsall RF, Kwak Y-S, Paulitz TC, Thomashow LS, Weller DM (2012) Accumulation of the antibiotic phenazine-1-carboxylic acid in the rhizosphere of dryland cereals. Appl Environ Microbiol 78:804–812

Mayak S, Tirosh T, Glick BR (2004a) Plant growth-promoting bacteria that confer resistance in tomato to salt stress. Plant Physiol Biochem 42:565–572

Mayak S, Tirosh T, Glick BR (2004b) Plant growth-promoting bacteria that confer resistance to water stress in tomato and pepper. Plant Sci 166:525–530

Nautiyal CS, Srivastava S, Chauhan PS, Seem K, Mishra A, Sopory SK (2013) Plant growth-promoting bacteria *Bacillus amyloliquefaciens* NBRISN13 modulates gene expression profile of leaf and rhizosphere community in rice during salt stress. Plant Physiol Biochem 66:1–9

Ngumbi E, Kloepper J (2016) Bacterial-mediated drought tolerance: current and future prospects. Appl Soil Ecol 105:109–125

Orozco-Mosqueda MC, Duan J, DiBernardo M, Zetter E, Campos-García J, Glick BR, Santoyo G (2019) The production of ACC deaminase and trehalose by the plant growth promoting bacterium *Pseudomonas* sp. UW4 synergistically protects tomato plants against salt stress. Front Microbiol 10:1392

Ort DR, Long SP (2014) Limits on yields in the corn belt. Science 344:484–485

Peng J, Wu D, Liang Y, Li L, Guo Y (2019) Disruption of *acdS* gene reduces plant growth promotion activity and maize saline stress resistance by *Rahnella aquatilis* HX2. J Basic Microbiol 2019:1–10

Perera I, Subashchandrabose SR, Venkateswarlu K, Naidu R, Megharaj M (2018) Consortia of cyanobacteria/microalgae and bacteria in desert soils: an unexplored microbiota. Appl Microbiol Biotechnol 102:7351–7363

Qin S, Zhang Y-J, Yuan B, Xu PY, Xing K, Wang J, Jiang J-H (2014) Isolation of ACC deaminase-producing habitat-adapted symbiotic bacteria associated with halophyte *Limonium sinense* (Girard) Kuntze and evaluating their plant growth-promoting activity under salt stress. Plant Soil 374:753–766

Qin Y, Druzhinina IS, Pan X, Yuan Z (2016) Microbially mediated plant salt tolerance and microbiome-based solutions for saline agriculture. Biotechnol Adv 34:1245–1259

Ravanbakhsh M, Kowalchuk GA, Jousset A (2019) Root-associated microorganisms reprogram plant life history along the growth-stress resistance tradeoff. ISME J 13:3093–3101

Rodriguez-Salazar J, Suarez R, Caballero-Mellado J, Iturriaga G (2009) Trehalose accumulation in *Azospirillum brasilense* improves drought tolerance and biomass in maize plants. FEMS Microbiol Lett 296:52–59

Salwan R, Sharma A, Sharma V (2019) Microbes mediated plant stress tolerance in saline agricultural ecosystem. Plant Soil 442:1–22

Saravankumar D, Samiyappan R (2007) ACC deaminase from *Pseudomonas fluorescens* mediated saline resistance in groundnut (*Arachis hypogea*) plants. J Appl Microbiol 102:1283–1292

Sergeeva E, Shah S, Glick BR (2006) Tolerance of transgenic canola expressing a bacterial ACC deaminase gene to high concentrations of salt. World J Microbiol Biotechnol 22:277–282

Siddikee MA, Glick BR, Chauhan PS, Yim W-J Sa T (2011) Enhancement of growth and salt tolerance of red pepper seedlings (*Capsicum annuum* L.) by regulating stress ethylene synthesis with halotolerant bacteria containing ACC deaminase activity. Plant Physiol Biochem 49:427–434

Stearns JC, Woody OZ, McConkey BJ, Glick BR (2012) Effects of bacterial ACC deaminase on *Brassica napus* gene expression. Mol Plant-Microbe Interact 25:668–676

Suarez R, Wong A, Ramirez M, Barraza A, Orozco MC, Cevallos MA, Lara M, Hernandez G, Iturriaga G (2008) Improvement of drought tolerance and grain yield in common bean by overexpressing trehalose-6-phosphate synthase in rhizobia. Mol Plant-Microbe Interact 21:958–966

Subramanian P, Krishnamoorthy R, Chanratana M, Kim K (2015) Expression of an exogenous 1-aminocycloproprane-1-carboxylate deaminase gene in psychrotolerant bacteria modulates ethylene metabolism and induced genes in tomato under chilling stress. Plant Physiol Biochem 89:18–23

Sugawara M, Cytryn EJ, Sadowsky MJ (2010) Functional role of *Bradyrhizobium japonicum* trehalose biosynthesis and metabolism genes during physiological stress and nodulation. Appl Environ Microbiol 76:1071–1081

Tewari S, Arora NK (2014) Multifunctional exopolysaccharides from *Pseudomonas aeruginosa* PF23 involved in plant growth stimulation, biocontrol and stress amelioration in sunflower under saline conditions. Curr Microbiol 69:484–494

Timmusk S, Paalme V, Pavlicek T, Bergquist J, Vangala A, Danilas T, Nevo E (2011) Bacterial distribution in the rhizosphere of wild barley under contrasting microclimates. PLoS One 6(3): e17968

Timmusk S, Abd El-Daim IA, Copolovici L, Tanilas T, Kannaste A, Behers L, Nevo E, Seisenbaeva G, Stenstrom E, Niinemets U (2014) Drought-tolerance of wheat improved by rhizosphere bacteria from harsh environments: enhanced biomass production and reduced emissions of stress volatiles. PLoS One 9(5):e96086

Tittabutr P, Piromyou P, Longtonglang A, Noisa-Ngiam R, Boonkerd N, Teaumroong N (2013) Alleviation of the effect of environmental stresses using co-inoculation of mungbean by *Bradyrhizobium* and rhizobacteria containing stress-induced ACC deaminase. Soil Sci Plant Nutr 2013:1–13. https://doi.org/10.1080/00380768.2013.804391

Trejo A, de Bashan LE, Hartmann A, Hernandez J-P, Rothballer M, Schmid M, Bashan Y (2012) Recycling waste debris of immobilized microalgae and plant growth-promoting bacteria from wastewater treatment as a resource to improve fertility of eroded desert soil. Environ Exp Bot 75:65–73

Xu J, Li X-L, Luo L (2012) Effects of engineered *Sinorhizobium meliloti* on cytokinin synthesis and tolerance of alfalfa to extreme drought stress. Appl Environ Microbiol 78:8056–8061

Yaish MW, Al-Harrasi I, Alansari AS, Al-Yahyai R, Glick BR (2016a) The use of high throughput DNA sequence analysis to assess the endophytic microbiome analysis of date palm roots grown under different levels of salt stress. Int Microbiol 19:143–155

Yaish MW, Al-Lawati A, Jana GA, Parankar HV, Glick BR (2016b) Impact of soil salinity on the structure of the bacterial endophytic community identified from the roots of Caliph Medic (*Medicago truncatula*). PLoS One 11(7):e0159007

Yoolong S, Kruasuwan W, Pham HTT, Jaemsaeng R, Jantasuriyarat C, Thamchaipenet A (2019) Modulation of salt tolerance in Thai jasmine rice (*Oryza sativa* L. cv KDML105) by *Streptomyces venezuelae* ATCC10712 expressing ACC deaminase. Sci Rep 9:1275

Zhou X, Smaill SJ, Clinton PW (2014) Methane oxidation needs less stressed plants. Trends Plant Sci 18:657–659

Mycorrhizal–Plant Interactions

9

Abstract

In addition to their relationship with a variety of different plant growth-promoting bacteria, the roots of >90% of all land plants form a mutualistic relationship with plant beneficial fungi called mycorrhizae. In this chapter, the different types of mycorrhizae are elaborated and the benefits that they provide to plants are discussed including the increased uptake of water and a range of different minerals from the soil. Typically, mycorrhizae and plant growth-promoting bacteria utilize different mechanisms to promote plant growth and numerous reports suggest that the two often act synergistically. However, most mycorrhizae cannot be grown in culture so that the commercial use of these beneficial fungi has so far been quite limited.

9.1   Overview of Mycorrhizae

Mycorrhizae are plant beneficial fungi that appear to form a symbiotic/mutualistic relationship with >90% of all land plants. The word mycorrhiza (singular) is derived from Greek in which "mykes" means fungus and "rhiza" means roots. In a mycorrhizal association, the fungus colonizes the host plant's roots, either intracellularly or extracellularly where energy and carbon compounds move from the plant to the fungus and inorganic resources (i.e., minerals) and water move from the fungus to the plant. It has been said that, "it is as normal for the roots of plants to be mycorrhizal as it is for the leaves to photosynthesize."

There are several different types of mycorrhizae; the two main types being ectomycorrhizae and endomycorrhizae. Ectomycorrhizae are widespread and extracellularly colonize the outside of the roots and are commonly found colonizing the root surfaces (Fig. 9.1) of approximately 3% of vascular plant families especially gymnosperms and other woody plants such as birch, oak, fir, pine, and rose. Ectomycorrhizae are readily manipulable in that they can easily be grown in culture. Many forest trees are highly dependent upon their association with ectomycorrhizae

© Springer Nature Switzerland AG 2020
B. R. Glick, *Beneficial Plant-Bacterial Interactions*,
https://doi.org/10.1007/978-3-030-44368-9_9

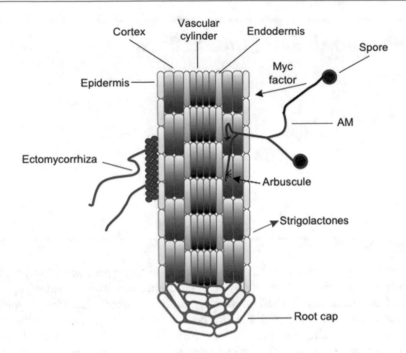

Fig. 9.1 Schematic representation of a plant root with ectomycorrhizae shown bound to the left side of the root and AM shown bound to the right side of the root. However, in practice, only one type of mycorrhiza will specifically bind to a particular plant root. Different types of root cells are shown in different colors. The ectomycorrhizae bind to the root surface while the AM, attracted by root-synthesized strigolactones, enter root cortical cells. Myc factors are signaling molecules produced by AM

which mobilize minerals (including phosphorus, sulfur, and zinc) from the soil and deliver them to the plant. These fungi become part of a sheath around the root tip forming a so-called hartig net which has been suggested to protect host trees from attack by parasites, nematodes and other soil pathogens. At the present time, it is estimated that there are about 5000–6000 different species of ectomycorrhizal fungi including mostly Basidiomycetes, some Ascomycetes, and a few Zygomycetes.

On the other hand, the much more common endomycorrhizae or arbuscular mycorrhiza (AM), formerly known as vesicular arbuscular mycorrhiza (VAM), colonizes roots intracellularly. With AM, the fungus enters the root cortical cell walls and forms highly branched intracellular fungal structures called arbuscules; the AM are obligate symbionts and cannot be grown independently of the plant (Fig. 9.1). The mutualistic relationship between this beneficial fungus and plants provides the fungus with fixed carbon and nitrogen, and in exchange the plant is provided with an increased ability to take up water and nutrients from the soil. The arbuscules are the main site of nutrient exchange between the AM fungus and the plant, and it is from here that nutrient transport to the rest of the plant is initiated as AM fungal hyphae extend into the soil and act as effective extensions of the roots.

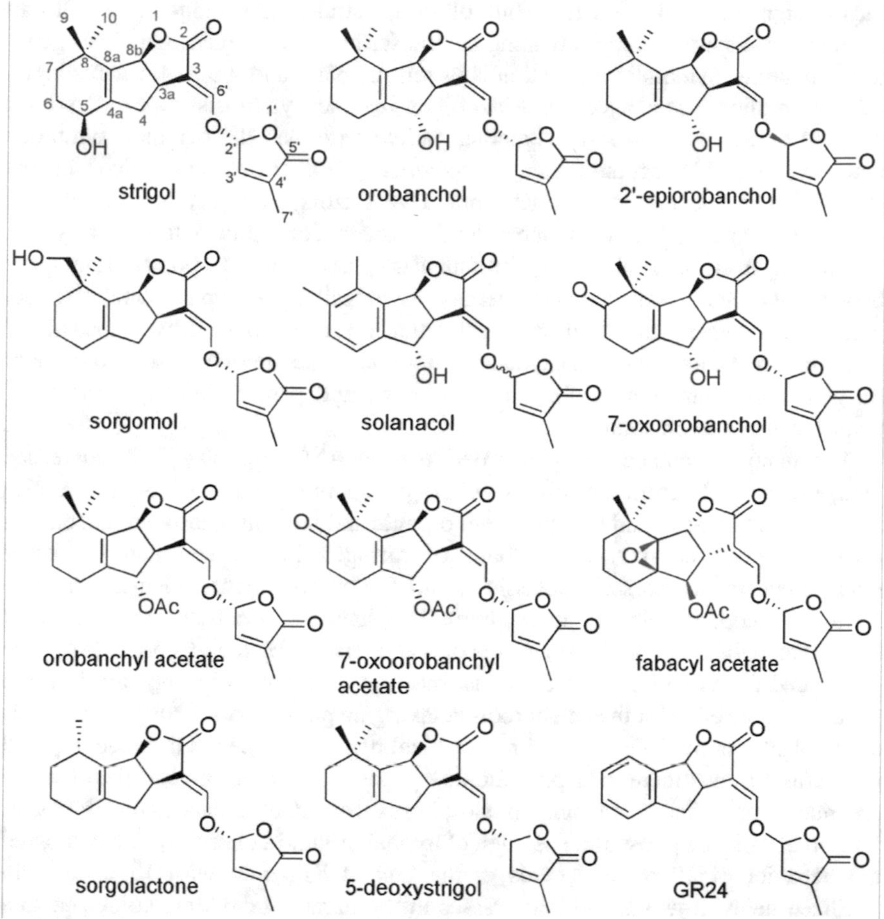

Fig. 9.2 Chemical structures of some plant synthesized strigolactones that are taken up by AM and alter its metabolism

Arbuscular mycorrhizae are exclusively formed by the phylum of fungi called *Glomeromycota*. In addition, both fossil evidence and DNA sequence data are consistent with this mutualistic relationship having begun around 400–460 million years ago, about the same time that the first vascular plants were colonizing the land. In fact, it has been suggested that it is precisely the symbiotic relationship between AM fungi and the descendants of freshwater algae that facilitated the development of land plants. Thus, it is thought that plant roots and the relationship with AM fungi may have co-evolved.

Plant hormones (synthesized from carotenoids) called strigolactones (molecular weight ~300; Fig. 9.2) stimulate spore germination in AM, as well as fungal metabolism and hyphal branching. In addition, strigolactones are known to modulate both shoot and root architecture of higher plants. In turn, the fungi release "Myc factor" signaling molecules (lipochito-oligosaccharides) that induce root cell calcium levels to spike thereby activating the transcription of the plant symbiosis genes to facilitate the symbiotic interaction between the plant and the AM. Interestingly, ~75–80 million years ago, the Rhizobium–legume symbiosis appears to have coopted elements of the same signaling pathway (termed the common symbiotic pathway) to enable legume plants to recognize rhizobial produced "Nod" factors (lipochito-oligosaccharides distinct from Myc factors) and facilitate nodulation, including receptors, from the much older AM interaction pathway to form a symbiotic interface. In the root cortex, the fungal hyphae enter the apoplast (the space between the cells) and branches laterally. Eventually, the hyphae enter the root cortical cells where they form arbuscules that are encompassed by a specialized membrane that allows nutrient exchange. New spores are then synthesized outside of the plant root. Thus, AM are obligate symbionts that depend entirely upon their host plant for their carbon sources.

The major advantage that plants derive from AM symbiosis is the increased uptake of nutrients from the soil, especially phosphorus, and water. AM have also been shown to uptake and transport both organic and inorganic nitrogen (depending upon the plant cultivar, the amount of nitrogen in the soil and the other microorganisms in the soil), potassium, and the micronutrients selenium, magnesium, zinc, copper, calcium, and sodium. This increase in nutrient uptake is most likely due to the additional surface area of soil contacted by the mycorrhizal hyphae (compared to the soil contacted by just the roots). In particular, mycorrhizae are generally more efficient than plant roots at taking up phosphorus. Phosphorus travels to the plant root by diffusion, and mycorrhizal hyphae reduce the distance required for diffusion, thus increasing phosphorus uptake. In addition, mycorrhizal hyphae are much more efficient at taking up phosphorus than are root hairs so that the major portion of a plant's phosphorus is often of hyphal origin. In one study that compared the behavior of 10 organic fields in the United Kingdom with 13 chemically fertilized fields, it was found that grasses in the organic field were colonized to a greater extent by AM with a much greater number of AM spores found in the soil coincident with the presence of much less phosphorus in the soil (Table 9.1). This observation is consistent with the notion that the presence of excess phosphorus (commonly added as a chemical fertilizer) can inhibit the interaction between plants and AM. In addition, those plants that were grown organically and were colonized to a greater extent by AM also grew to a greater size despite the lack of phosphorus fertilizer. It is argued that the increased use of AM should lead to a decrease in the use of phosphorus fertilizers resulting in less fertilizer runoff from agricultural fields and a decrease in harmful algal blooms in nearby lakes.

In sub-Saharan Africa, Bambara groundnut is one the most important grain legumes. It is highly caloric, rich in minerals and protein, grows underground, is resistant to the high temperatures that are common in the region and can grow on marginal soils. In one series of experiments, researchers demonstrated that AM (actually a mixture of *Gigaspora margarita* and *Acaulogpora tubercolata*) significantly enhanced both Bambara groundnut plant growth and phosphorus content when the plants were grown in the field (Table 9.2). This is important since many

Table 9.1 A comparison of the behavior of organic versus chemically fertilized fields with regard to AM colonization, the presence of spores in the soil, the amount of P in the soil and the response of clover plants to the presence of P and AM

Measured trait	Type of grassland field	
	Chemically fertilized	Organic
Soil extractable P (mg kg^{-1})	5.4	2.0
Colonization by AM, %	40.4	63.6
Number of spores/gram dry soil	12.1	34.4
Clover shoot dry weight, mg	9.4	10.3
Clover shoot P content	7.6	8.2
Clover colonization by AM, %	55	63

Table 9.2 The effect of the addition of soluble phosphorus and an AM mixture on the biomass and leaf phosphorus content of Bambara groundnut plants grown in the field

Phosphorus added, mM	± AM	Shoot fresh weight, g	Root fresh weight, g	Leaf phosphorus, %	Number of nodules per plant
0	−	9.48	10.73	0.067	21
0	+	11.70	16.94	0.083	65
1	−	13.73	14.88	0.087	34
1	+	18.98	23.56	0.128	101

The germinated seeds were sown together with the AM inoculum. Plants were watered with a nutrient solution every second day and were harvested 2 months after sowing

sub-Saharan soils are quite poor in terms of phosphorus content. The results of this experiment indicate that (i) the addition of phosphorus fertilizer significantly increases both shoot and root growth as well as leaf P content while (ii) the addition of AM also increases these three parameters. In this study, the combination of both added phosphorus and added AM yielded the healthiest plants with by far the largest number of root nodules, with the expectation that these plants will produce the greatest yield of groundnut.

The extent to which plants depend on AM colonization is largely a function of the environmental conditions where the plant is growing. In addition to functioning as the major mechanism that most plants use to acquire phosphorus from the soil, AM fungi also greatly facilitate the uptake of minerals and water from the soil. In soils that are deficient in various minerals (or water in the case of moderate drought), mycorrhizal plants have a significant selective advantage over non-mycorrhizal plants. Interestingly, plants can have several different AM fungal partners and mycorrhiza can interact with more than a single plant at the same time. This often results in so-called common mycorrhizal networks (Fig. 9.3). In the case of the few plants that do not contain chlorophyll (and therefore are not photosynthetic) those plants can obtain all of their resources, including carbon compounds, from those common mycorrhizal networks. Thus, those plants that lack chlorophyll can only exist as parasites depending on taking up carbon-containing exudates from chlorophyll-containing plants.

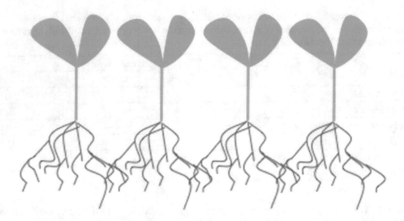

Fig. 9.3 Schematic representation of mycorrhizal underground networks. Mycorrhizae bound to roots can interact with mycorrhizae from nearby plants sometimes exchanging metabolites

Plants typically sequester about 25% of the anthropogenic (originating from human activity) carbon dioxide thereby slowing the rate of climate change. However, in addition to increased photosynthesis, plant growth also requires an increase in available nitrogen. Furthermore, in instances with low levels of available nitrogen in the soil, ectomycorrhizal plants show a much greater biomass increase then AM plants. Thus, in terms of ameliorating carbon dioxide caused climate change, trees whose roots are associated with ectomycorrhizae should be a key environmental consideration.

9.1.1 Selecting New Arbuscular Mycorrhizae

Cassava is a plant with tuberous roots that is rich in starch and is a food staple to more than 800 million people living in developing countries. In 2014 ~270 million tons of cassava were produced worldwide with nearly 70% of this production in Nigeria, Brazil, Thailand, Indonesia, and the Democratic Republic of Congo. Unfortunately, the yield of cassava in some countries is only about half of the global average of 13 tons per hectare. This low productivity has been attributed to low soil fertility, drought and the inhibition of cassava growth by root-knot nematodes. In an attempt to significantly increase the cassava yield in those countries with low productivity, researchers have sought to isolate and characterize strains of AM that might help to overcome these limitations. Specifically, the researchers wanted to identify AM strains that were superior to the existing commercial strains of AM in their ability to confer upon cassava plants better (i) plant growth and yield, (ii) tolerance to low levels of water, and (iii) resistance/tolerance to root-knot nematodes. To do this, researchers first collected soil samples from the rhizospheres of cassava plants growing in different locations during the dry season (Fig. 9.4). The collected soils were mixed with sand and sterilized soil, and then placed in 10 L pots

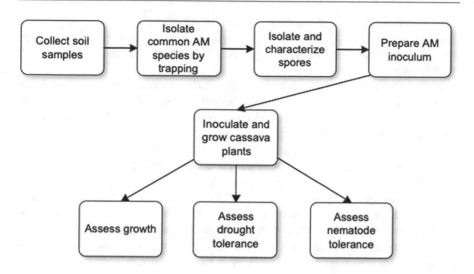

Fig. 9.4 Flow chart of the isolation and characterization of new and more effective strains of AM

in which cassava was planted and allowed to grow for 4 months. Spores were then extracted from the soil and the AM strain was by determined by PCR. The most prevalent strains (i.e., the common ones) were grown to produce an inoculum that was used to inoculate cassava plants. After four additional months, the biomass, drought tolerance and nematode tolerance of the mycorrhizal plants was assessed. In this way, researchers identified two AM strains that, in the laboratory, promoted cassava growth and increased drought tolerance. And, one of these two strains also conferred bioprotective effects to cassava plants against root-knot nematodes. Importantly, the "best" new AM strain significantly outperformed the commercial inoculant in the field.

9.1.2   Growing and Formulating Mycorrhizae

Since AM, unlike ectomycorrhizae, cannot be grown in culture, it is a somewhat difficult and labor-intensive process to produce AM inoculants for crops. Nevertheless, there are approximately 200 small companies worldwide that produce and sell AM inoculants for use in agriculture. The most common methods for producing AM involve the use of cultivation of the AM in pots with sterilized soil, aeroponics, hydroponics, or greenhouse-based in vivo methods followed by harvesting of the roots. Unfortunately, these techniques suffer from problems of cross-contamination in the inoculum production so that the production of high-quality inoculants using these approaches remains a major challenge. More recently, some researchers have utilized transformed root organ cultures to produce AM propagules, without microbial contaminants, under strictly controlled sterilized conditions after pure AM fungi are inoculated into the transformed root organ. However, this is still a slow, difficult and painstaking process so that the production of AM inoculants is generally limited to high valued crops. Moreover, despite a large number of successful field trials with AM, the conditions for successful large-scale field application of AM are not well elaborated.

Since most mycorrhizae lack a high degree of host-specificity, it is necessary to establish the desired plant–mycorrhizal interaction during the very early stages of plant growth. Inoculation of plants with mycorrhizae are typically done in one of five ways: broadcasting, in-furrow application, seed dressing, root dipping, or seedling inoculation. With broadcasting, seeds and the AM inoculum are spread over the soil surface and then mixed with the top few centimeters of soil. This can be done either manually or mechanically. The advantage of this procedure is that it does not require any special equipment to cover a large area. The disadvantage is that this procedure requires a large amount of inoculum since seeds will only be colonized by the inoculum to which they are immediately adjacent in the soil. For in-furrow application, the inoculum is placed in a small trench or furrow, and the seeds are placed over the inoculum and then covered with soil. With seed dressing, pellets of the inoculum are mixed with a sticky substance such as gum acacia before mixing them with the seeds. The treated seeds are then dried and subsequently planted. The advantage of this procedure is that the inoculum readily colonizes the germinating seeds. With root dipping, plants are first grown in the absence of mycorrhizae in a nursery and once they reach a certain size the roots are dipped for several minutes into the inoculum (mixed with a sticky carrier). This technique is often used with nursery-raised plants. Seedling inoculation involves growing seedlings in nursery beds containing mycorrhizae and then transplanting the seedlings to the field. This approach is generally used to inoculate forest tree species. In addition to these more established techniques of formulating mycorrhizae, one group of researchers reported encapsulating ectomycorrhizae (*Pisolithus microcarpus*) in ~4 mm Ca alginate beads before using the beads to treat Eucalyptus cuttings. In this case, in the presence of low levels of available phosphate, the mycorrhizal inoculation resulted in significantly increased shoot height and dry mass as well as greater root mass, suggesting that this is an effective method for rooting tree cuttings in a nursery setting.

9.2   Phytohormones

As previously discussed, phytohormones regulate various aspects of plant growth and development. Moreover, there is not always a simple relationship between plant growth and development and the concentrations of various phytohormones, especially since many phytohormones affect the expression and functioning of other phytohormones. Nevertheless, given the intimate and long-standing relationship between plants and mycorrhizae, it is not surprising that various phytohormones should affect and modulate the colonization of plants roots by mycorrhizae.

The plant hormone gibberellin, which has a positive impact on many aspects of plant development including stem elongation and leaf expansion, plays an important

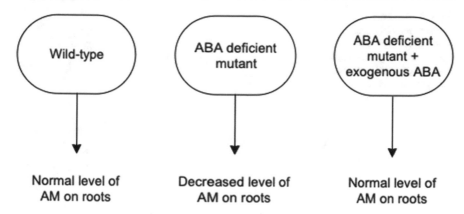

Fig. 9.5 Overview of an experiment designed to demonstrate that abscisic acid (ABA) positively affects the colonization of tomato plant roots by AM

role in limiting AM colonization. Application of gibberellins inhibits the formation of arbuscular mycorrhizae across species, and this is supported by studies with mutant and transgenic plants that have been disrupted in gibberellin biosynthesis or signaling. Moreover, this effect of low levels of gibberellins on mycorrhizal colonization appears to be independent of plant ethylene levels. In contrast, researchers recently demonstrated that cytokinins had a stimulatory role in the development of the symbiosis between AM and pea plants. These experiments included treating plants with the synthetic cytokinin 6-benzylaminopurine, or the cytokinin degradation inhibitor 2-chloro-6-(3-methoxyphenyl)-aminopurine, or the cytokinin receptor antagonist 6-(2-hydroxy-3-methylbenzyl)-aminopurine. The effects of the abovementioned synthetic compounds on the colonization of pea plants by AM was assessed after approximately 1 month's time.

Similar to the results observed with cytokinin, scientists have observed that abscisic acid (ABA) is essential for full AM colonization of (tomato) plants (Fig. 9.5). In this case, wild-type tomato plants were compared with ABA-deficient mutants of the same variety of tomatoes. In addition, it was shown that exogenously adding ABA to the mutant tomato plants fully restored the ability of the mutant plants to form a normal level of AM on the plant roots. A subsequent experiment using a different ABA-deficient mutant was consistent with the notion that ABA deficiency enhances plant ethylene levels. This latter work concluded that while ABA deficiency did inhibit AM colonization, this effect was exacerbated by the ethylene inhibition of AM root colonization. Unfortunately, experiments using ethylene overproducing mutants and ethylene insensitive mutants of tomato yielded somewhat ambiguous results in terms of the inhibition of AM colonization by ethylene. However, as discussed in Sect. 9.3, experiments with ethylene lowering PGPB clearly demonstrated that ethylene was inhibitory to AM colonization of plants.

While it has been known for some time that strigolactones stimulate hyphal branching and growth (which leads to increased contact with plant roots), until recently it has not been clear whether or not strigolactones also stimulate early

Fig. 9.6 Schematic
representation of bacteria
bound to the roots or
mycorrhizae of plants. Plant
roots are depicted in brown
and bacteria in red

steps of root colonization. More recently, it has become clear that strigolactones act directly on the fungi and thus do not play a role as a plant hormone per se. This is demonstrated by the fact that strigolactone deficient mutants form fewer arbuscular mycorrhizae but they do not display altered development of the nutrient exchange unit, the arbuscule, and mutant deficient in a strigolactones receptor do not form fewer arbuscules.

9.3   Mycorrhizae and Bacteria

In the soil, mycorrhizal fungi interact with a large number of different microorganisms, many of which can affect the mycorrhizal–plant symbiosis. Among the many soil microorganisms, a number of bacteria, sometimes called mycorrhiza helper bacteria (MHB) actively bind to both plant roots and mycorrhizal hyphae and promote the mycorrhizal symbiosis (Fig. 9.6). Different MHB can affect mycorrhiza in a variety of ways including (i) stimulating spore germination and mycelial extension; (ii) increasing mycorrhizal root colonization; (iii) reducing the impact of adverse environmental conditions, e.g., by inhibiting microbes that are antagonistic to mycorrhizal fungi; (iv) facilitating nutrient mobilization from soil minerals; (v) contributing to mycorrhizal nutrition by providing nitrogen (in the case of diazotrophs) or iron (from bacterial siderophores); or (vi) by directly lowering plant stress. Table 9.3 summarizes the results of some studies where the combination of both AM and PGPB have been found to enhance plant growth.

Some scientists have suggested that of the three major types of bacteria that are found to be associated with ectomycorrhizae, i.e., *Pseudomonas*, *Bacillus*, and *Streptomyces*, some typically have positive while other have negative effects on this fungus. That is, it is thought that *Pseudomonas* are generally helper bacteria, *Bacillus* generally have little or no effect on the binding of the AM fungus and *Streptomyces* tends to inhibit the fungus. While this generalization may be true in some instances, whether a bacterium has a positive, neutral or negative effect on the

Table 9.3 The effect of the inoculation of both AM and PGPB on various plants

Plant species	AM applied	PGPB applied	Result
Lycopersicon esculentum (Tomato)	Glomus intraradices	Pseudomonas fluorescens, Enterobacter cloaceae	Reduced Fusarium infection
Glycine max (Soybean)	Glomus clarum	Bradyrhizobium japonicum	Increased number of nodules
Trifolium repens (White clover)	Glomus mosseae	Brevibacillus brevis	Increase shoot and root dry weight
Cicer arietinum (Chickpea)	Glomus intraradices	Rhizobium sp. and Pseudomonas straita	Reduced nematode population and number of galls
Capsicum chinense (Bonnet pepper)	Mixed AM inoculum	Azotobacter chroococcum and Azospirillum brasilense	Increased number of leaves, plant height and fresh weight
Cucumis sativum (Cucumber)	Gi. rosea	Pseudomonas putida	Increased root and shoot fresh weight and leaf area
Capsicum annuum (Sweet pepper)	Mixed AM inoculum	Methylobacterium oryzae	Increased root and shoot dry weight and N, P, K content
Carthamus tinctorius (Safflower)	Glomus intraradices	Azotobacter chroococcum	Increased root dry weight and grain yield
Sesamum indicum (Sesame seeds)	G. fasciculatum, Acaulospora laevis	Pseudomonas straita	Increased root and shoot length and P content
Phaseolus vulgaris (Common bean)	Glomus mosseae, Glomus intraradices	Rhizobium tropici	Increased shoot and pod dry weight
Lactuca sativa (Lettuce)	Glomus mosseae	Pseudomonas mendocina	Increased root and shoot dry weight
Triticum aestivum (Wheat)	Glomus mosseae	Paenibacillus polymyxa and Paenibacillus brasilensis	Increased shoot length and dry weight
Oryza sativa (Rice)	Glomus intraradices	Azospirillum brasilense	Increased shoot length and photosynthetic efficient

functioning of ectomycorrhizae is largely a consequence of the physiology and biochemistry of the bacterium and is not necessarily related to its genus and species.

In one series of experiments, *Capsicum annum* plants (producing spice peppers) were treated with either the AM *Glomus deserticola*, the PGPB *Azospirillum brasilense*, or both of these organisms. When the plants were grown in the field in nutrient deficient soil, the plants that were treated with both the AM and the PGPB grew to the biggest size, took up more phosphorus and nitrogen from the soil (although some of the increased nitrogen may have been from nitrogen fixation by the *Azospirillum*), and produced more fruit (peppers) than the plants treated with

Table 9.4 Influence of AM (*Glosmus deserticola*) and PGPB (*Azospirillum brasilense*) on the growth of pepper plants (*Capsicum annum*) in the field

Treatment	Plant height, cm	Total plant P, mg/g dry weight	Total plant N, mg/g dry weight	Fruit dry weight, g per fruit
No additions	60	0.2	2.02	1.62
Glosmus deserticola	70	0.58	4.46	1.75
Azospirillum brasilense	66.4	0.38	3.15	1.89
G. deserticola + A. brasilense	74.1	0.64	6.01	2.02

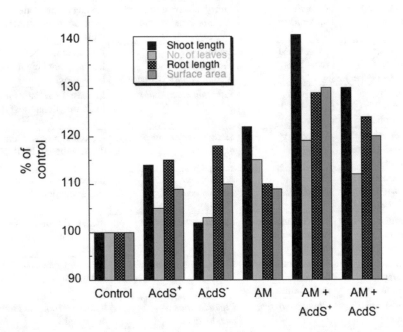

Fig. 9.7 Effect of mycorrhiza (AM) and PGPB on the growth of cucumber plants. In all cases the values are expressed as the percentage of the control value. The AM is *Gigaspora rosea* while the PGPB is *Pseudomonas* sp. UW4. The wild-type strain (with ACC deaminase) is indicated by the symbol AcdS$^+$ while the mutant strain (minus ACC deaminase) is indicated by the symbol AcdS$^-$

only one (or neither) of the AM and PGPB (Table 9.4). This is a good demonstration of how the synergism between these two types of organisms may be used to facilitate plant growth.

In one series of experiments, the PGPB *Pseudomonas* sp. UW4 was shown to facilitate the interaction between the AM fungus *Gigaspora rosea* and cucumber plants. These experiments tested the ability of both the wild-type bacterium and an ACC deaminase minus mutant of this strain to enhance plant growth (Fig. 9.7). These experiments clearly showed that for every parameter that was measured, including plant shoot length, number of leaves, root length and leaf surface area, maximal plant growth was achieved when both the AM fungi and the wild-type

PGPB bacterial strain were present together from the outset of the experiment. Interestingly, the next best condition in terms of promoting plant growth was when the AM fungi and the ACC deaminase minus mutant were present together. This result suggests that while the ACC deaminase from strain UW4 significantly contributes to the observed effects on plant growth, other bacterial activities (such as IAA production) are probably also important in promoting plant growth. Thus, strain UW4, which can facilitate plant growth under a range of stressful conditions can act as an MHB. This may occur by decreasing the small amount of stress incurred by the plants as a consequence of mycorrhizal infection but may also include other bacterial mechanisms. The abovementioned experiment, and many other experiments with AM and MHB have been performed under controlled greenhouse conditions where the plants were grown under near optimal conditions. However, it would certainly be of interest to ascertain that the affect that this and other MHB have on mycorrhizal infection, and ultimately on plant growth, is reproducible under field conditions.

In agriculture, what growers are most concerned about is not how to increase the biomass of the nonedible parts of the plant but rather how to increase the amount and quality of the edible product (e.g., fruit or grain) that the plant produces. Having previously ascertained that AM and PGPB (both separately and together), using different mechanisms, are beneficial for plant growth and development, one group of researchers examined in detail the effect of AM and PGPB on the growth and nutritive value of tomatoes (Table 9.5). In the first instance, these researchers compared tomato growth and nutritional content between tomatoes fertilized in the standard way with tomatoes fertilized using only 70% of the normal amount of fertilizer. Next, tomatoes were grown with 70% of the normal amount of fertilizer plus a commercial mixture of five different AM strains together with a PGPB strain. In the first instance, the collar diameter of the tomato fruit, the number of flowers per plant, and the total number of fruits were significantly decreased when the standard fertilizer was replaced by the 70% fertilizer. However, these values were partially restored in the presence of the AM and PGPB. Fructose is the most important sugar for tomato sweetness, and it was found to be at its highest level in the presence of the AM and PGPB. As a result of this study, it was concluded that the inoculation of tomato plants with AM and PGPB can significantly reduce the use of chemical fertilizers while generally maintaining or even improving the tomato fruit yield and quality.

Finally, it is interesting to note that some researchers have observed that in addition to augmenting crop yields in mycorrhiza treated plants, some MHB can also stimulate anthocyanin levels in some plants. Thus, in strawberry fruits, the presence of AM increases the concentration of the antioxidant anthocyanin cyaniding-3-glucoside (responsible for the fruit's red color) and the addition of an MHB further increases the amount of this compound.

9.4 Mycorrhizae and Stress

Not surprisingly, a large number of studies have revealed that many different abiotic stresses can inhibit mycorrhizal spore germination and hyphae elongation with the result that mycorrhizal root colonization is decreased. These abiotic stresses include

Table 9.5 The effect of AM (a commercial mixture of five different strains), and a PGPB (*Pseudomonas fluorescens* C7) on the size and nutritional content of tomato fruits

Parameter	Standard fertilization	70% fertilization	70% fertilization + AM + PGPB
Average fruit weight, g	64.3	64.4	71.3
Average fruit length, cm	5.81	5.49	6.05
Average fruit diameter, cm	4.62	4.24	4.78
% Water	94.79	94.71	94.34
% Dry biomass	5.20	5.29	5.66
Glucose, g/kg	10.45	11.00	10.66
Fructose, g/kg	10.77	11.14	11.91
Malate, mg/100 g	18.68	17.31	19.04
Ascorbate, mg/100 g	5.47	7.12	4.30
Citrate, g/100 g	0.15	0.14	0.16
Nitrate, mg/kg	41.62	63.27	46.85
pH	4.3	4.4	4.5
β-Carotene, μg/100 g FW	2.224	2.178	2.829
Lycopene, μg/100 g FW	2799.9	2889.3	2408.2
Luteine, μg/100 g FW	1.381	1.377	1.501

the presence in the soil of petroleum products, polycyclic aromatic hydrocarbons, fungicides, metals, salinity, drought, flooding, cold temperature, and high calcium carbonate concentrations. These negative effects may result from either a direct effect on the AM or from alterations to plant roots. To protect themselves from inhibition by reactive oxygen species (produced as a consequence of abiotic stress) including superoxide anion radical, hydrogen peroxide, hydroxyl radicals and singlet oxygen, AM synthesize a number of small molecules that act as antioxidants as well as chaperone proteins to prevent protein misfolding or aggregation. Like many PGPB and plants, AM have also been shown to synthesize trehalose in response to heat shock. Moreover, as a consequence of significantly extending the region of soil covered by plant roots, AM mitigate plant stress that might otherwise result from insufficient levels of water and high levels of salt. This is in part because AM increase the availability of water and nutrients from the soil. In addition, AM partially exclude the influx of sodium ions under salt stress conditions thereby decreasing the build-up of toxic sodium ions within plant tissues.

The resilience of a wide range of host plants to both drought and salt stress has been shown to result from the presence of AM. In overcoming drought and salt stress, the presence of AM has been correlated with accumulation of proline; phosphorus acquisition; increased levels of potassium, zinc, copper, magnesium, and calcium; an increase in antioxidant enzymes (such as superoxide dismutase, catalase, peroxidase, and ascorbate peroxidase); lowered abscisic acid levels; higher strigolactones levels; and increased water levels. The host plants that AM has been shown to protect include (but are not limited to): pepper, watermelon, muskmelon, lettuce, cabbage, eggplant, pepper, cabbage, tomato, corn, mungbean, onion, basil, clover, sorghum, zucchini squash, carrot, cucumber, garlic, broccoli, mint, radish, and spinach. In addition, based on numerous reports, colonization of the roots a number of different plants (i.e., both C_3 and C_4 plants) by AM results in a much healthier plant that is much more resistant to drought and salt stress.

9.4.1 Mycorrhizae and Viruses

In addition to pathogenic fungi and bacteria with which mycorrhizae work together with PGPB to protect plants, plant viruses generally induce a disease syndrome in infected plants and potentially cause large losses in plant productivity and quality. Since insects are generally the main way that viruses are spread between plants, and climate change favors insect colonization of various new habitats, many viral diseases may be considered to be an emerging problem in modern agriculture. Currently, one of the main approaches that is employed to limit plant viral diseases is through the use of transgenic plants that express various viral components (e.g., viral coat protein) that interfere with viral assembly or functioning. However, this approach is not effective when dealing with uncharacterized emerging (new) viral pathogens. Thus, one suggested method for dealing with these new pathogens is through the use of AM.

In a small number of instances, it has been shown that prior AM colonization of plants reduced the effectiveness of infecting viruses (in causing disease in plants). This prophylactic effect of AM on plant virus infection is currently thought to be a consequence of the induction of hormone biosynthesis by the AM. However, other literature reports indicate that mycorrhizal colonization of some plants enhances virus multiplication. In this instance, it has been suggested that this effect might be a consequence of the increased phosphorus availability that is found in mycorrhizal plants. Despite the limited scientific literature regarding the interactions between mycorrhizae, viruses, and plants, it has been reported that in addition to mycorrhizae affecting a plant's interaction with pathogenic viruses, viral infection can sometimes negatively impact the ability of mycorrhizae to colonize a plant's roots.

Questions
1. What are the differences between ectomycorrhizae and arbuscular mycorrhizae?
2. How are mycorrhizae helpful for plant growth?
3. What are strigolactones?
4. How do various phytohormones affect the colonization of plant roots by mycorrhizae?
5. How would you prove that ethylene is inhibitory to the colonization of plant roots by mycorrhizae?

6. How would you select and characterize new strains of mycorrhizae that facilitate plant growth, protect plants against salt stress and protect plants against inhibition by root-knot nematodes?
7. Suggest five different methods for inoculating plants with mycorrhizae.
8. How do different phytohormones affect the interaction between mycorrhizae and plants?
9. How can some mycorrhizae protect plants against various types of abiotic stress?
10. How does the presence of various levels of phosphorus fertilizer in the soil affect the interaction between plant roots and mycorrhizae?
11. What are some advantages of using both AM and PGPB?
12. What are mycorrhizae helper bacteria?
13. How do mycorrhizae and viruses affect one another's interactions with plants?

Bibliography

Altuntas O, Kustal IK (2018) Use of some bacteria and mycorrhizhae as biofertilizers in vegetable growing and beneficial effects in salinity and drought stress conditions. In: El-Esawi MA (ed) Physical methods for stimulation of plant and mushroom development. IntechOpen, London. https://doi.org/10.5772/intechopen.69094

Bona E, Cantamessa S, Massa N, Manassero P, Marsano F, Copetta A, Lingua G, D'Agostino G, Gamalero E, Berta G (2017) Arbuscular mycorrhizal fungi and plant growth-promoting pseudomonads improve yield, quality and nutritional value of tomato: a field study. Mycorrhiza 27:1–11

Bona E, Todeschini V, Cantamessa S, Cesaro P, Copetta A, Lingua G, Gamalero E, Berta G, Massa N (2018) Combined bacterial and mycorrhizal inocula improve tomato quality at reduced fertilization. Sci Hortic 234:160–165

Bonfante P, Anca I-A (2009) Plants, mycorrhizal fungi, and bacteria: a network of interactions. Annu Rev Microbiol 63:363–383

Brito I, Goss MJ, Alho L, Brígido C, van Tuinen D, Félix MR, Carvalho M (2019) Agronomic management of AMF functional diversity to overcome biotic and abiotic stresses – the role of plant sequence and intact extraradical mycelium. Fungal Ecol 40:72–81

Brundrett MC (2002) Coevolution of roots and mycorrhizas of land plants. New Phytol 154:275–304

Campos C, Nobre T, Goss MJ, Faria J, Barrulas P, Carvalho M (2019) Transcriptome analysis of wheat roots reveals a differential regulation of stress responses related to arbuscular mycorrhizal fungi and soil disturbance. Biology 8:93

Chandrasekaran M, Chanratana M, Kim K, Seshadri S, Sa T (2019) Impact of arbuscular mycorrhizal fungi on photosynthesis, water status, and gas exchange of plants under salt stress-a meta-analysis. Front Plant Sci 10:457

Chen M, Arato M, Borghi L, Nouri E, Reinhardt D (2018) Beneficial services of arbuscular mycorrhizal fungi – from ecology to application. Front Plant Sci 9:1270

Da Costa LS, Grazziotti PH, Silva AC, Fonseca AJ, Gomes ALF, Grazziotti DCFS, Rossi MJ (2019) Alginate gel entrapped ectomycorrhizal inoculum promoted growth of cuttings of *Eucalyptus* clones under nursery conditions. Can J For Res 49:978–985

Evelin H, Devi TS, Gupta S, Kapoor R (2019) Mitigation of salinity stress in plants by arbuscular mycorrhizal symbiosis: current understanding and new challenges. Front Plant Sci 10:470

Foo E, Ross JJ, Jones WT, Reid JB (2013a) Plant hormones in arbuscular mycorrhizal symbioses: an emerging role for gibberellins. Ann Bot 111:769–779

Foo E, Yoneyama K, Hugill C, Quittenden L, Reid JB (2013b) Strigolactones and the regulation of pea symbioses in response to nitrate and phosphate deficiency. Mol Plant 6:76–87

Frey-Klett P, Garbaye J, Tarkka M (2007) The mycorrhiza helper bacteria revisited. New Phytol 176:22–36

Gamalero E, Berta G, Massa N, Glick BR, Lingua G (2008) Synergistic interactions between the ACC deaminase-producing bacterium *Pseudomonas putida* UW4 and the AM fungus *Gigaspora rosea* positively affect cucumber plant growth. FEMS Microbiol Ecol 64:459–467

Goh DM, Cosme M, Kisiala AB, Mulholland S, Said ZMF, Spíchal L, Emery RJN, Declerck S, Guinel FC (2019) A stimulatory role for cytokinin in the arbuscular mycorrhizal symbiosis of pea. Front Plant Sci 10:262

Gomez-Roldan V, Fermas S, Brewer PB, Puech-Pagés V, Dun EA, Pillot JP, Letisse F, Matusova R, Danoun S, Portais JC, Bouwmeester H, Bécard G, Beveridge CA, Rameau C, Rochange SF (2008) Strigolactone inhibition of shoot branching. Nature 455:189–194

Gutjahr C, Gobbato E, Choi J, Riemann M, Johnston MG, Summers W, Carbonnel S, Mansfield C, Yang SY, Nadal M, Acosta I (2016) Rice perception of symbiotic arbuscular mycorrhizal fungi requires the karrikin receptor complex. Science 350:1521–1524

Lenoir I, Fontaine J, Sahraoui AL-H (2016) Arbuscular mycorrhizal fungal responses to abiotic stresses: a review. Phytochemistry 123:4–15

Lingua G, Bona E, Manassero P, Marsano F, Todeschini V, Cantamessa S, Copetta A, D'Agostino G, Gamalero E, Berta G (2013) Arbuscular mycorrhizal fungi and plant growth-promoting pseudomonads increases anthocyanin in strawberry fruits (*Fragaria x ananassa* var. Selva) in conditions of reduced fertilization. Int J Mol Sci 14:16207–16225

McGuiness PN, Reid JB, Foo E (2019) The role of gibberellins and brassinosteroids in nodulation and arbuscular mycorrhizal associations. Front Plant Sci 10:269

Miozzi L, Vaira AM, Catoni M, Fiorilli V, Accotto GP, Lanfranco L (2019) Arbuscular mycorrhizal symbiosis: plant friend or foe in the fight against viruses? Front Microbiol 10:1238

Parihar M, Meena VS, Mishra PK, Rakshit A, Choudhary M, Yadav RP, Rana K, Bisht JK (2019) Arbuscular mycorrhiza: a viable strategy for soil nutrient loss reduction. Arch Microbiol 201:723–735

Parniske M (2008) Arbuscular mycorrhiza: the mother of plant root endosymbiosis. Nat Rev Microbiol 6:763–775

Ramasamy K, Joe MM, Kim K, Lee S, Shagol C, Gangasamy A, Chung J, Islam MR, Sa T (2011) Synergistic effects of arbuscular mycorrhizal fungi and plant growth promoting rhizobacteria for sustainable agricultural production. Korean J Soil Sci Fert 44:637–649

Recchia GH, Konzen ER, Cassieri F, Caldas DGG, Tsai SM (2018) Arbuscular mycorrhizal symbiosis leads to differential regulation of drought-responsive genes in tissue-specific root cells of common bean. Front Microbiol 9:1339

Séry DJ-M, Kouadjo ZGC, Voko BRR, Zézé A (2016) Selecting native arbuscular mycorrhizal fungi to promote cassava growth and increase yield under field conditions. Front Microbiol 7:2063

Temegne NC, Foh TDN, Taffouo VD, Wakem G-A, Youmbi E (2018) Effect of mycorrhization and soluble phosphate on growth and phosphorus supply of Voandzou [*Vigna subterranea* (L.) Verdc.]. Legum Res 41:879–884

Terrer C, Vicca S, Hungate BA, Phillips RP, Prentice IC (2016) Mycorrhizal association as a primary control of the CO_2 fertilization effect. Science 353:72–74

Vyas M, Vyas A (2014) Field response of *Capsicum annum* dually inoculated with AM fungi and PGPR in Western Rajasthan. Int J Res Stud Biosci 2:21–26

Phytoremediation

10

Abstract

The use of plant growth-promoting bacteria need not be limited to agriculture and horticulture. In this chapter, the use of these bacteria to facilitate the remediation of polluted soil and water is considered. For organic pollutants, plant growth-promoting bacteria in concert with plants can help to break down the pollutants rendering them harmless. In metal and metaloid contaminated soils, plant growth-promoting bacteria aid in the uptake of the metallic contaminants by the plants. In both types of phytoremediation, organics and metals, the mechanisms used by the bacteria to aid this process are examined in some detail. While neither of these approaches is as yet commercially viable, phytoremediation of organic compounds works quite well in some field experiments so that this means of cleaning the environment shows a great deal of promise.

10.1 Problem Overview

The problem of toxic waste disposal has accelerated dramatically since the beginning of the industrial revolution. The problems caused by this waste are enormous. It has been estimated by the United States Environmental Protection Agency that more than 40 million tons of hazardous waste are produced each year in the United States. Of the many chemicals found in hazardous waste sites in the United States the ATSDR (Agency for Toxic Substance and Disease Registry lists a number of chemicals as priority substances) (Table 10.1). Not shown in this table, cyanide is number 35 on the ATSDR list, cobalt is number 51, nickel is number 57, zinc is number 75 and asbestos is number 94. Altogether, the ATSDR lists 275 compounds as priority substances, all of which are considered to pose a significant threat to human health. Here it is important to point out that the priority list does not rank substances based solely on their toxicity, but rather on a combination of their

© Springer Nature Switzerland AG 2020
B. R. Glick, *Beneficial Plant-Bacterial Interactions*,
https://doi.org/10.1007/978-3-030-44368-9_10

Table 10.1 The United
States Department of
Health and Human
Services, ATSDR list of
Priority Toxic Substances
for 2017

Rank	Substance
1	Arsenic
2	Lead
3	Mercury
4	Vinyl chloride
5	Polychlorinated biphenyls (PCBs)
6	Benzene
7	Cadmium
8	Benzo(A)pyrene
9	Polycyclic aromatic hydrocarbons (PAHs)
10	Benzo(B)fluoranthene
11	Chloroform
12	Arochlor1260 (a specific PCB)
13	Dichlorodiphenyltrichloroethane (DDT)
14	Arochlor 1254 (a specific PCB)
15	Dibenzo(A,H)anthracene
16	Trichloroethylene
17	Hexavalent chromium
18	Dieldrin (a chlorinated hydrocarbon)
19	White phosphorus
20	Hexachlorobutadiene

toxicity, their frequency in the environment, and for the possibility that humans will
be exposed to them.

Scientists and engineers have developed several technologies and methods to
remove toxic compounds from polluted environments. The majority of these
methods include the physical removal of soil to landfill sites or extraction of
contaminated soil or water through chemical or physical means. Unfortunately, the
majority of these procedures are prohibitively expensive. Moreover, many
contaminated environments contain mixtures of contaminants that are not readily
dealt with in a simple manner. Thus, many soils are contaminated with one or more
metals, other inorganic compounds, radioactive material or various organic
compounds. The metals may include lead, zinc, cadmium, selenium, chromium,
cobalt, copper, nickel, or mercury; the other inorganic compounds might include
arsenic, sodium, nitrate, ammonia, or phosphate; the radioactive compounds may be
uranium, cesium, or strontium; and the organic compounds may include trichloroethylene, trinitrotoluene (TNT); benzene, toluene, and xylene (BTX); polycyclic
aromatic hydrocarbons (PAHs); polychlorinated biphenyls (PCBs); and pesticides
such as atrazine and bentazon.

Some organic compounds can be broken down by bacteria that may either be
naturally present in or added to the soil, in the absence of plants; however, this
process is usually quite slow and inefficient, in part as a consequence of the relatively
low concentration of these degradative bacteria in soil. On the other hand, the use of
plants to facilitate the remediation of polluted soils, i.e., phytoremediation, is a

Table 10.2 Overview of phytoremediation technologies

Technology	Definition
Phytodegradation	Biodegradation of contaminants is facilitated
Phytostimulation	Plant roots stimulate soil microbial communities in plant root zones to break down contaminants; also called rhizodegradation
Rhizofiltration	Plant roots acquire contaminants from solution
Phytoextraction	Plant accumulate contaminants from soil
Phytostabilization and Phytoimmobilization	Roots and their exudates immobilize contaminants within the root zone, and thus prevent their spreading
Phytovolatilization	Contaminants are taken up by the roots through the plants to the leaves and are volatized through stomata
Phytotransformation	Contaminants are transformed to a less harmful substance

potentially clean, effective and relatively inexpensive technology that is likely to be readily accepted by a concerned public. However, this approach also has its drawbacks, as few plant species can tolerate high concentrations of environmental contaminants. Some organic compounds can be directly degraded and completely mineralized by plant enzymes through phytodegradation; many plants produce, and often secrete to the environment, enzymes that can degrade a wide range of organic compounds. However, inorganic pollutants cannot be degraded. Inorganic pollutants may be either stabilized in the soil to make them less bioavailable (i.e., phytostabilization); or extracted, transported, and accumulated in plant tissues (i.e., phytoextraction); or transformed into volatile forms (i.e., phytovolatilization). Table 10.2 summarizes the possible fates of various environmental contaminants that occur as a consequence of phytoremediation (Fig. 10.1).

Given the fact that phytoremediation is still a relatively new and not completely proven technology, it is useful to review some of the advantages and disadvantages of this approach. In this regard, the perceived advantages of phytoremediation technology include the following (Table 10.3):

- The cost of using a phytoremediation approach as compared to soil washing or soil removal or other more traditional methods of cleaning contaminated environmental sites has been estimated to be from 4 to 20 times less expensive.
- Phytoremediation is a passive solar-driven process that does not disturb the environment to any great extent.
- Compared to other more intrusive methods, phytoremediation has a relatively high measure of public acceptance.
- Phytoremediation does not disturb or remove topsoil.
- In the field, phytoremediation does not generate any secondary wastes.

On the other hand, phytoremediation is also considered to have a number of potential drawbacks including the following:

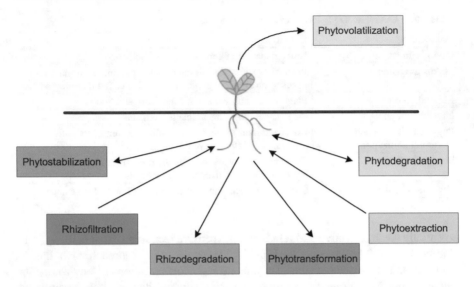

Fig. 10.1 Schematic view of various phytoremediation technologies. These technologies are briefly defined in Table 10.2. The direction of the arrow indicates where the phytoremediation most likely takes place

Table 10.3 Some advantages and disadvantages of phytoremediation

Advantages	Disadvantages
1. Less costly than mechanical methods	1. Only accesses shallow contaminants
2. Solar driven	2. Much slower than mechanical methods
3. Good public acceptance	3. Contaminants may be toxic to plants
4. Topsoil is retained	4. Some biodegradative products may be toxic
5. Generates less secondary waste	5. Contaminants could enter the food chain

- Effective phytoremediation may often be limited to dealing with contaminants that are found within the top few cm of the soil because of the fact that except for trees, the roots of most plants are not especially deep.
- To be effective and remove a sufficient amount of a particular contaminant from soil (or water) and thereby satisfy government mandated contamination limits, several seasons of phytoremediation are likely to be required. In practice, this could take years with the contaminated land being taken out of use for an extended period of time.
- Many environmental contaminants are toxic to plants with the result that plants intended to be part of a phytoremediation protocol may grow only very slowly or to a very limited extent in the presence of the target contaminant.
- In the process of breaking down some organic contaminants occasionally bio-degradation products will be produced that are as toxic (or even more toxic) to plant and animal life than the original contaminant. Thus, it is possible that phytoremediation might actually make things worse.

- With phytoremediation, there may be a real chance that the targeted contaminants can enter the food chain. For example, photovolatilization could result in a particular contaminant becoming airborne, eventually contaminating a food crop in a farmer's field that is some distance from the original contaminated site. In addition, contaminants that are tightly bound to soil particles could become more bioavailable (and soluble) and subsequently spread to nearby farmer's fields as runoff following a heavy rainstorm.

10.2 Metals

Until now, the standard approach to cleaning metal-contaminated sites has been physical removal of the contaminants followed by landfilling. However, this relatively expensive option merely moves the contaminants from one location to another. Thus, it is essential that alternative methods of cleaning metal-contaminated sites, such as phytoremediation, be developed.

Plants that are exposed to relatively high concentrations of metals (depending on the metal involved) may exhibit a decrease in both root and shoot development, some level of iron deficiency, damage to the plant resulting from the presence of reactive oxygen species, lower than normal levels of chlorophyll, an increase in the expression of stress proteins and an increase in stress ethylene production. Only a small number of plant species can naturally tolerate/accumulate metals to any significant extent. Unfortunately, many of the plants that are most effective at removing metals from the soil (metal hyperaccumulators) are both small in size and slow growing. Thus, many of these so-called metal hyperaccumulating plants are not especially useful as a component of phytoremediation protocols. In addition, to be more effective for soil remediation, plants should be tolerant to one or more pollutants (often including several different metals), highly competitive, fast growing and produce as high a level of biomass as possible. One way to improve the usefulness of many plants in phytoremediation protocols is to include specific soil bacteria along with the plants. These bacteria may contain one or more of the traits discussed below.

To date, a few practical problems have prevented the more rapid development of metal phytoremediation technology. First, in many metal contaminated soils, the metals are tightly bound to soil particles so that they are not readily taken up by growing plants. These metal contaminants are said to not be bioavailable. Researchers are trying to address this issue by employing chemical compounds (either directly added to the soil or through the use of bacteria that produce these compounds) that can solubilize or release the bound metals thereby making them more bioavailable for uptake by growing plants. The second problem in the development of practical metal phytoremediation strategies is the fact that many, if not most, of the plants that take up metals from the soil localize those metals preferentially within the plant roots and only translocate a small fraction of that metal to the rest of the plant. Thus, to remove a metal from soil, after the metal has been taken up by a plant it is necessary to remove the entire plant including the roots from the soil, a

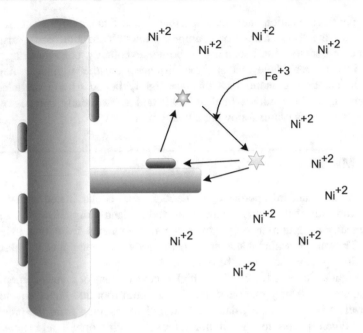

Fig. 10.2 Schematic representation of how a siderophore-producing PGPB strain (red ovals) bound to a plant root or root hair can secrete siderophores (purple star), sequester trace levels of Fe^{+3} in the presence of a very large molar excess of Ni^{+2} (the iron–siderophore complex is shown as a yellow star) and provide some of the bound Fe^{+3} to the plant

difficult and labor-intensive task. A third problem in developing metal phytoremediation strategies relates to the fact that some added bacteria while highly effective at improving the growth of plants in contaminated environments do not increase the uptake of metals into the plants, and in fact may even prevent plant uptake of some metals.

10.2.1 Siderophores

When a large amount of metal (such as nickel) is present in the soil, plants become highly stressed, not only from the damage directly caused by the presence of the contaminating metals, but also because they cannot acquire sufficient iron for all of their growth and developmental needs. However, as discussed in Chap. 4, the siderophores secreted by some PGPB have an extremely high affinity for iron, forming an iron–siderophore complex that can be taken up by the plant where this complex will be cleaved, the iron released and then used by the plant. Thus, siderophores are capable of taking up sufficient Fe^{+3}, even in the presence of enormous molar excesses of contaminating metals (such as nickel), so that the plant will not become starved for iron (Fig. 10.2). Bacterial siderophores may also sometimes form metal–siderophore complexes facilitating the uptake of a contaminating metal into plants. While this may be detrimental for plant growth, it may act to increase the amount of metal contaminant that a plant–PGPB combination can take up from the environment.

In one series of experiments, researchers selected for a naturally occurring mutant of a PGPB strain that overproduced siderophores from a large population of bacteria (Fig. 10.3). This procedure selects for the genetic variation that normally exists in a bacterial population (these mutations naturally occur in about one in every million or so bacteria). Thus, it is not necessary to mutagenize bacterial cells chemically or with radiation in order to find siderophore overproducing cells. In this regard, it is likely that siderophore overproduction occurs as a result of a mutation in a siderophore gene(s) regulatory region.

In one experiment, tomato plants were grown either with no added PGPB, a wild-type PGPB or a siderophore overproducing PGPB whose selection is described above, all in the presence of 1 mM nickel added to the soil (Fig. 10.4). Following several weeks of growth in the presence of nickel, the plants were harvested and various growth-associated parameters, including stem length, plant wet weight, and leaf chlorophyll content, were measured. For all of the parameters that were measured, the presence of the wild-type PGPB strain stimulated plant growth compared to the absence of any added bacterium. Moreover, for each of the measured parameters, the mutant PGPB strain promoted growth to a greater extent than the wild-type. Since none of the traits of the PGPB strain other than siderophore production (which was increased several fold) were altered, this data provides direct evidence for the role of bacterial siderophores in facilitating plant growth in the presence of inhibitory amounts of metals. And, the bottom line here is: more plant growth means that more metal will be taken up from the soil by the plant.

10.2.2 IAA

It is well established that PGPB-produced IAA can both directly promote plant growth and protect plants from various environmental stresses such as the presence of high salt concentrations. Therefore, it is not surprising that a number of literature reports indicate that numerous IAA-producing bacteria have been found to either facilitate the uptake of metal contaminants by plants or to protect plants from growth inhibition by the presence of metals in the environment. Most of these studies are focused on identifying bacteria that will make metal phytoremediation work better and, while they suggest that a number of bacterial traits that may be involved, they generally do not provide any definitive proof for the involvement of one or more specific traits. Thus, the evidence for the involvement of IAA in facilitating metal phytoremediation is based on circumstantial and indirect evidence. However, from what is known about IAA's ability to facilitate plant growth, it is quite likely that it plays a key role in facilitating metal phytoremediation.

In one study, *Bacillus subtilis* strain SJ-101 was utilized to facilitate the nickel accumulation in Indian mustard (*Brassica juncea*). Researchers chose to use Indian mustard because it could accumulate moderate levels of nickel while producing a

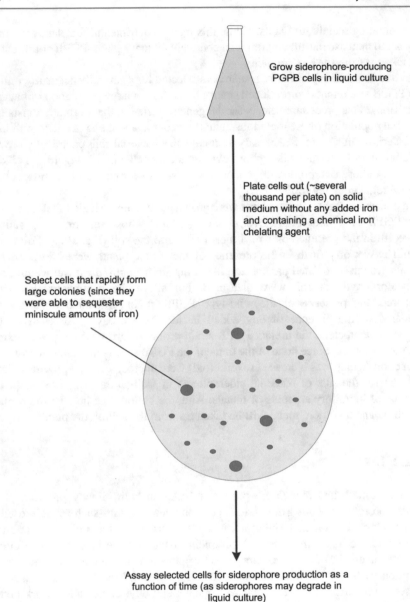

Grow siderophore-producing PGPB cells in liquid culture

Plate cells out (~several thousand per plate) on solid medium without any added iron and containing a chemical iron chelating agent

Select cells that rapidly form large colonies (since they were able to sequester miniscule amounts of iron)

Assay selected cells for siderophore production as a function of time (as siderophores may degrade in liquid culture)

Fig. 10.3 A scheme for selecting a siderophore-overproducing mutant of a PGPB from a large, non-mutagenized, population of bacteria. After growth in liquid culture the cells are plated onto solid medium that does not contain any added iron and also contains a chemical chelating agent that is intended to bind all of the available iron making it less available to the bacteria. Any cells that grow rapidly and form large colonies probably overproduce siderophores, however, this needs to be checked

Fig. 10.4 Several normalized measures of tomato plant growth in the presence of nickel in the soil. All parameters (plant stem length, wet weight and chlorophyll content) were normalized with the value in the absence of any added PGPB being set to 100%

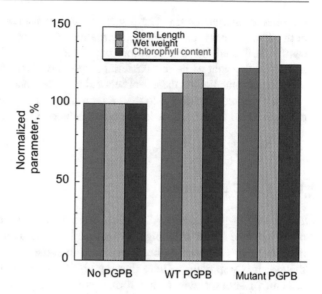

Fig. 10.5 Growth of Indian mustard plants, shoot fresh weight (FW), in the presence or absence of an added IAA-producing PGPB in soil that contains varying amounts of nickel (in µM)

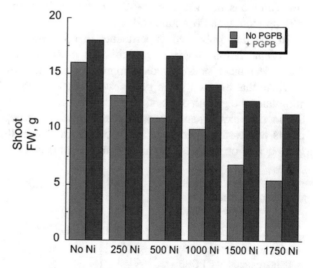

reasonably high level of biomass. Moreover, Indian mustard is relatively efficient at translocating metals from roots to shoots. Strain SJ-101 was used because it is nickel resistant and produces both IAA and phosphate solubilizing activity. As expected, the presence of strain SJ-101 facilitated the growth of plants (as measured by shoot fresh or dry weight, root fresh or dry weight, shoot length, or root length) to a significant extent, especially in the presence of high levels of nickel in the soil (Fig. 10.5). Importantly, the results showed a similar trend regardless of whether the comparison is of shoot fresh or dry weight, root fresh or dry weight, shoot length, or root length. In addition, at all levels of nickel that were tested, the amount of nickel taken up by the plant was increased significantly (typically by around 50%) by the presence of the bacterium. The result of these measurements is that by increasing both plant biomass and metal concentration, metal phytoremediation was significantly facilitated by the addition of strain SJ-101. The one caveat to these experiments is that while this bacterial strain played an important role in enhancing metal phytoremediation, these workers did not prove that IAA per se was responsible for the observed results. To do that, it would have been necessary to create mutants of strain SJ101 that either under- or overproduced IAA and test their efficacy in this system.

10.2.3 ACC Deaminase

The ability of the enzyme ACC deaminase to facilitate plant growth by lowering plant ethylene levels that might otherwise increase as a result of both biotic and abiotic stresses has already been discussed in some detail (see Chaps. 5–8). Nevertheless, it is important to note that the ability of ACC deaminase to decrease the amount of stress ethylene that forms as a consequence of the presence of high concentrations of metals in the soil permits plants to grow larger and therefore take up more metal from the soil.

In the first published experiments that included a PGPB to facilitate metal phytoremediation, it was shown that the presence of a bacterium that contained ACC deaminase both lowered (approximately two- to threefold) the amount of ethylene that evolved when plants were treated with a 2 mM nickel solution and stimulated the growth of plants treated with 1 mM nickel (Fig. 10.6 and Table 10.4). Subsequent experiments from the same research group compared the growth of plants that were subjected to metal contamination in the presence of either wild-type PGPB or an ACC deaminase minus mutant of the wild-type strain. These

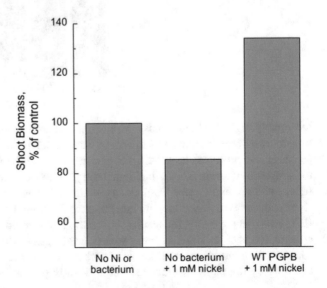

Fig. 10.6 Growth of Indian mustard plants, shoot dry weight percentage of control (no nickel or bacterium added), in the presence or absence of an ACC deaminase-producing PGPB in the soil. The nickel concentration added periodically was 1 mM

Table 10.4 Effect of an ACC deaminase-containing PGPB on the growth of canola plants in the presence of 2 mM nickel

Treatment	Shoot length (cm)	Shoot fresh weight (g)	Root fresh weight (g)
− PGPB	12.2	163	16.7
+ PGPB	16.1	290	37.0

experiments clearly demonstrated that the ACC deaminase minus mutants provided the plants with only a very small benefit (compared to the benefit provided by the wild-type strain) in terms of growth in the presence of metal contaminants. Most likely, this small benefit reflected the fact that the ACC deaminase minus mutant could still synthesize both IAA and siderophores. Nevertheless, this experiment clearly demonstrates that one of the major mechanisms used by PGPB to facilitate metal phytoremediation includes the lowering of stress ethylene levels by the enzyme ACC deaminase.

Given the success of many researchers in demonstrating that soil bacteria that contain both ACC deaminase and resistance to the metal of interest, it has become common practice in recent studies for researchers to use these traits as the starting point for the isolation of new PGPB that will be useful in metal phytoremediation. Typically, researchers first isolate a few hundred bacterial strains from the rhizosphere soil of a host plant growing in the presence of metal contamination (notwithstanding that the rhizosphere may contain a mixture of both rhizospheric and endophytic bacteria, both of which may be useful). Here, it should be noted that many metal-contaminated soils are also salt-affected. In this case, the selection of appropriate PGPB, of necessity, must include the isolation of PGPB that are halophytic as well as metal resistant. These strains are then tested for their ability to proliferate on minimal medium with ACC as the sole nitrogen source, this is commonly around 10–40% of the bacterial strains originally isolated. Next, bacterial strains are tested for the ability to synthesize IAA and siderophores and to solubilize inorganic phosphate. In this way, the original several hundred strains may be winnowed down to 5–25 potentially useful strains. These strains are then tested under controlled conditions for the ability to facilitate the growth of the host plant in the presence of high levels of the target metal contaminant. This usually allows researchers to settle on two or three strains which are then subjected to field trials. Minor variations on this strategy have been repeated successfully in dozens of labs throughout the world so that there are currently a large number of laboratories worldwide at the field-testing stage and a few laboratories that have commercialized this approach, albeit on a small scale.

In some instances, metals such as cadmium can pose a human and animal health risk at levels that are not at all toxic to plants. Thus, it may be desirable in those circumstances to find, if possible, conditions in which plant cadmium uptake is decreased. One group of researchers tested whether cadmium uptake into three different plants (*Rumex palustris*, *Alcea aucheri* and *Arabidospsis thaliana*) might be affected by the ethylene level in those plant. They grew the plants on cadmium-spiked soil in the presence of a rhizospheric ACC deaminase-producing PGPB and

Fig. 10.7 Growth of corn plants, shoot dry weight normalized to the control (C) in which no bacteria were added, in the presence of (from left to right) two different PGPB strains that naturally contained ACC deaminase (+ ACC-D), two PGPB strains that were able to fix nitrogen (+N Fixn), and two strains that could both fix nitrogen and contained ACC deaminase (+ ACC-D, + N-Fixn)

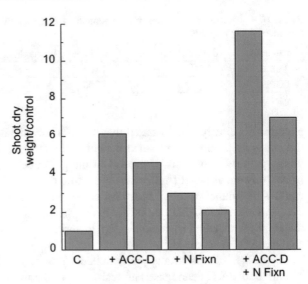

an ACC deaminase minus mutant of the same bacterium. With all three plants, the wild-type PGPB decreased cadmium accumulation within the plant shoots (by as much as 45%) while the mutant PGPB increased cadmium accumulation (by as much as 30%). Thus, higher levels of plant ethylene (produced in the presence of the ACC deaminase minus mutant) appear to result in higher levels of cadmium in plant shoots while the lower level of ethylene that is present in the plants treated with the wild-type bacterium results in much less cadmium being taken up and translocated to the shoots. Whether this effect is also the case for various crop plants and whether this approach can reduce cadmium levels in plants to levels below those that are toxic to humans and animals remains to be determined. However, based on other reports regarding cadmium uptake and translocation within plants, it appears that factors other than ACC deaminase, such as IAA, may be involved in cadmium uptake and translocation.

10.2.4 Nitrogen Fixation

Recently, one group of workers compared six different PGPB for the ability to promote corn plant growth in the presence of lead in the soil. Two of the PGPB strains contained ACC deaminase, two were nitrogen fixing and two contained ACC deaminase and could also fix nitrogen (Fig. 10.7). In these experiments, corn plants were grown in soils containing different amounts of added lead. Then, shoot and root biomass, shoot and root lead content, as well as (shoot) chlorophyll and (shoot and root) protein were measured. In all instances, the plants derived a small benefit from the ability of the bacteria to fix nitrogen, a larger benefit from the presence of ACC deaminase activity and a very large benefit from the ability of the added PGPB to *both* fix nitrogen and produce ACC deaminase. Similarly, a different group of

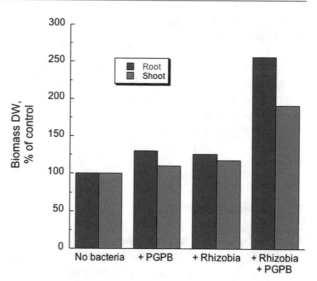

Fig. 10.8 Growth of *Lotus edulis* plants in heavy metal polluted mine waste. The PGPB is a strain of *Variovorax paradoxus* that produces a high level of ACC deaminase activity. The rhizobium strain is *Mesorhizobium loti*. Plant biomass dry weight (DW) is expressed as the percentage of the control values (in the absence of added bacteria) for roots and shoots under greenhouse conditions following 6 weeks of growth

researchers showed that lotus plants that were grown in the presence of mine wastes that contained lead, zinc, and cadmium produced slightly elevated levels of root and shoot biomass when the lotus plant seeds were first incubated with either an ACC deaminase containing PGPB or a nitrogen fixing strain of rhizobia (Fig. 10.8). In this case, the PGPB was expected to lower the plant ethylene levels that occur as a consequence of the metal stress while the rhizobia was expected to increase plant growth because of the presence of additional fixed nitrogen (i.e., a fertilizer effect). Interestingly, it was only when the plants were treated with both the PGPB and the rhizobia strain that dramatic increases in plant growth were observed.

The low level of rainfall in arid and semiarid regions of the world make plants in these zones readily affected by the presence of toxic metals such as lead, cadmium, nickel, cobalt, iron, zinc, chromium, silver, titanium, and copper. In one study, researchers found that *Prosopis laevigata* trees (*Prosopis* spp. or mesquite are small leguminous trees native to the southwestern US and Mexico) that normally grow in metal polluted soils in Mexico can hyperaccumulate aluminum, iron, titanium and zinc but not chromium. They then isolated 88 bacterial strains from the nitrogen fixing nodules found on the roots of young plants growing in metal-contaminated soil. When they were tested, none of these bacterial strains was able to induce nodulation, however, several of these strains were able to tolerate high levels of chromium (as well as other metals). One strain in particular was able to tolerate up to 15,000 mg/L chromium. When this *Bacillus* sp. strain was added to *P. laevigata* seedlings, it enabled the plants to grow on relatively high levels of chromium. In essence, the coexistence of a chromium resistant *Bacillus* sp. with nitrogen-fixing rhizobial bacteria in the nodules of this plant increased the fitness of the plant so that it could be used for the phytoremediation of chromium together with other metals.

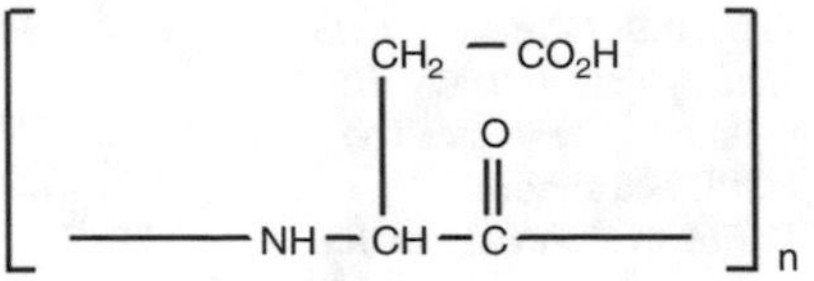

Fig. 10.9 Chemical structure
of polyaspartic acid

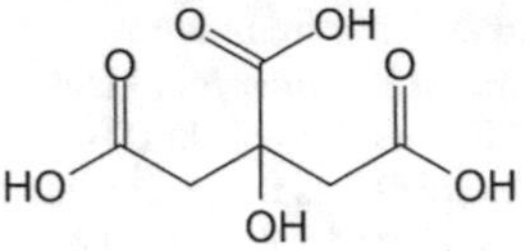

Fig. 10.10 Chemical
structure of citric acid

10.2.5 Biosurfactants

As pointed out earlier, in metal contaminated soils the metals are often tightly bound to soil particles and therefore are not bioavailable. Researchers are trying to address this issue by employing chemical compounds (either directly added to the soil or through the use of bacteria that produce these compounds) that can solubilize or release the bound metals thereby making them more bioavailable for uptake by growing plants. A number of chelating agents, such as ethylenediaminetetraaceticacid (EDTA), trans-1,2-diaminocyclohexane-N,N,N',N'-tetraacetic acid (CDTA), ethyleneglycol-bis (b-aminoethyl ether), N,N,N',N-tetraacetic acid (EGTA), ethylenediamine-N,N'-bis (2-hydroxyphenyacetic acid) (EDDHA), nitrilotriacetic acid (NTA), and organic acids such as citric, malic, acetic and oxalic acid have been used in an effort to desorb metals from soil particles. However, because of their high solubility and persistence in the soil, some chelating agents can cause metal leaching, thereby posing an environmental risk for groundwater quality. In addition, some chelating agents are toxic to plants and animals. For example, EDTA is considered to be toxic based on invertebrate toxicity tests and it has been found to reduce growth and cause leaf abscission in some plants. In the search for safer and environmentally friendlier means of making metal contaminants more bioavailable, researchers have tested various compounds including polyaspartic acid (Fig. 10.9) which possesses several carboxylic groups able to coordinate various metals, is not toxic, is rapidly biodegraded, and efficiently forms complexes with different metals. When polyaspartic acid was tested in the lab using metal-spiked soil, Cu and Zn uptake in plant roots was improved to the same extent as when the best chemical chelating agent was used. Since polyaspartic acid does not impair either plant or bacterial functioning, this suggests that this highly biodegradable and nontoxic poly-amino acid may be a good choice for enhancing the phytoremediation of some metals. Moreover, the large-scale production of this compound is already a well-established and cost-effective industrial process.

In addition to using polyaspartic acid as a means of making metal contaminants more bioavailable, some researchers have utilized citric acid (Fig. 10.10). In one series of experiments, researchers used soybean as the host plant, *Kochuria rhizophilia* (a Gram-positive soil bacterium that contained ACC deaminase activity, IAA production, siderophore synthesis and phosphorus solubilizing activity) as the

PGPB and citric acid (with its three carboxylic acid groups) as a means of solubilizing metals from various soils. In these experiments, researchers measured both plant growth and the ability of the plants to take up Cd, Cr, Cu and Ni from the soil in the presence of (i) no additions, (ii) citric acid, (iii) the PGPB strain, and (iv) both citric acid and the PGPB strain. While the individual application of either citric acid or the PGPB strain led to an increase in both plant growth and metal uptake, the largest effects were a consequence of adding both citric acid and the PGPB strain (where the latter led to an increase in biomass of ~40% and an increase in the uptake of the four metals tested of ~40–60% compared to non-inoculated and non-treated plants). Given that citric acid is relatively inexpensive and is itself nontoxic, it is possible that its addition may help to make the phytoremediation of some metals (from some soils) a more practical endeavor.

Another way of assuring that soil contaminants are bioavailable is through the use of biosurfactant-producing bacteria. Biosurfactants are molecules that can be produced either extracellulary or attached to a cell membrane. They are amphiphilic compounds, i.e., they contain both hydrophobic groups (tails) and hydrophilic groups (heads) that can reduce surface and interfacial tensions in both aqueous solutions and hydrocarbon mixtures. Biosurfactants have some advantages over synthetic surfactants in that they are generally both high specificity and biodegradable. At the present time, the commercial use of biosurfactants is somewhat limited because they are relatively expensive to produce. In this regard, optimization of the bacterial growth medium may be key to reducing the costs of some biosurfactants. A number of labs have reported preliminary experiments in which bacteria that actively produce biosurfactants, including some PGPB, have been used as adjuncts of phytoremediation processes (Fig. 10.11). For phytoremediation, it is advantageous to utilize bacteria that produce biosurfactants rather than purified biosurfactants so that the biosurfactants do not have to be added continuously. Biosurfactants are useful for metal phytoremediation because they readily form complexes with metals, as a result releasing them from the surface of soil particles (Fig. 10.12). Anionic biosurfactants form nonionic bonds with metals while cationic biosurfactants replace the charged metal ions by competition for some of the negatively charged surfaces. Eventually, the metal–biosurfactant complex is desorbed from the surface of the soil particle and the metal is incorporated into a micelle. In fact, biosurfactants have been used to increase the bioavailability of both metals and organic contaminants that are bound to soil particles.

In one preliminary experiment, a strain of *Pseudomonas aeruginosa* that produced a high level of di-rhamnolipid biosurfactant (Fig. 10.13) was found to successfully remove chromium, lead, cadmium, and copper from soil. In this case a 0.1% di-rhamnolipid biosurfactant solution was used to remove the metals. Notwithstanding the success of this experiment, two caveats need to be kept in mind. First, not all of the metals were removed from the soil to the same extent so that while this procedure may facilitate phytoremediation by making the metals more available, it does not completely remove all metals from all soils. Second, while *Pseudomonas aeruginosa* is commonly found in many environments, it is a disease-causing bacterium in both animals and humans so that it not likely to be applied as an

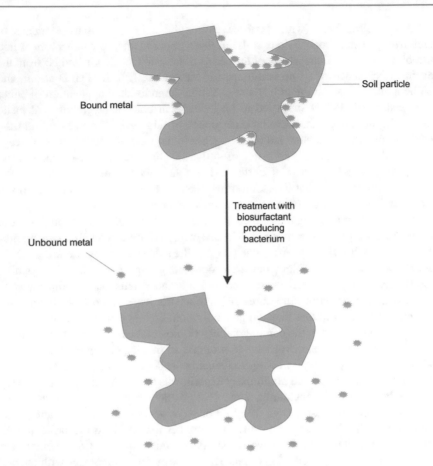

Fig. 10.11 Schematic representation of a biosurfactant-producing bacterium making soil bound metals more bioavailable (for phytoremediation)

adjunct to phytoremediation protocols in the environment. In a separate study, other researchers found that a biosurfactant-producing strain of the nontoxic PGPB *Bacillus edaphicus* increased the amount of water-soluble (i.e., bioavailable) lead from contaminated soil by approximately 40–60%.

To get around the problems of using the pathogen *Pseudomonas aeruginosa* in the environment, in one study this bacterium was grown for 4 days on distillery wastewater before the di-rhamnolipid biosurfactant was isolated from the collapsed foam that was produced as a byproduct of the fermentation process using activated carbon as an adsorbent. The biosurfactant (free of any bacterial cells) was then used to wash metal contaminated soil with the result that up to 90% of the metal was removed from the soil (Table 10.5). Whether it is possible to increase the efficiency of metal removal by further optimizing this process and whether this process can be scaled up to a sufficient extent, so that it is practical, remains to be seen.

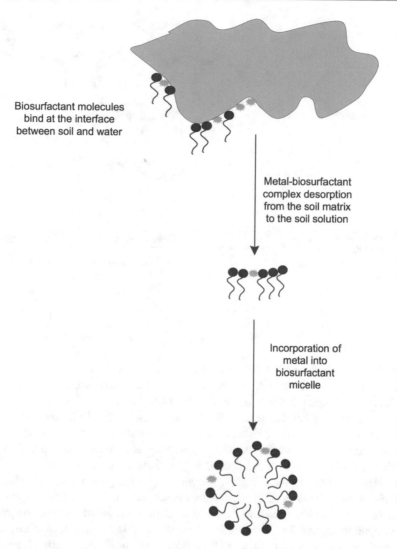

Biosurfactant molecules bind at the interface between soil and water

Metal-biosurfactant complex desorption from the soil matrix to the soil solution

Incorporation of metal into biosurfactant micelle

Fig. 10.12 Schematic representation of a biosurfactant binding to a soil particle, forming a complex with bound metals and eventually incorporating the metal into a biosurfactant micelle

Certainly, regarding the use of biosurfactant-producing bacteria as a component of metal phytoremediation processes, initial results indicate that this may become an important approach in which the problem of contaminant bioavailability may be successfully addressed.

For a large number of metal phytoremediation studies, no single mechanism has been shown to be central to the ability of a PGPB to facilitate plant growth and metal uptake in metal contaminated soils. Rather, many of the studies that have been reported have utilized bacteria that have a number of traits that may together

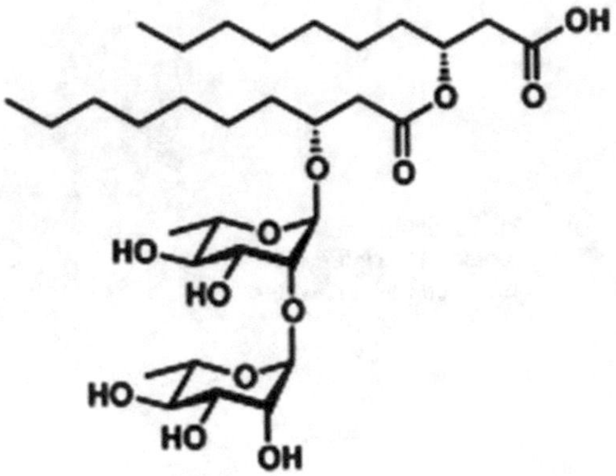

Fig. 10.13 Chemical structure of a di-rhamnolipid

Table 10.5 Metal removed from soil by rhamnolipid biosurfactant treatment

Metal	Before treatment	After treatment
Chromium	940.5	80.0
Lead	900.4	120.5
Cadmium	430.4	40.5
Nickel	880.6	220.5
Copper	480.6	60.5

The metal concentration is in parts per million (ppm)

contribute to the success of this process. As indicated above, these traits include siderophore production, IAA synthesis, ACC deaminase activity, biosurfactant production, nitrogen fixation activity, and phosphate solubilization activity. In addition, it is essential that the PGPB that are used in metal phytoremediation be resistant to a range of different potentially inhibitory metals (and organic contaminants) and be competitive with existing soil bacteria so that they may efficiently colonize the host plants used in the phytoremediation process.

Recently, following the complete sequencing of the genome of a PGPB strain, *Pseudomonas rhizophila* S211, researchers discovered that in addition to encoding proteins that have been shown to be involved in plant growth promotion such as ACC deaminase and IAA biosynthesis enzymes, this bacterium also encoded a rhamnolipid biosurfactant. In fact, this bacterium produced maximal levels of rhamnolipid following its growth on olive mill wastewater, a low-cost substrate. Thus, it may be advantageous to include this and similar bacterial strains in metal phytoremediation protocols.

10.2.6 Consortia

The Tinto river estuary in southern Spain is an area that is highly polluted by a number of different metals. To determine whether this area might be amenable to clean up by phytoremediation, scientists first isolated several metal-resistant PGPB (from the rhizospheres of plants growing in this estuary) and tested them for a variety

Table 10.6 Distribution of metals in contaminated Tinto river soil (in %) in the absence and presence of an added bacterial consortium

Metal or metaloid	Control (no added bacteria)		Added bacteria	
	Plant	Soil	Plant	Soil
As	1.30	98.70	2.20	97.80
Cd	1.92	98.08	3.00	97.00
Cu	1.46	98.54	2.60	97.40
Ni	0.55	99.45	1.20	98.80
Pb	0.43	99.57	0.86	99.14
Zn	1.00	99.00	1.97	98.03

of known functions in plant growth promotion including nitrogen fixation, phosphate solubilization, biofilm-forming capacity, synthesis of siderophores, ACC deaminase activity and IAA biosynthesis. Based on the presence of these activities, four strains were selected to be part of a phytoremediation consortium, they included *Bacillus methylotrophicus* SMT38, *B. aryabhattai* SMT48, *B. aryabhattai* SMT50 and *B. licheniformis* SMT51. The plant that was selected for these experiments was *Spartina maritima* (small cordgrass), a perennial plant that is native to the coastal regions of southern Europe and can grow to up to 70 cm in height. Clumps of this plant along with soil were obtained from the Tinto river marshes and were planted in large plastic pots in a greenhouse maintained at 21 °C–25 °C. The plants were kept in the greenhouse for 2 weeks and watered with tap water prior to the start of the experiment when 10^9 cells of each of the four bacterial strains were added to each pot. A few plants were harvested and analyzed at the start of the experiment with the remainder being harvested and analyzed after 30 days of growth. The addition of the four-strain consortium increased both plant root growth and plant metal uptake (Table 10.6). Interestingly, in the presence of the bacterial consortium, the uptake of all of the metal contaminants by the plant tissues increased. While this experiment lasted only 30 days, it would be interesting to know whether a longer experiment might result in increased metal uptake. Moreover, to ensure that the bacterial consortium had remained intact during the course of the experiment and to test whether additional bacterial aliquots might yield better results, experiments that included multiple additions of the bacterial consortium over the 30-day incubation were performed. Unfortunately, multiple aliquots of the bacterial consortium did not significantly alter the amount of metal taken up by the plants. While the results of this experiment are encouraging, they are nevertheless preliminary. It remains to be seen whether this bacterial consortium will be effective in the natural environment and, importantly, how many seasons will be required to completely remediate this highly contaminated estuary.

Metal uptake and translocation are significantly influenced by the range of microorganisms in the soil. In one study, scientists compared the effects the native soil microbial community with the effects of this community on cadmium and zinc accumulation in the plant *Arabidopsis helleri* following gamma irradiation of the soil. The gamma irradiation did not alter either the plant growth or the metal content of the soil, however, it significantly altered the abundance of various bacteria within the soil. As a result, the plants accumulated much more cadmium and zinc when grown on the untreated soil as compared to the gamma-irradiated soil. While the mechanism(s) behind this altered behavior is unknown at the present time, what is clear from this experiment is the fact that, in the absence of adding specific bacteria to facilitate phytoremediation, the endogenous soil bacteria appear to play a significant role in this process.

10.3   Arsenic

Admittedly, most of the emphasis in phytoremediation is related to the ability of plants to take up and remove contaminants from soil, with the plants themselves not being used as a food source. However, in some instances, treatment of plants with specific bacteria may decrease the translocation of the contaminant from the roots to the edible (seed) portion of the plant thereby making it possible to consume some crops that have been grown in contaminated soils.

Despite the fact that it is toxic to both animals and plants, the level of arsenic in many soils is high and increasing. Not only does arsenic inhibit plant growth, but by entering the edible parts of the plant, it may make some plants unsafe to eat.

Chickpea (*Cicer arietinum* L.) is a rich source of seed protein and is consumed worldwide. Chickpea is major source of protein in some countries such as India (which accounts for >70% of the world's chickpea production) so that it is essential to ensure that arsenic from contaminated soils does not either decrease the yield of chickpeas or make them contaminated and therefore inedible.

In one series of experiments, an arsenic-resistant PGPB strain identified as *Acinetobacter schindleri*, a Gram-positive rod that actively produced IAA, ACC deaminase activity, and siderophores, and had the ability to solubilize phosphate, was used to treat the roots of chickpea plants grown in the presence of 10 mg arsenic per kg of soil. The treated plants were compared in terms of their shoot length, root length, plant biomass, and chlorophyll content (Fig. 10.14). For each parameter that was measured, the presence of arsenic significantly inhibited the plant's normal value. Moreover, in every instance, the addition of the aforementioned PGPB strain increased the measured value to a level equal to or greater than the level observed in the absence of added arsenic. Importantly, not only did the presence of the PGPB facilitate plant growth in the presence of arsenic, it resulted in an increased level of arsenic in the plant roots and a decreased level of arsenic in the shoots and leaves. Thus, the net result of treating chickpea plants with this PGPB strain was that the edible part of the plant continued to have a concentration of arsenic that was significantly below the level that is deemed to be permissible for human consumption.

In another series of experiments, it was shown that pretreatment of rice plants with the PGPB *Pseudomonas chlororaphis* decreased both the toxicity of those plants from arsenic as well as their susceptibility to the fungal pathogen *Magnaporthe oryzae*, the causative agent of rice blast disease (Fig. 10.15).

Fig. 10.14 Growth of chickpea plants in the presence of arsenic compared to growth in the absence of arsenic (control). Growth is inhibited by the presence of arsenic. However, even in the presence of arsenic, growth is stimulated by the presence of the PGPB, *Acinetobacter schindleri*

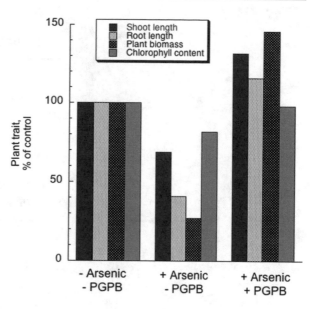

Moreover, an examination of the rice genes whose expression was altered indicated that the PGPB primed and then upregulated the plant's defensive mechanisms enabling these rice plants to simultaneously cope with the two different stresses. The results of this experiment suggest that microbes such as the particular strain of *Pseudomonas chlororaphis* that was used here might, by imparting a measure of resistance to arsenic toxicity to various plants, be a useful and key component in phytoremediation procedures to remove arsenic from soils. Of course, for this approach to be useful it will first be necessary to define what particular traits this bacterium contains that allows to function in this manner.

10.4   Organics

The development of effective phytoremediation schemes for removing organic contaminants from the environment depends upon a number of factors. First, it is necessary to identify plants that can tolerate relatively high concentrations of the target contaminant. This generally means that the plant is both able to grow in the presence of the contaminant and can, to some extent, degrade the contaminant (typically to harmless environmentally benign compounds). Second, the contaminant(s) must be bioavailable. Third, ideally a plant that is used for phytoremediation of organic contaminants should grow relatively rapidly and attain a high to moderate level of biomass. Fourth, the plant should be capable of interacting synergistically with one or more soil microorganisms that can facilitate the breakdown of the organic contaminant. Fifth, the phytoremediation process should occur within a reasonable time frame. For example, it may be acceptable to develop a phytoremediation process that requires 3 years to reduce the level of a particular

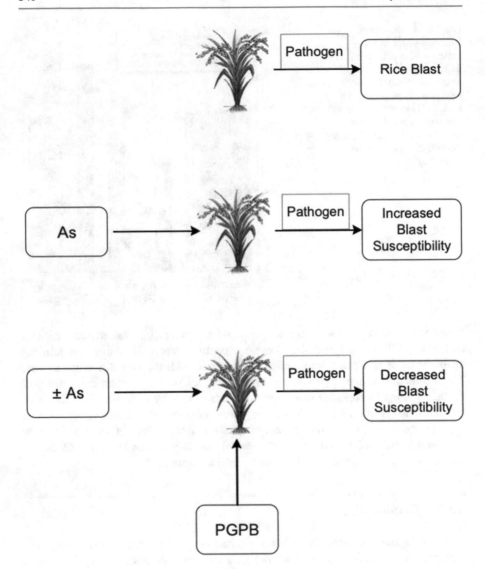

Fig. 10.15 Flowchart of an experiment showing the effect of a fungal pathogen and added As on rice blast in the presence and absence of an added rhizospheric PGPB

contaminant in soil despite the fact that traditional soil removal strategies (that are much more expensive to implement) might occur in a much shorter time frame. However, if the land that needs to be remediated is needed for a particular use, it is generally not acceptable to depend upon a process that could take up to 20 years to implement.

Polycyclic aromatic hydrocarbons (PAHs) are a particularly recalcitrant group of organic contaminants and are persistent in the environment (Fig. 10.16). They are

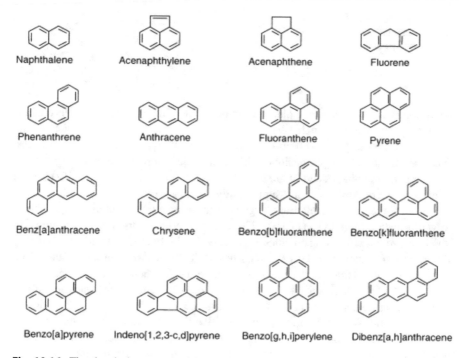

Fig. 10.16 The chemical structures of 16 priority aromatic hydrocarbons (PAHs)

difficult to degrade and it is both expensive and time consuming to remediate them from soils, and the techniques that are used are relatively inefficient, primarily degrading the smaller PAHs while leaving the larger molecules. There are numerous sources for PAH contamination in soils including creosote from the burning of wood and coal, fossil fuel processing and steel production. Some scientists have attempted to degrade PAHs in bioreactors; however, in this case it is still necessary to transport the contaminated soils to the reactor for the cleanup, an expensive and time-consuming process that often damages the natural structure and texture of the soil. Other scientists have attempted in situ bioremediation of PAHs. In this case, it is often extremely difficult to generate sufficient biomass in natural soils to achieve an acceptable rate of movement of the hydrophobic PAHs (generally tightly bound to soil particles) to the soil microbes where they can be degraded. In addition, most soil microbes are unable to degrade larger PAHs. The net result is that microbial biodegradation is, by itself, too slow to be a feasible approach to deal with these contaminants.

Phytoremediation (i.e., the degradation of organic compounds by plants) alone is generally not significantly faster than bioremediation (i.e., the biodegradation of the organic compounds by microorganisms) for removal of PAHs that include three rings or less. However, it is possible to address the relatively slow rate of biodegradation of PAHs by plants by adding both degradative and plant growth-promoting bacteria to the plant rhizosphere. As discussed earlier, by lowering a portion of the

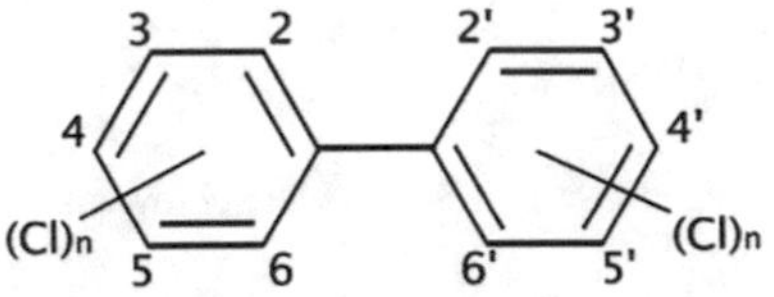

Fig. 10.17 The general
chemical structure of
polychlorinated biphenyls
(PCBs)

stress that is imposed upon the plant by the presence of the PAHs, the PGPB enables seeds to germinate to a much greater extent than they would in the absence of the PGPB, and then to grow well and thereby accumulate a large amount of biomass. The more numerous, larger, healthier plants, in conjunction with any biodegradative bacteria that are either added or are already present in the soil, degrades PAHs (and other organic contaminants) at a significantly increased rate.

Another group of recalcitrant organic environmental contaminants includes polychlorinated biphenyls (PCBs). These are synthetic organic compounds in which one to ten chlorine molecules are attached to biphenyl, a molecule consisting of two benzene rings (Fig. 10.17). PCBs were widely used in a variety of industrial applications as solvents and insulators prior to 1979 when their production was banned in the U.S. PCBs been shown to cause cancer in animals and humans as well as endocrine disruption and neurotoxicity. The toxicity of PCBs is a function of not only the number of chlorine atoms but also their position within the benzene rings. PCBs are considered to be a persistent organic pollutant, soluble in lipids and organic solvents, but not water, and are still found in many former industrial sites throughout the world.

In one series of laboratory experiments, canola plants were grown in soil that had been previously mixed with creosote, a common PAH. To possibly facilitate the phytoremediation of the creosote, a PGPB strain and its derivatives was used to inoculate the canola seeds when they were planted (Fig. 10.18). In these experiments, the various treatments included (i) no added PGPB, (ii) addition of the PGPB *Pseudomonas asplenii* previously isolated from PAH-contaminated soil, (iii) addition of the aforementioned *P. asplenii* transformed with an ACC deaminase gene (and its regulatory region) isolated from a different *Pseudomonas* strain and (iv) addition of the transformed *P. asplenii* strain encapsulated in alginate beads. The data from this experiment clearly indicates that the added PGPB strain facilitates both root and shoot growth compared to the control; the ACC deaminase transformed strain facilities growth to a greater extent (the wild-type bacterium does not contain ACC deaminase); and the alginate encapsulated bacterium stimulated growth to the greatest extent of all of the treatments. This result may be explained as follows. In the first instance, the presence of the PGPB strain facilitates plant growth, even in the presence of plant inhibitory levels of creosote. In this case, the observed plant growth promotion is likely a consequence of the bacterial IAA that is provided to the plant by the PGPB. Second, as expected based on the behavior of other PGPB, the presence of ACC deaminase activity lowers the level of inhibitory stress ethylene in the plant. Third, it is assumed that as the alginate slowly dissolves in the soil, the encapsulated PGPB are steadily released, with the consequence that there is a continuous source of the bacterium in the plant rhizosphere

Fig. 10.18 Canola shoot and root biomass following growth in soil containing 6 g/ kg PAHs (creosote) for 25 days. Control has no added PGPB. The PGPB is a strain of *Pseudomonas asplenii*. The abbreviation ACC-D indicates that the PGPB strain has been transformed to express ACC deaminase activity. The notation "+ Alginate" indicates that the bacterium has been encapsulated in alginate

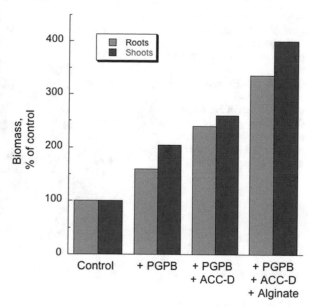

potentially allowing for more extensive colonization of the plant roots by the PGPB. In addition to the demonstrated advantage of encapsulating the abovementioned PGPB in alginate, other researchers have shown that alginate encapsulation may provide some bacterial strains with long-term stability thereby facilitating their use on a commercial scale. While work of the sort described above is an important step in developing protocols for the removal of PAHs from contaminated environments, it is essential to test whether the success seen in the lab can be reproduced in a field situation.

In a different laboratory/greenhouse experiment, researchers have shown that the addition of an ACC deaminase containing PGPB to Kentucky blue grass growing in the presence of varying amounts of added PAHs caused the plants to grow to a significantly greater size than when the PGPB was not added (Fig. 10.19). In this experiment, plants were grown for 120 days before their growth was assessed. In addition to facilitating an increase of plant biomass in these experiments, when the PGPB were present not only were PAHs degraded to a greater extent, but more of the recalcitrant larger molecules were degraded than in the absence of the PGPB. Moreover, experiments with different grasses and several different PGPB that all contained ACC deaminase have shown that the results seen in the laboratory are also seen in the field. In one experiment, with the addition of these bacteria, 30–55% of either PAHs or total petroleum hydrocarbons (Fig. 10.20) have been observed to be removed from a contaminated soil in a single growing season.

The insecticide cypermethrin (Fig. 10.21) is a synthetic pyrethroid (natural pyrethrins are produced by the flowers of pyrethrums) that has been used extensively as a commercial insecticide since the 1980s. It functions as a relatively fast-acting neurotoxin of insects (both target and beneficial insects are killed) but, in high doses,

Fig. 10.19 Kentucky blue
grass biomass following
120 days of growth in a
greenhouse with and without
an ACC deaminase-
containing PGPB in the
presence of soil
concentrations of PAH from
0.5 to 3.0 g/kg. In this
experiment, the control refers
to plants that were grown in
absence of any PAH

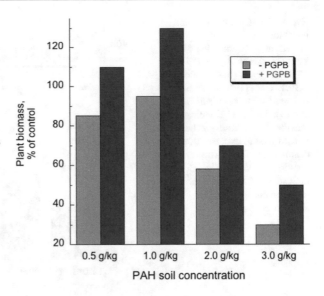

Fig. 10.20 Decrease of total
petroleum hydrocarbons in a
field soil following the growth
of annual ryegrass with or
without an ACC deaminase-
containing PGPB

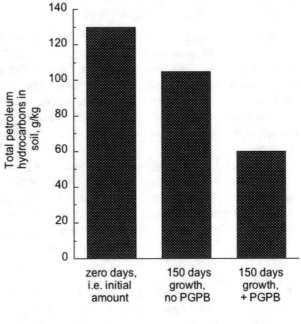

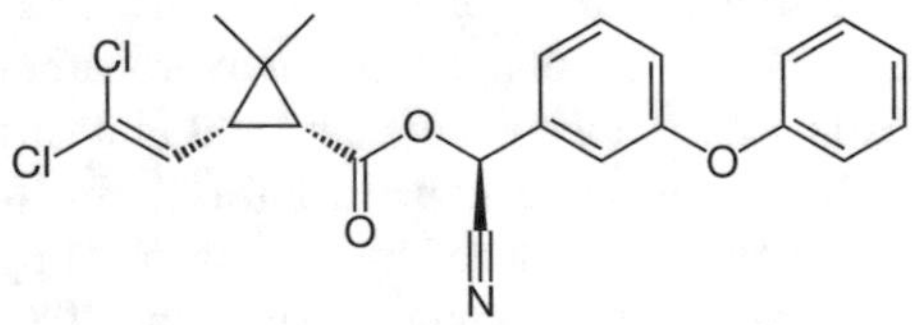

Fig. 10.21 The chemical
structure of the pyrethroid
insecticide cypermethrin

is also toxic to humans. This notwithstanding, cypermethrin continues to be used widely. One group of researchers isolated three bacterial strains, *Acinetobacter calcoaceticus*, *Brevibacillus parabrevis* and *Sphingomonas* sp., from agricultural soil, that were able to efficiently degrade cypermethrin and other pyrethroids. In addition, these three bacterial strains displayed some plant growth-promoting traits such as IAA synthesis and phosphate solubilization. Thus, one or more of these bacteria might be a useful means of removing cypermethrin from contaminated soils.

Contaminated soils in the environment often contain both metal and organic contaminants complicating the task of remediation. In one recent series of experiments, researchers sought to remove both pyrene (a PAH that consists of four benzene rings; see Fig. 10.16) and nickel from contaminated soil. Pyrene is commonly produced from the incomplete combustion of organic compounds such as the burning of gasoline in automobile engines. To perform these experiments, researchers used the plant *Scirpus triqueter* (a sedge or bulrush) in conjunction with a newly isolated PGPB strain (that was not characterized to the genus and species level) that contained ACC deaminase activity and the ability to synthesize IAA, siderophores, and solubilize phosphorus. Moreover, this bacterium was also able to partially degrade pyrene. In this case, the addition of the PGPB increased the tolerance of the plant to both nickel and a mixture of nickel and pyrene. The effectiveness of this PGPB is especially important since the mixture of nickel and pyrene was much more toxic to soil microorganisms than either of these contaminants separately. Notwithstanding this encouraging result, it remains to be determined how this PGPB behaves under field conditions.

In one study, researchers used a combination of rhamnolipids and a rhamnolipid-producing strain of the bacterium *Shewanella* sp. to remediate petroleum hydrocarbon-contaminated soil (soil spiked with 10% crude oil). This treatment, in concert with the indigenous soil microbial community, resulted in more than 75% of the contaminant being removed following a 120-day incubation. Again, while these results are very encouraging, it remains to be seen how well this approach works under field conditions.

10.5 Endophytes

Interestingly, the vast majority of the studies where PGPB have been used to facilitate phytoremediation have employed bacterial endophytes rather than rhizospheric bacteria. In one study, researchers transferred a biodegradative plasmid from a rhizospheric bacterium to an endophytic strain and found that the newly engineered endophyte significantly improved the phytoremediation of certain compounds (Fig. 10.22). Thus, when a rhizospheric bacterium with a large plasmid encoding a biochemical pathway for the breakdown of toluene was present in the rhizosphere of yellow lupin plants, the plants were able to partially breakdown toluene that was present in the soil. Despite the fact that some of the toluene was degraded, a significant fraction of the toluene was volatilized so that this combination of plant and bacteria resulted in the transfer of a good deal of the soil toluene to

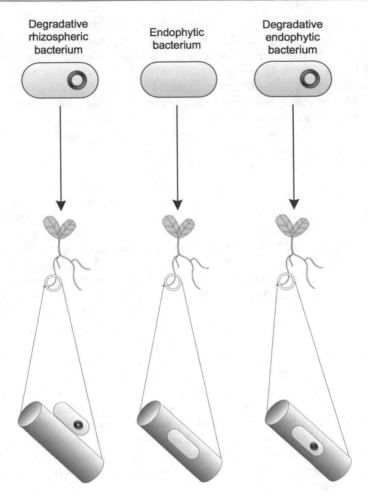

Fig. 10.22 Schematic representation of toluene degradation by a protein encoded by a degradative plasmid (shown in red) in conjunction with yellow lupin plants. The degradative rhizospheric bacterium, carrying the degradative plasmid, binds to the surface of plant roots (a root segment is shown) and partially degrades toluene in the soil. The endophytic bacterium is localized inside of the plant tissues but cannot degrade the toluene. The endophytic bacterium that is carrying the degradative plasmid is localized inside of the plant tissues and can degrade large amounts of the toluene that is taken up into the plant

the air, a problematic outcome. On the other hand, when the plasmid was transferred, by conjugation, to a bacterial endophyte, the degradative endophytic bacterium, when it was inside of the plant, largely degraded the toluene with a 50–70% reduction in evapotranspiration through the plant leaves. In addition, at all of the toluene concentrations tested, plants achieved the highest level of biomass in the presence of the endophytic strain carrying the degradative plasmid (Fig. 10.23). In this experiment, both the rhizosphere-binding bacterium and the endophyte were

Fig. 10.23 Plant biomass of yellow lupins grown in the presence of different bacteria and 500 mg toluene per liter of soil. The control does not contain any added bacteria. The plasmid encodes a biochemical pathway for the degradation of toluene

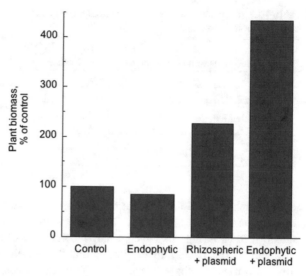

strains of *Burkholderia* spp. Since the biodegradative plasmid was transferred by conjugation and not by any so-called unnatural form of genetic engineering, the scientists who performed these experiments expect that such an engineered strain will be acceptable to various regulatory agencies for use in the environment. These workers anticipate that this approach may eventually be utilized with (fast growing) poplar trees to phytoremediate groundwater that is contaminated with water-soluble organic compounds. Moreover, to make this system more efficient, it may be possible to introduce degradative plasmids into endophytic PGPB strains that can promote plant growth and lower plant stress thereby further increasing the effective plant biomass.

In a more recent experiment, researchers added a suspension of a strain of *Pseudomonas putida* that had previously been shown to act as a root endophyte of poplar trees, to the roots of small poplar trees. *P. putida* is a very common soil bacterium and is considered to not be harmful to either animals or humans. Prior to inoculating trees with the selected endophytic bacterium, a naturally occurring plasmid that encoded the ability to degrade trichloroethylene was transferred into the *P. putida* strain by bacterial conjugation. Trichloroethylene, often found as a contaminant of groundwater, is an industrial solvent that is considered to be a human carcinogen. Unfortunately, some anaerobic soil bacteria can reductively dehalogenate trichloroethylene to produce vinyl chloride, which is an even more toxic compound. These researchers sought to test whether the in planta bioaugmentation of phytoremediation with endophytic bacteria would enhance the remediation of trichloroethylene. The behavior of three uninoculated poplar trees was compared with three poplar trees that were inoculated with the transformed *P. putida* endophytic strain containing the trichloroethylene degrading plasmid. The trees were all planted in soil where the groundwater contained trichloroethylene in concentrations up to 100 mg per liter. Two years after the trees were planted,

Table 10.7 Evapotranspiration of poplar trees grown at a site where the groundwater contains a high level of trichloroethylene

Tree #	± Transformed endophyte	Trichloroethylene transpiration (10^{-2} ng cm^{-2} h^{-1})	
		Initial value	Final value
1	−	8.9	7.8
2	−	7.1	7.2
3	−	8.4	6.8
4	+	7.9	0.9
5	+	8.1	0.9
6	+	7.5	0.5

researchers measured the amount of trichloroethylene that was evapotranspirating through the leaves of these trees. The three trees that had been incubated with the transformed bacterium displayed a much lower level of evapotranspiration than the trees without this bacterium suggesting that the bacterially treated trees had broken down the trichloroethylene while the untreated trees continued to transfer the trichloroethylene from the groundwater to the air (Table 10.7). Not only are these results highly encouraging, in addition, according to current European Union legislation, the transformed *P. putida* endophytic strain is considered to be a nongenetically modified organism (i.e., a non-GMO) with the result that it can be employed in field applications. This is because the naturally occurring plasmid that encoded the ability to degrade trichloroethylene was transferred into the endophytic *P. putida* strain by bacterial conjugation, which is not considered to be genetic manipulation. It is interesting to note that in the presence of trichloroethylene, the introduced bacterium became the dominant endophytic strain in the three treated poplar trees. However, when the transformed *P. putida* strain was grown in the absence of trichloroethylene, the introduced bacterium lost its degradative plasmid in ~20 generations.

One endophytic PGPB strain was not only able to facilitate the phytoremediation of petroleum-contaminated soil, it did so in the presence of very high levels of salt. The bacterium, a strain of *Bacillus safensis*, degraded petroleum hydrocarbons, produced biosurfactants, synthesized IAA and siderophores, and solubilized phosphate. This PGPB strain worked in conjunction with the halophytic plant *Chloris vergata* (feather fingergrass), a plant that is common in subtropical and tropical regions. This bacterium was able to degrade C_{12}-C_{32} n-alkanes of diesel oil as well as common PAHs, all in the presence of 6% NaCl. In a pot trial, after 120 days, ~63% of the added petroleum was degraded; this compares to ~37% of the petroleum being degraded in the absence of the endophytic PGPB. According to the authors of this study, this PGPB has an enormous potential for facilitating phytoremediation in the field. However, this remains to be demonstrated.

Scientists have discussed for many years whether PGPB prefer to bind to specific plants. While no one would argue that PGPB show the same level of plant specificity that is exhibited by various species of *Rhizobia*, some studies indicate that certain PGPB strains are more effective in promoting the growth of certain plants compared to other plants. In one study bacterial endophytes were isolated from the stem tissues of plants growing in soil contaminated with petroleum hydrocarbons and the five most effective PGPB strains, *Curtobacterium herbarum*, *Paenibacillus taichungensis*, *Rhizobium selenitireducens*, and two strains of *Plantibacter flavus*, were tested for the ability to promote the growth of *Arabidopsis*, lettuce, basil and bok choy plants. The two *P. flavus* strains increased total biomass and root length of *Arabidopsis* plants by ~ fivefold over control plants and also promoted the shoot and root growth of lettuce, basil and bok choy plants. The other three bacteria were most efficient at promoting host-specific plant growth. Thus, in this instance some endophytic bacterial strains were able to promote the growth of a range of different plants and some were more narrowly limited in their ability to promote plant growth.

Scientists have developed so-called "floating wetlands" as a means of treating polluted wastewater. In this system, plants exist in a floating mat while their roots are extended down, often through around 1 m of water. The roots take up both nutrients and pollutants, and the pollutants are degraded by a combination of enzymes provided by the plant and the root endophytes. A variety of plants have been used in these systems with the key determinant being the ability of the plants' fibrous roots having the ability to withstand a particular pollutant/toxicant. Depending upon the locale where this technology is applied, there are often significant seasonal differences in the extent of pollutant degradation that occurs. This is because of the fact that there are typically significant seasonal differences in both water and air temperature and the extent of solar radiation. Although this system has only been used to a limited extent, for certain types of pollutants and industrial wastes, this may be a very attractive means of remediating waterborne pollutants.

According to some estimates, the aerial surfaces of all of the plants in the world plants is $\sim 4 \times 10^8$ km^2, and these leaves are estimated to contain around 10^{26} bacterial cells (both on their surface and internally as endophytes). These plants' leaves are therefore able to absorb and degrade a wide range of air pollutants into less toxic or even harmless molecules. These air pollutants may include particulate matter, O_3 (ozone), NO_2 (nitrogen dioxide), SO_2 (sulfur dioxide), CO (carbon monoxide), volatile PAHs, volatile organic compounds (e.g., benzene, toluene, etc.), formaldehyde, radon, asbestos, and tobacco smoke. In general, plant leaves together with their associated microbes can absorb and remediate a significant portion of these air pollutants. In this regard, it is important to remember that just as not all bacteria can remediate all pollutants, not all plants can thrive in the presence of all pollutants.

10.6 Phytoremediation and OMICS

A large number of variables can affect the success of using phytoremediation to clean up various environmental contaminants. These variables include the plant (genus, species and cultivar), the nature of the contaminants (metals or organics), the type of soil (and its ability to bind the contaminants), the presence of compounds (other than the target contaminants) in the soil that affect plant growth, the microorganisms present in the soil (and their metabolic capabilities) and the weather. For a phytoremediation process to be as effective as possible, it is important to understand how these variables might affect plant growth and development. One way to gain some insight into how these variables affect plant growth is to employ the techniques of transcriptomics, proteomics and/or metabolomics. In addition, it is also possible to utilize the above mentioned "omics" techniques to document changes to PGPB that are used to facilitate the phytoremediation process that are caused by various environmental contaminants.

10.6.1 Changes to Plants

In one study, researchers isolated and characterized a PGPB, *Pseudomonas* sp. TLC 6-6.5-4, from lake sediment that promoted the growth of corn (maize) plants in the presence of high levels of copper. This bacterium is free-living, resistant to high concentrations of a number of different metals, produces IAA and siderophores, and solubilizes inorganic phosphate. In the first instance, it was observed in this study that the way in which plants were inoculated with the PGPB significantly affected plant growth and metal uptake. Thus, the greatest level of plant biomass was obtained when plant roots were immersed in a suspension of the PGPB prior to planting. On the other hand, when the PGPB was mixed with the soil prior to sowing seeds or the seeds were coated with the PGPB prior to planting, a significantly greater level of copper uptake resulted.

In the sandy soil that was used for these experiments, the amount of available phosphorus was quite low. Fortunately, the *Pseudomonas* sp. strain that was utilized was able to solubilize inorganic phosphate from the environment and provide it to the plants. In addition, the PGPB slightly, but significantly, increased the translocation of the copper taken up by the roots to the shoots. As well, the amount of soluble protein in the plant grown in the presence of the PGPB was increased by approximately 20% compared to the absence of the PGPB. Importantly, at the end of a 45 day growing period in copper-contaminated soil, maize plants not only produced more biomass in the presence of the PGPB, they also took up considerably more copper from the soil (Fig. 10.24).

Based on a proteomic analysis of the 45-day-old plants grown either in normal soil or in copper-containing soil, with or without PGPB, 85 maize proteins were observed to show significant changes in response to PGPB inoculation. In this case, a significant change was defined as an increase in protein expression ≥ 2 or a decrease in expression ≤ 0.5. The identities of many of the 85 proteins that changed significantly were determined by MALDI-TOF (matrix-assisted laser desorption/ionization—time of flight) mass spectrometry analysis. In normal soil, the main host proteins altered by PGPB inoculation are related to plant development and photosynthesis. In addition, maize proteins involved in regulation and signal transduction, cellular metabolism, and protein folding and degradation also changed in response to PGPB inoculation. In copper-contaminated soil, treatment with PGPB affected proteins that impact the plant's cellular metabolism and stress response. Thus, in

Fig. 10.24 Normalized maize root and shoot biomass and copper concentration following 45 days of growth compared to maize plants grown without the addition of a PGPB

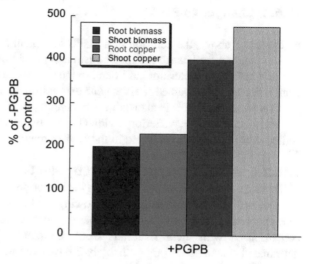

the presence of copper, proteins involved in DNA repair, methionine biosynthesis, malate metabolic process, photosynthesis, and carbon fixation were upregulated while the synthesis of several antioxidant enzymes (such as glutathione S-transferase, catalase and superoxide dismutase) was significantly reduced as a result of PGPB inoculation. Plant phytochelatins play an important role in plant development and metal detoxification. The enzyme γ-glutamylcysteine synthetase (GCL), which plays a key role in phytochelatin synthesis, was significantly upregulated in contaminated soil compared with maize grown in uncontaminated soil. GCL expression was downregulated in the presence of the PGPB in contaminated soil and upregulated in the presence of PGPB in uncontaminated soil. These results indicate that the addition of this PGPB alleviates the stress of host plants caused by high concentration of copper.

Analysis of the metabolites present in maize plants treated or untreated with PGPB and grown in copper-contaminated soil indicated a number of significant changes in these plants. This analysis revealed that the pathways that were increased by the presence of the PGPB included those encoding glutathione, proline, and ascorbate metabolism, aminoacyl-tRNA biosynthesis, the TCA cycle, galactose metabolism, and carbon fixation. In the contaminated soil, the presence of the PGPB led to the downregulation in maize of glutathione and proline metabolism as well as significant decreases in antioxidant compounds such as phenolics. Overall, the metabolomics results matched the results obtained by proteomics. That is, the PGPB stimulated plant growth and development and decreased the expression of plant defense proteins and metabolites, presumably because the PGPB alleviated a significant portion of the stress caused by the presence of the copper.

10.6.2 Changes to PGPB

As discussed above, the addition of a PGPB to a plant can significantly change the gene expression, and hence the metabolism and physiology of its plant host regardless of whether the bacterium has bound to the root surface or is localized within the plant's tissues. At the same time, the gene expression, and hence the metabolism and physiology of the PGPB is also subject to change, in this case as a consequence of the interaction of the bacterium with (1) the plant and (2) the wide range of compounds and conditions found in the environment.

10.6.2.1 Bacterial Changes Caused by Plants

The changes in a PGPB, especially when the bacterium is found primarily within the plant rhizosphere, may be examined following the treatment of the PGPB with plant root exudates. In this way, bacterial proteins, nucleic acids and metabolites are not contaminated with these materials from the plant. One group of researchers germinated canola seeds (approximately 2 days) and then grew them for a total of 7 days in growth pouches in water in the absence of soil. The remaining liquid was filtered using a 2 μm filter to remove any particulate matter or microorganisms. PGPB (*Pseudomonas* sp. UW4) cells were then grown in minimal medium containing filtered canola root exudate before the protein contents of the bacteria were analyzed. The experiments included both the wild-type and an ACC deaminase minus mutant of the PGPB. Out of 1700–1800 proteins that were detected on the two-dimensional gels, the expression levels of ~300 proteins changed. About 60% of the proteins whose expression was altered had increased expression while the remaining 40% of proteins showed decreased expression. Of the proteins with altered expression, 72 proteins could be identified by mass spectrometry. This data indicates that treatment of the wild-type PGPB with canola root exudates causes (i) an increase in nutrient utilization, (ii) an increase in membrane protein expression, with these proteins possibly involved in the bacterial colonization of roots, (iii) an increase in several soluble proteins previously implicated in interacting with plants, and (iv) a decrease in bacterial communication (chemotaxis and quorum sensing) proteins.

Four of the identified proteins (whose expression levels changed dramatically) were selected for additional analysis. These proteins included one upregulated protein (outer membrane protein F, OmpF) and three downregulated proteins (peptide deformylase, Pdf; transcription regulator Fis; and a previously uncharacterized protein, Hyp). The genes encoding these four proteins were amplified by PCR using primers whose DNA sequences were based on analogue proteins from other *Pseudomonas* strains. These four genes were cloned into a *Pseudomonas–Escherichia* shuttle expression vector and each cloned gene was used to construct a PGPB strain that overproduced one of the selected proteins. In addition, the genes encoding three of these four proteins were disrupted to produce mutant strains of the PGPB. The strain with the Pdf minus mutant (the fourth gene) was unable to survive this procedure, presumably because this mutation was lethal. The canola root elongation promoting ability of the wild-type, the four overexpressing mutants, and the three

Fig. 10.25 Normalized canola root elongation in the presence of no PGPB (control), wild-type PGPB, strains of the PGPB that overexpress one particular protein, and strains of the PGPB where the expression of one particular protein has been prevented (knockout mutant). The target proteins include peptide deformylase (Pdf), a transcription regulator (Fis), a previously uncharacterized protein (Hyp) and outer membrane protein F (OmpF). The *pdf* knockout mutant strain was not viable and therefore could not be tested

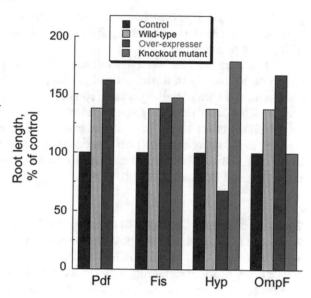

disruption mutant (nonexpressing) strains were tested (Fig. 10.25). This assay revealed that (i) the Fis protein overexpressing and knockout strains showed no significant difference compared with the wild-type, (ii) the strains overexpressing the Pdf and the OmpF proteins and the Hyp protein knockout strain promoted root length more than the wild-type, and (iii) the OmpF protein knockout strain and the Hyp protein overexpresser lost their root-length-promoting activity. In fact, overexpression of Hyp inhibited root growth relative to the untreated plants, suggesting that high levels of this protein were detrimental to plant growth.

Based on the results summarized in Fig. 10.22 the plant-growth-promoting activity of strain UW4 depends on the outer membrane protein OmpF, an integral membrane protein that forms a nonspecific transport channel responsible for the passive transport of small molecules across the outer membrane of Gram-negative bacteria. Thus, knocking out the production of OmpF may disrupt the transport of ACC from plant cells into attached bacterial cells, because ACC deaminase is localized and functions in the bacterial cytosol. In addition, the OmpF protein may be involved in bacterial cell adhesion to the plant. Second, Pdf is involved in protein maturation after translation through its activity to remove the *N*-formyl group of *N*-formyl methionine. The fact that the Pdf-protein-disrupted strain was not viable suggests that this protein is essential in bacterial cells. Downregulation of Pdf in response to plant exudates could reflect a plant mechanism to inhibit bacterial growth and limit colonization by bacterial pathogens. The strain overexpressing the Pdf protein promoted root length to a greater extent than the wild-type, possibly by overcoming the inhibitory effects of a putative plant compound. The protein Hyp is predicted, based on its homology to proteins of known function, to be a potential DNA binding protein. It is possible that the Hyp protein has a role in regulation of genes related to virulence in this PGPB strain, causing detrimental effects on plant growth. Thus, the Hyp-protein-overexpressing strain may have transformed this bacterium into a deleterious organism that caused a decrease in root length compared with control plants.

What is clear from the work that has been reported up until now is that when a bacterium interacts with a plant, the physiology of both of them is altered. This means that there are potentially a large number of bacterial proteins, many of them that have not been previously associated with plant growth promotion that can impact on the effectiveness of a particular PGPB. It is hoped that a better understanding of the many bacterial activities that affect the interaction between bacteria and plants will lead to an increased ability to select and utilize new and more efficacious PGPB strains.

10.6.2.2 Bacterial Changes Caused by Environmental Contaminants

In addition to changes to PGPB physiology that are brought about by its interaction with a host plant, to be effective PGPB need to be able to function in the presence of a wide range of soil components, some of them contaminants. The simplest way of assessing how various contaminants affect bacterial metabolism is to compare the differences in mRNA production, protein expression or metabolite levels in PGPB in the absence and presence of one or more contaminants.

PGPB containing certain traits have been shown to be able to facilitate plant growth, and increase contaminant uptake by plants and/or contaminant degradation. PGPB that possess one or more of these traits are more effective at facilitating phytoremediation; however, the limited number of known bacterial traits that facilitate phytoremediation are only part of the story.

In one series of experiments, the PGPB *Pseudomonas* sp. UW4 was grown in the presence of high levels of nickel and the changes in the protein profile of the bacterium were determined. In total, the levels of 35 proteins were found to change significantly. Several proteins involved in amino acid synthesis and protein synthesis, DNA replication and cell division were down regulated. On the other hand, several transport proteins (both import and export) were all up regulated. The other upregulated proteins included general stress proteins and anti-oxidative proteins. This preliminary result indicates that the functioning of a PGPB is affected by both the soil environments and the interaction of the bacterium with plants. These biochemical changes in turn can alter the impact that the PGPB has on plant growth and development.

Questions
1. What is phytoremediation?
2. Briefly define phytovolatilization, phytodegradation, phytoextraction, phytotransformation, rhizodegradation, rhizofiltration, and phytostabilization.
3. How do siderophores facilitate phytoremediation?
4. How does the presence of IAA-producing bacteria affect metal phytoremediation?
5. How does bacterial ACC deaminase facilitate phytoremediation of metals?

6. How does metal bioavailability affect phytoremediation? How can problems of limited metal bioavailability be overcome?
7. What environmentally friendly chelating agents are able to facilitate metal bioavailability?
8. What are biosurfactants and how might they be used to improve phytoremediation?
9. How does the presence of nitrogen-fixing bacteria affect plant growth in the presence of metal contaminants?
10. How might bacterial consortia affect metal phytoremediation?
11. How can plants grown in the presence of arsenic be acceptable for human consumption?
12. What are PAHs and PCBs? Why are they difficult to remove from contaminated soils?
13. How might alginate encapsulation of a PGPR strain affect its behavior?
14. Why might it be advantageous to use endophytes rather than rhizospheric bacteria in some phytoremediation protocols?
15. Can a PGPB strain also have biodegradative activity?
16. Is it possible to simultaneously phytoremediate metals and organics?
17. What is a floating wetland and how does it work?
18. How can bacterial endophytes that carry a plasmid from another bacterium that encodes a trichloroethylene degradation pathway be considered to the non-GMOs?
19. How might the addition of a PGPB affect the expression of specific plant proteins in plants grown in contaminated soil the presence and absence of exogenous PGPB?
20. How would you determine how a particular plant affects the physiology/metabolism of a PGPB with which it interacts?
21. How would you identify bacterial proteins that respond to plants and affect the ability of a bacterium to promote plant growth?

Bibliography

Akbar S, Sultan S, Kertesz M (2015) Determination of cypermethrin degradation potential of soil bacteria along with plant growth-promoting characteristics. Curr Microbiol 70:75–84

Balseiro-Romero M, Gkorezis P, Kidd PS, Van Hamme J, Weyens N, Monterroso C, Vangreonsveld J (2017) Use of plant growth promoting bacterial strains to improve *Cytisus striatus* and *Lupinus luteus* development for potential application in phytoremediation. Sci Total Environ 571-572:676–688

Barac T, Taghavi S, Borremans B, Provoost A, Oeyen L, Colpaert JV, Vangronsveld J, van der Lilie D (2004) Engineered endophytic bacteria improve phytoremediation of water-soluble, volatile, organic pollutants. Nat Biotechnol 22:583–588

Becerra-Castro C, Kidd P, Kuffner M, Prieto-Fernández A, Hann S, Monterroso C, Sessitsch A, Wenzel W, Puschenreitere M (2013) Bacterially induced weathering of ultramafic rock and its applications for phytoextraction. Appl Environ Microbiol 79:5094–5103

Bell TH, Joly S, Pitre FE, Yergeau E (2014) Increasing phytoremediation efficiency and reliability using novel omics approaches. Trends Biotechnol 32:271–280

Brígido C, Glick BR (2015) Phytoremediation using rhizobia. In: Ansari AA, Gill SS, Gill R, Lanza GR, Newman L (eds) Phytoremediation: management of environmental contaminants, vol 2. Springer, New York, pp 95–114

Burd GI, Dixon DG, Glick BR (1998) A plant growth promoting bacterium that decreases nickel toxicity in plant seedlings. Appl Environ Microbiol 64:3663–3668

Burd GI, Dixon DG, Glick BR (2000) Plant growth-promoting bacteria that decrease heavy metal toxicity in plants. Can J Microbiol 46:237–245

Cabello-Conejo MI, Becerra-Castro C, Prieto-Fernández A, Monterroso C, Saavedra-Ferro A, Mench M, Kidd PS (2014) Rhizobacterial inoculants can improve nickel phytoextraction by the hyperaccumulator *Alyssum pintodasilvae*. Plant Soil 379:35–50

Chen L, Luo S, Li X, Wan Y, Chen J, Liu C (2014) Interaction of Cd-hyperaccumulator *Solanum nigrum* L. and functional endophyte *Pseudomonas* sp. Lk9 on soil heavy metals uptake. Soil Biol Biochem 68:300–308

Chen X, Liu X, Zhang X, Cao L, Hu X (2017) Phytoremediation effect of *Scirpus triqueter* inoculated plant-growth-promoting bacteria (PGPB) on different fractions of pyrene and Ni in co-contaminated soils. J Hazard Mater 325:319–326

Cheng Z, Duan J, Hao Y, McConkey BJ, Glick BR (2009a) Identification of bacterial proteins mediating the interactions between *Pseudomonas putida* UW4 and *Brassica napus* (canola). Mol Plant Microbe Interact 22:686–694

Cheng Z, Wei Y-YC, Sung WWL, Glick BR, McConkey BJ (2009b) Proteomoic analysis of the response of the plant growth-promoting bacterium *Pseudomonas putida* UW4 to nickel stress. Proteome Sci 7:18. https://doi.org/10.1186/1477-5956-7-18

Crombie AT, Larke-Mejia NL, Emery H, Dawson R, Pratscher J, Murphy GP, McGenity TJ, Murrell JC (2018) Poplar phyllosphere harbors disparate isoprene-degrading bacteria. Proc Natl Acad Sci USA 115:13081–13086

Doty SL (2008) Enhancing phytoremediation through the use of transgenics and endophytes. New Phytol 179:318–333

Doty SL, Freeman JL, Cohu CM, Burken JG, Firrincieli A, Simon A, Khan Z, Isebrands JG, Lukas J, Blaylock MJ (2017) Enhanced degradation of TCE on a superfund site using endophyte-assisted poplar tree phytoremediation. Environ Sci Technol 51:10050–10058

Fang Q, Fan Z, Xie Y, Wang X, Li K, Liu Y (2016) Screening and evaluation of the remediation potential of Cu/Zn-resistant autochthonous *Acinetobacter* sp. FQ-44 from *Sonchus oleraceus* L. front. Plant Sci 7:1487

Feng N-X, Yu J, Zhao H-M, Cheng Y-T, Mp C-H, Cai Q-Y, Li Y-W, Li H, Wong M-H (2017) Efficient phytoremediation of organic contaminants in soils using plant-endophyte partnerships. Sci Total Environ 583:352–368

Gamalero E, Glick BR (2010) Bacterial ACC deaminase and IAA: interactions and consequences for plant growth in polluted environments. In: Golubev IA (ed) Handbook of phytoremediation. Nova Science Publishers, New York, pp 763–774

Gamalero E, Glick BR (2012) Plant growth-promoting bacteria and metal phytoremediation. In: Anjum NA, Pereira ME, Ahmad I, Duarte AC, Umar S, Khan NA (eds) Phytotechnologies: remediation of environmental contaminants. Taylor & Francis, Boca Raton, FL, pp 359–374

Gamalero E, Lingua G, Berta G, Glick BR (2009) Beneficial role of plant growth promoting bacteria and arbuscular fungi on plant responses to heavy metal stress. Can J Microbiol 55:501–514

Germaine KJ, Keogh E, Ryan D, Dowling DN (2009) Bacterial endophyte-mediated naphthalene phytoprotection and phytoremediation. FEMS Microbiol Lett 296:226–234

Gkorezis P, Daghio M, Franzetti A, Van Hamme JD, Sillen W, Vangronsveld J (2016) The interaction between plants and bacteria in the remediation of petroleum hydrocarbons: an environmental perspective. Front Microbiol 7:1836

Glick BR (2003) Phytoremediation: synergistic use of plants and bacteria to clean up the environment. Biotechnol Adv 21:383–393

Glick BR (2010) Using soil bacteria to facilitate phytoremediation. Biotechnol Adv 28:367–374

Gurska J, Wang W, Gerhardt KE, Khalid AM, Isherwood DM, Huang X-D, Glick BR, Greenberg BM (2009) Three year field test of a plant growth promoting rhizobacteria enhanced phytoremediation system at a land farm for treatment of hydrocarbon waste. Environ Sci Technol 43:4472–4479

Gurska J, Glick BR, Greenberg BM (2015) Gene expression of *Secale cereale* (fall rye) grown in petroleum hydrocarbon (PHC) impacted soil with and without plant growth-promoting rhizobacteria (PGPR), *Pseudomonas putida*. Water Air Soil Pollut 226:308–327

Hassan W, Bano R, Bashir F, David J (2014) Comparative effectiveness of ACC-deaminase and/or nitrogen-fixing rhizobacteria in promotion of maize (*Zea mays* L.) growth under lead pollution. Environ Sci Pollut Res Int 21:10983–10996

Hassan W, Neifar M, Cherif H, Najjari A, Chouchane H, Driouich RC, Salah A, Naili F, Mosbah A, Souissi Y, Raddadi N, OuzariHI FF, Cherif A (2018) *Pseudomonas rhizophila* S211, a new plant growth-promoting rhizobacterium with potential in pesticide-bioremediation. Front Microbiol 9:34

Huang X-D, El-Alawi Y, Penrose DM, Glick BR, Greenberg BM (2004) Responses of three grass species to creosote during phytoremediation. Environ Pollut 130:453–463

Hussain A, Amna KMA, Javed MT, Hayat K, Farooq MA, Ali N, Ali M, Manghwar H, Jan F, Chaudhary HJ (2019) Individual and combinatorial application of *Kochuria rhizophila* and citric acid on phytoextraction of multi-metal contaminated soils by *Glycine max* L. Environ Exp Bot 159:23–33

Imperato V, Portillo-Estrada M, McAmmond BM, Douwen Y, Van Hamme JD, Gawronski SW, Vangronsveld J, Thijs S (2019) Genomic diversity of two hydrocarbon-degrading and plant growth-promoting Pseudomonas species isolated from the oil field of Bóbrka (Poland). Genes 10:443

Joe MM, Gomathi R, Benson A, Shalini D, Rengasamy P, Henry AJ, Truu J, Truu M, Sa T (2019) Simultaneous application of biosurfactant and bioaugmentation with Rhamnolipid-producing *Shewanella* for enhanced bioremediation of oil-polluted soil. Appl Sci 9:3773

Juwarkar AA, Dubey KV, Nair A, Singh SK (2008) Bioremediation of multi-metal contaminated soil using biosurfactant – a novel approach. Ind J Microbiol 48:142–146

Kong Z, Glick BR (2017a) The role of bacteria in phytoremediation. In: Yoshida T (ed) Applied bioengineering: innovations and future directions. Wiley-VCH Verlag GmbH & Co, Weinheim, pp 315–341

Kong Z, Glick BR (2017b) The role of plant growth-promoting bacteria in metal phytoremediation. In: Poole RK (ed) Advances in microbial physiology, vol 71. Academic Press, Oxford, pp 97–132

Kong Z, Glick BR, Duan J, Ding S, Tian J, McConkey B, Wei G (2015a) Effects of 1-aminocyclopropane-1-carboxylate (ACC) deaminase-overproducing *Sinorhizobium meliloti* on plant growth and copper tolerance of *Medicago lupulina*. Plant Soil 391:383–398

Kong Z, Mohamad OA, Deng Z, Liu X, Glick BR, Wei G (2015b) Rhizobial symbiosis effect on the growth, metal uptake, and antioxidant responses of *Medicago lupulina* under copper stress. Environ Sci Pollut Res 22:12479–12489

Kong Z, Deng Z, Glick BR, Wei G, Chu M (2017) A nodule endophytic plant growth-promoting *Pseudomonas* and its effects on growth, nodulation and metal uptake in *Medicago lupulina* under copper stress. Ann Microbiol 67:49–58

Kong Z, Wu Z, Glick BR, He S, Huang C, Wu L (2019) Co-occurrence patterns of microbial communities affected by inoculants of plant growth-promoting bacteria during phytoremediation of heavy metal-contaminated soils. Ecotoxicol Environ Saf 183:109504

Lakshmanan V, Cottone J, Bais HP (2016) Killing two birds with one stone: natural rice rhizospheric microbes reduce arsenic uptake and blast infection in rice. Front Plant Sci 7:1514

Li K, Pidatala V, Shaik R, Datta R, Ramakrishna W (2014) Integrated metabolomics and proteomic approaches dissect the effect of metal-resistant bacteria on maize biomass and copper uptake. Environ Sci Technol 48:1184–1193

Lingua G, Todeschini V, Grimaldi M, Baldantoni D, Proto A, Cicatelli A, Biondi S, Torrigiani P, Castiglione S (2014) Polyaspartate, a biodegradable chelant that improves the phytoremediation potential of poplar in a highly metal-contaminated agricultural soil. J Environ Manag 132:9–15

Liu J, Zhang Z, Sheng Y, Gao Y, Zhao Z (2018) Phenanthrene-degrading bacteria on root surfaces: a natural defense that protects plants from phenanthrene contamination. Plant Soil 425:335–350

Maier RM, Neilson JW, Artiola JF, Jordan FL, Glenn EP, Descher SM (2001) Remdiation of metal-contaminated soil and sludge using biosurfactant technology. Int J Occup Med Environ Health 14:241–248

Mayer E, de Quandros PD, Fulthorpe R (2019) *Plantibacter flavus, Curtobacterium herbarum, Paenibacillus taichungensis*, and *Rhizobium selenitireducens* endophytes provide host-specific growth promotion of *Arabidopsis thaliana*, basil, lettuce, and bok choy plants. Appl Environ Microbiol 85:e00383–e00319

Mesa J, Rodríguez-Llorente ID, Pajuelo E, Piedras JMB, Caviedes MA, Redondo-Gómez S, Mateos-Naranjo E (2015) Moving closer towards restoration of contaminated estuaries: bioaugmentation with autochthonous rhizobacteria improves metal rhizoaccumulation in native *Spartina maritima*. J Hazard Mater 300:263–271

Mesa-Marín J, Del-Saz NF, Rodríguez-Llorente ID, Redondo-Gómez S, Pajuelo E, Ribas-Carbó M, Mateo-Naranjo E (2018) PGPR reduce root respiration and oxidative stress enhancing *Spartina martima* root growth and heavy metal rhizoaccumulation. Front Plant Sci 9:1500

Muehe EM, Weigold P, Adaktylou IJ, Planer-Friedrich B, Kraemer U, Kappler A, Behrens S (2015) Rhizosphere microbial community composition affects cadmium and zinc uptake by the metal-hyperaccumulating *Arabidopsis helleri*. Appl Environ Microbiol 81:2173–2181

Nie L, Shah S, Burd GI, Dixon DG, Glick BR (2002) Phytoremediation of arsenate contaminated soil by transgenic canola and the plant growth-promoting bacterium *Enterobacter cloacae* CAL2. Plant Physiol Biochem 40:355–361

Pacwa-Plociniczak M, Plaza GA, Piotrowska-Seget Z, Cameotra SS (2011) Environmental applications of biosurfactants: recent advances. Int J Mol Sci 12:633–654

Pilon-Smits E (2005) Phytoremediation. Annu Rev Plant Biol 56:15–39

Rajkumar M, Ae N, Prasad MNV, Fritas H (2010) Potential of siderophore-producing bacteria for improving heavy metal phytoextraction. Trends Biotechnol 28:142–149

Ramírez V, Baez A, López P, Bustillos R, Villalobos MA, Carreño R, Contreras J, Muñoz-Rojas J, Fuentes LE, Martínez J, Munive JA (2019) Chromium hyper-tolerant Bacillus sp. MH778713 assists phytoremediation of heavy metals by mesquite trees (*Prosopis laevigata*). Front Microbiol 10:1833

Raskin I, Ensley BD (eds) (2000) Phytoremediation of toxic metals: using plants to clean up the environment. Wiley, New York

Ravanbakhsh M, Kowalchuk GA, Jousset A (2019) Optimization of plant hormonal balance by microorganisms prevents plant heavy metal accumulation. J Hazard Mater 37:120787

Reed MLE, Glick BR (2005) Growth of canola (*Brassica napus*) in the presence of plant growth-promoting bacteria and either copper or polycyclic aromatic hydrocarbons. Can J Microbiol 51:1061–1069

Rodriguez H, Vesely S, Shah S, Glick BR (2008) Isolation and characterization of nickel resistant *Pseudomonas* strains and their effect on the growth of non-transformed and transgenic canola plants. Curr Microbiol 57:170–174

Ryan RP, Germaine K, Franks A, Ryan DJ, Dowling DN (2008) Bacterial endophytes: recent developments and applications. FEMS Microbiol Lett 278:1–9

Safranova VI, Piluzza G, Zinovkina NY, Kimeklis AK, Belimov AA, Bullitta S (2012) Relationships between pasture legumes, rhizobacteria and nodule bacteria in heavy metal polluted mine waste of SW Sardinia. Symbiosis 58:149–159

Salt DE, Blaylock M, Kumar NPBA, Dushenkov V, Ensley BD, Chet I, Raskin I (1995) Phytoremediation: a novel strategy for the removal of toxic metals from the environment using plants. Biotechnology 13:468–474

Sharma RK, Archana G (2016) Cadmium minimization in food crops by cadmium resistant plant growth promoting rhizobacteria. Appl Soil Ecol 107:66–78

Sheng XF, Jiang CY, He LY (2008) Characterization of a plant growth-promoting *Bacillus edaphicus* NBT and its effect on lead uptake by Indian mustard in a lead-amended soil. Can J Microbiol 54:417–422

Singh T, Singh DK (2019) Rhizospheric *Microbacterium* sp. P27 showing potential of lindane degradation and plant growth promoting traits. Curr Microbiol 76:888–895

Singh R, Glick BR, Rathore D (2018) Biosurfactants: development of biological tools to increase micronutrient availability in soil. Pedosphere 28:170–189

Srivastava S, Singh N (2014) Mitigation approach of arsenic toxicity in chickpea grown in arsenic amended soil with arsenic tolerant plant growth promoting *Acinetobacter* sp. Ecol Eng 70:146–153

Taghavi S, Barac T, Greenberg B, Borremans B, Vangronsveld J, van der Lelie D (2005) Horizontal gene transfer to endogenous endophytic bacteria from poplar improves phytoremediation of toluene. Appl Environ Microbiol 71:8500–8505

Thijs S, Weyens N, Sillen W, Gkorezis P, Carleer R, Vangronsveld J (2014) Potential for plant growth promotion by a consortium of stress-tolerant 2,4-dinitrotoluene-degrading bacteria: isolation and characterization of a military soil. Microbiol Biotechnol 7:294–306

Wei X, Lyu S, Yu Y, Wang Z, Liu H, Pan D, Chen J (2017) Phylloremediation of air pollutants: exploiting the potential of plant leaves and leaf-associated microbes. Front Plant Sci 8:1318

Wenzel WW (2009) Rhizosphere processes and management in plant-assisted bioremediation (phytoremediation) of soils. Plant Soil 321:385–408

Weyens N, van der Lelie D, Artois T, Smeets K, Taghavi S, Newman L, Carleer R, Vangronsveld J (2009) Bioaugmentation with engineered endophytic bacteria improves contaminant fate in phytoremediation. Environ Sci Technol 43:9313–9318

Wu T, Xu J, Xie W, Yao Z, Yang H, Sun C, Li X (2018) *Pseudomonas aeruginosa* L10: a hydrocarbon-degrading biosurfactant-producing, and plant-growth-promoting endophytic bacterium isolated from a reed (*Phragmites australis*). Front Microbiol 9:1087

Wu T, Xu J, Liu J, Guo W-H, Li X-B, Xia J-B, Xie W-J, Yao Z-G, Zhang Y-M, Wang R-Q (2019) Characterization and initial application of endophytic *Bacillus safensis* strain ZY16 for improving phytoremediation of oil-contaminated saline soils. Front Microbiol 10:991

Zaidi S, Usmani S, Singh BR, Musarrat J (2006) Significance of *Bacillus subtilis* strain SJ-101 as a bioinoculant for concurrent plant growth promotion and nickel accumulation in *Brassica juncea*. Chemosphere 64:991–997

Zhang J, Yin R, Lin X, Liu W, Chen R, Li X (2010) Interactive effect of biosurfactant and microorganism to enhance phytoremediation for removal of aged polycyclic aromatic hydrocarbons from contaminated soils. J Health Sci 56:257–266

Issues Regarding the Use of PGPB

11

Abstract

The use and projected use of plant growth-promoting bacteria on a large scale requires the examination of a number of issues. A number of those issues are addressed in this chapter. First, of necessity, the deliberate large-scale release of bacteria into the environment requires a detailed consideration of the environmental consequences of such a release. Moreover, eventually arguments will be made to deliberately release genetically engineered bacteria into the environment. Thus, these considerations are discussed from several different perspectives. Finally, commercial considerations regarding the use of plant growth-promoting bacteria are also discussed in that companies involved in this technology will wish to protect their intellectual property.

11.1 PGPB in the Environment

Irrespective of how well PGPB perform to facilitate plant growth under controlled laboratory conditions, their ultimate utility depends upon how effective they are under greenhouse and field conditions. A variety of considerations can have an impact on the use of PGPB in situations outside of the laboratory including whether or not the bacterium has been genetically modified, the type of plant being cultivated, temperature and, soil conditions including the presence of other microorganisms, pH, compaction, and mineral content. In addition, it is essential to consider not only those factors that permit an introduced PGPB to be effective in the field, but also any governmentally imposed guidelines that regulate the deliberate release of these microorganisms to the environment. In some jurisdictions only the release of genetically engineered organisms is controlled by legislation while in many others even the release of native unmodified microorganisms is subject to regulatory oversight.

The deliberate environmental release of PGPB is not a new or recent development. In the 1930s through the 1950s, a number of different bacteria including

© Springer Nature Switzerland AG 2020
B. R. Glick, *Beneficial Plant-Bacterial Interactions*,
https://doi.org/10.1007/978-3-030-44368-9_11

Azotobacter spp. and *Bacillus megaterium* were used in the field on a large scale in an attempt to provide crops with either fixed nitrogen or solubilized phosphate (or both)—*Azotobacter* spp. are diazotrophic while *Bacillus megaterium* is well known for its ability to solubilize phosphate. The results of these early experiments, while initially promising, were highly variable causing this approach to be discarded. To a large extent, the variability that was observed was a consequence of a lack of an in depth understanding of PGPB and the multiplicity of mechanisms that they employed, notwithstanding the fact that this work was undertaken several years prior to the existence of the term PGPB. With the fact that bacterial fertilizers were seemingly unreliable, this approach was largely abandoned as worldwide agricultural practice began to emphasize the use of inexpensive, reliable, and effective chemical fertilizers and pesticides. However, with the realization, some years later, that many agricultural chemicals can cause or contribute to a variety of environmental problems, interest in using PGPB to facilitate plant growth was renewed beginning in the 1980s and 1990s. Moreover, many developing countries lack the funds to purchase chemical fertilizers and pesticides so that bacterial inoculants could play a significant role in improving agricultural productivity in these countries. In addition, with the advent of recombinant DNA technology, many individuals in society became concerned about what they mistakenly perceived to be the looming dangers of this new and "unnatural" biological technology and the possibility that it might become a major and pervasive component of modern agriculture.

Regardless of whether the microorganisms that are deliberately introduced into the environment are native unmodified strains or genetically engineered microorganisms, it is essential that they are not harmful to the environment (especially plants and animals) in any way, either in the short or long term. Thus, prior to the deliberate release of PGPB to the environment a number of factors need to be thoroughly considered (Fig. 11.1). Specific knowledge regarding the behavior of the PGPB is required ahead of time including (i) how well the organism is able to survive in the environment, (ii) whether it will be able to grow and proliferate in different environments, (iii) what the probability is that DNA from the released PGPB will be transferred to other microorganisms in the environment, (iv) what the likelihood is that the PGPB will be widely disseminated throughout the environment, and (v) whether the PGPB is harmful to any other organisms in the environment.

Generally, the number of introduced PGPB in the environment declines after their introduction, sometimes after a brief period of proliferation during which time the number of introduced organisms may increase. Typically, the number of introduced PGPB in the bulk soil decline rapidly; however, once they become established in the rhizosphere, they may persist for several months with little or no further decline in their number.

A wide range of different factors can affect the survival and proliferation of PGPB that are introduced into the environment. (i) Bacterial survival varies considerably in different soil types with organisms generally surviving to a much greater extent in nutrient rich soils and in soils with a high clay content. (ii) PGPB that are bound to the roots of plants can persist for months while a population of the same organism in the bulk soil may decrease by five or six orders of magnitude in the same period of

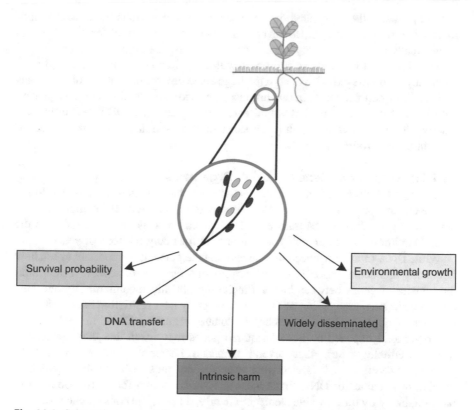

Fig. 11.1 Schematic representation of the knowledge needed before deliberately releasing PGPB to the environment

time. (iii) Endophytic PGPB, once they become established within a plant's tissues, can persist for prolonged periods of time (sometimes years), often related to the life of the plant host. (iv) Bacterial survival is influenced by several soil parameters such as pH, temperature, compaction, and oxygen content. (v) In general, PGPB that have been genetically modified are less able to survive and proliferate in the environment than are the non-transformed forms of these organisms. In most cases, this is probably a consequence of the metabolic load (see below) that is imposed on genetically transformed cells by the expression of foreign DNA. In addition, laboratory culturing of PGPB, especially in rich growth media, may result in some strains losing genes that previously made them more competitive/effective in the environment.

11.1.1 Environmental Concerns

To date the deliberate release of PGPB has taken place mainly on a controlled, small to moderate scale, and no incidents have occurred to suggest that the release of bacteria, genetically engineered or otherwise, poses any serious threat to the environment or to human health. However, since there is the potential for these bacteria to eventually be used on a very large scale, considering the huge area of agricultural land worldwide, it is necessary to proceed with considerable caution. This caution is based largely on the experience with the large-scale environmental use of chemicals such as the insecticide DDT. In that instance, the successful small-scale tests gave no indication of the problems that would eventually arise when DDT was used on a massive scale worldwide, resulting in substantial accumulation of toxic levels of DDT in the environment.

11.1.1.1 Transfer of Genes to Other Organisms
Bacterial genes expressed in a bacterial strain, whether the genes are native to the host bacterium or introduced by genetic engineering, can readily be transferred to other organisms. In fact, the transfer of DNA from one organism to another in the natural environment has probably been going on for many millions of years.

While DNA transfer between organisms can take place, it is important to ask how frequently this actually occurs in the environment. For a start, several factors limit the transfer of genes between bacteria including plasmid incompatibility and host range restrictions, recognition of DNA as foreign by host restriction endonucleases, and lack of physiological competency to donate or take up DNA. In addition, the metabolic load imposed by newly acquired genes may mean that they are readily lost. Nevertheless, now that several thousand bacterial genomes have been completely sequenced, it is clear that many, if not most, soil bacteria contain a number of stretches of DNA that appear to have been transferred from other soil bacteria, likely by transposable elements. Clearly, all possible consequences of gene transfer cannot be predicted nor may the potential for harm be obvious.

With the sequencing of the genomes of a large number of bacteria, it has become increasingly clear that horizontal gene transfer is a key mechanism of bacterial evolution. Thus, while some pathogens may acquire genes that confer upon them the ability to promote plant growth (while still being pathogenic), other bacteria (apparently beneficial bacteria) have been observed to acquire virulence factors. This observation makes it essential that prior (and even subsequent to) environmentally releasing a PGPB that it is quite stable and cannot readily switch to virulence under different environmental conditions.

Even without gene transfer, introduced organisms may have consequences other than those intended. For example, secondary metabolites may affect nontarget organisms, especially if the production of these has been enhanced by genetic manipulation. For example, the overproduction of antibiotics by biocontrol PGPB, may be deleterious to other beneficial or benign microbes or to the host plant, as well as to the target phytopathogens.

11.1.1.2 Monitoring PGPB
In monitoring the effects of an introduced PGPB, it is essential to be able to separate out the effects attributable to the introduced bacterium from the effects caused by indigenous soil bacteria. To do this, the introduced bacterium may be detected by

using either enzyme linked immunosorbent assays (ELISA) of expressed proteins, nucleic acid hybridization—either with or without PCR amplification—or by using molecular markers. Marking or "tagging" a microorganism may be defined as, "the intentional introduction of genes conferring distinctive phenotypic properties which enable 'tracking' of the marked organism after introduction into the environment." For a start, the phenotype expressed by the marker gene should not be expressed by the indigenous microbial population. In addition, as a consequence of metabolic load, marker genes are considerably more stable when they are integrated into the chromosomal DNA of the organism being marked than when they are introduced on plasmids. At the present time, for all field applications, there is a strong reluctance of most regulatory agencies worldwide to permit the introduction of any exogenous genes into PGPB strains, thus limiting the usefulness of this approach.

Only a handful of genes have been used as molecular markers for the various microbial strains that are released into the environment. These genes include antibiotic resistance genes, which regulators discourage researchers from employing for fear that these genes will spread to other soil bacteria thereby increasing the proliferation of antibiotic resistant bacteria in the soil. However, it is often possible to select, in the laboratory, rifampicin resistant mutants that occur naturally within populations of many bacteria because of a minor alteration in the host RNA polymerase. In the lab including the greenhouse and growth chamber, a number of different marker genes have been tested successfully including the *Escherichia coli lacZY* genes encoding β-galactosidase and lactose permease—cells that express these genes can be detected on solid medium containing a chromogenic substrate which turns a blue color; the *Pseudomonas putida xylE* gene which codes for catechol 2,3-dioxygenase, an enzyme that converts colorless catechol to a yellow color; a *Pseudomonas* gene encoding the enzyme 2,4-dichlorophenoxyacetate monooxygenase which converts phenoxyacetate to phenol, and can produce a red color in a coupled reaction; the *Vibrio fischeri* or *V. harveyi lux*, or bioluminescence, genes encoding bacterial luciferase—detecting the activity of these genes involves the oxidation of a long-chain aliphatic aldehyde and reduced flavin mononucleotide with the liberation of free energy in the form of blue-green light; the *P. syringae* ice nucleation gene, *inaZ*, can be used to monitor bacterial activity on plant surfaces; and the gene for the green fluorescent protein, *gfp*–green fluorescent protein exhibits bright green fluorescence when exposed to blue to ultraviolet light.

In one study, the PGPB *Kluyvera ascorbata* SUD165/26 was labeled separately with either *lux* or *gfp* genes. Both labels enabled researchers to sensitively detect the presence of the labeled bacteria. Nevertheless, it was important to ascertain that the expression of these labels did not interfere with the biological activity of the PGPB strain. Thus, the labeled strains were compared to the native, unlabeled, strain in their ability to promote the growth of 25-day-old canola seedlings in a growth chamber in the presence of 3 mM nickel in the soil. As can be seen in Table 11.1, the labeled strains both promoted plant growth to statistically the same extent as the unlabeled strain, indicating that the expression of either of these two labels did not interfere with the functioning of the PGPB. In addition, over a period of 6 weeks, it

Table 11.1   The effect of labeling the PGPB *Kluyvera ascorbata* SUD165/26 with either *lux* or *gfp* genes on the ability of the bacterium to promote the growth of canola seedlings in soil in the presence of 3 mM nickel

Treatment	Fresh weight g/plant
No Ni	2.55
3 mM Ni	0.83
3 mM Ni + SUD165/26	1.36
3 mM Ni + SUD165/26-*lux*	1.25
3 mM Ni + SUD165/26-*gfp*	1.27

was observed that the labeled strains persisted to the same extent in soil as the unlabeled strain.

11.1.1.3 Some Environmental Hazards

As mentioned above, nearly all soil bacteria, including PGPB, appear to exchange DNA with other soil bacteria. In addition, in two separate studies of the phylogeny of ACC deaminase genes it was concluded that genes of bacterial origin had, on numerous occasions been transferred to the genomes of some alga and some fungi. This is just one example of how bacterial genes can be transferred to nontarget organisms that may be quite different from the original bacterium, and indicates that any deliberate release of PGPB into the environment must be done with caution, possibly including extensive long-term monitoring of the fate of the introduced organism and its unique genes.

In addition to the possibility of gene transfer from a released PGPB strain, there is a small but very real possibility for some PGPB to either act as human pathogens or to be contaminated with human pathogens. This situation is exacerbated by the fact that since PGPB are used to facilitate plant growth, they may become associated with the crop plant whose growth they have facilitated thereby potentially contaminating human food.

11.1.1.4 Nonreplicating Transgenic Bacteria

Several research teams have constructed bacteria that are able to proliferate in the lab (provided that the appropriate media components are added) but are unable to replicate once they are released to the natural environment. These bacteria have been engineered to contain one or more unnatural amino acids that allow their proteins to be as functional as their natural counterparts; however, once they have performed their (short term) function in the environment, these bacteria are unable to replicate in the absence of the unnatural amino acid (Fig. 11.2). The reason that natural amino acids cannot replace the unnatural amino acid is that the unnatural amino acid is incorporated into proteins at sites that would normally signify the rarest of the stop codons. Using such a bacterium in the environment would likely create problems for regulators despite the fact that the bacterium would be unable to proliferate outside of the lab. Up until now, bacteria of this sort have only been used in the lab. It remains to be seen whether this approach will become acceptable to

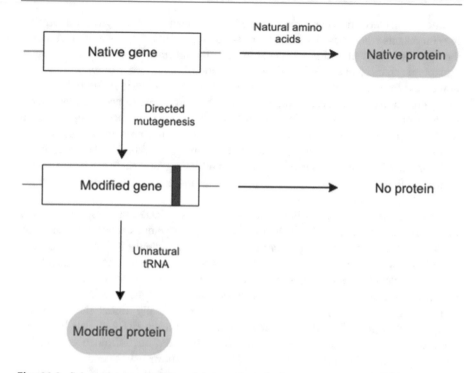

Fig. 11.2 Schematic representation of the creation of a bacterium with a modified gene that is unable to synthesize the encoded protein and therefore unable to replicate outside of the laboratory environment. An essential native gene is modified so that it contains a rare stop codon in the middle of the coding portion of the gene (shown in red). In the laboratory environment an unnatural tRNA that recognizes that stop codon and inserts an amino acid instead of stopping protein synthesis is added. The modified protein may contain the same (or an altered) amino acid at the site of the inserted stop codon as the native protein. The protein that was made functional by the addition of the unnatural tRNA enables the bacterium to proliferate (in the laboratory but not in the environment)

regulators and effective in the environment. In a variant of this procedure, scientists have redesigned some essential enzymes so that they metabolically depend on nonstandard amino acids for their survival. The microorganisms that carry these modified enzymes are therefore unable to proliferate outside of defined laboratory conditions.

11.1.1.5 Releasing Transgenic Rhizobia

Currently, Rhizobia are by far the most important species of bacterial inoculants worldwide, having been marketed for more than 100 years. In addition to being used on a relatively large scale, rhizobia have been extensively tested. Thus, prior to the development of genetic engineering, there were numerous field studies of the efficacy, competitive ability, and survival of different rhizobial strains.

It is important to monitor gene transfer in rhizobia because the symbiotic genes are typically located on the pSym, one of several large (>100 kb) plasmids that are present in most rhizobial species. Usually these plasmids are quite stable and their detailed characterization can provide an effective means of identifying (fingerprinting) inoculants when these bacteria are re-isolated from soil. This is notwithstanding the fact that some plasmids are capable of conjugal transfer thereby resulting in phenotypic changes in the recipient cell. In practice, both the possibility of transgene movement from the engineered inoculant strain to other bacteria in the field as well as the possibility that individual inoculant rhizobia might lose or acquire plasmids in the environment must be considered. In this regard, naturally occurring strains of *R. leguminosarum* typically have three to six plasmids while strains of *Sinorhizobium* spp. carry a smaller number of large plasmids. On the other hand, in *Bradyrhizobium* and *Mesorhizobium* the symbiotic genes are located on specific regions of the chromosome (rendering these genes more readily transmissible) rather than on large plasmids.

At the present time, according to European Union (EU) rules, "any organisms altered by recombinant DNA techniques and . . . . techniques involving the direct introduction into an organism of heritable material prepared outside the organism including micro-injection, macro-injection and micro-encapsulation" is subject to a very high level of government as well as community scrutiny before it can be used in the environment (http://europa.eu/legislation_summaries/agriculture/food/l28130_en.htm). Among other things, this restriction dramatically limits the deliberate release of transgenic PGPB in the environment within the EU and also limits the importation of transgenic PGPB into the EU. However, according to current EU rules, genetic modification does *not* include bacterial strains whose construction was by conjugation, transduction, or transformation (with naturally occurring plasmids). Nor does the EU definition of genetic modification include traditional mutagenesis. As a result, bacterial strains that are marked (so that they can be monitored in the environment) by the insertion of a transposon are classified as not genetically modified. Table 11.2 summarizes some of the information that is required in the EU before a genetically modified organism can be released into the environment. In addition to detailed information about the transgenic organism itself, the EU also requires information regarding the details of the proposed release including the date and place of the release, the amount of the transgenic organism that will be released, physical and chemical information about the soil at the release site, a detailed plan for ongoing monitoring the presence of the transgenic organism in the environment, etc. Here, it should be emphasized that (i) these requirements may be continually updated (especially as new knowledge is gained) and (ii) the guidelines/requirements established by the EU are similar to the approach taken in many other countries around the world.

Over the past 25 years a number of transgenic rhizobial strains have been released to the environment in both the United States and the European Union (Table 11.3). Most of these studies have been done with the objective of monitoring either (i) the environmental functioning of potentially improved strains (i.e., strains that appeared to be better than non-transgenic wild-type strains under laboratory conditions) or (ii) the environmental fate of marked strains. In one study with transgenic *S. meliloti*, when the indigenous soil population of *S. meliloti* was low, the transgenic strain occupied more than 90% of the nodules on host alfalfa plants. On the other hand,

Table 11.2 Brief overview of the information required by the EU about a transgenic organism when applying to release it to the environment

I. Characteristics of donor and recipient organism(s):

1. Scientific name; taxonomy

2. Phenotypic and genetic markers

3. Degree of relatedness between donor and recipient

4. Description of identification and detection techniques

5. Sensitivity, reliability, and specificity of detection and identification techniques

6. Description of the geographic distribution and of the natural habitat of the organism

7. Potential for genetic transfer and exchange with other organisms

8. Genetic stability of the organisms and factors affecting it

9. Pathological, ecological, and physiological traits including antibiotic resistance

10. Indigenous vectors: sequence; frequency of mobilization; specificity; presence of genes which confer resistance

11. History of previous genetic modifications

II. Characteristics of the vector:

1. Nature and source of the vector

2. Sequence of transposons, vectors and other non-coding genetic segments

3. Frequency of mobilization of inserted vector and/or genetic transfer capabilities;

III. Characteristics of the modified organism:

1. Information relating to the genetic modification including methods used for the modification; methods used to construct and introduce the insert(s) into the recipient or to delete a sequence; description of the insert and/or vector construction; sequence, functional identity and location of the altered/inserted/deleted nucleic acid segment(s) in question.

2. Information on the final transgenic organism: (a) genetic trait(s) or phenotypic characteristics and in particular any new traits and characteristics which may be expressed or no longer expressed; (b) structure and amount of any vector and/or donor nucleic acid remaining in the final construction of the modified organism; (c) genetic stability of the organism; (d) rate and level of expression of the new genetic material; (e) activity of the expressed protein(s); (f) techniques for identification and detection of the inserted sequence and vector; (g) history of previous releases or uses of the transgenic organism; (h) health considerations including allergenic effects

The rules regarding release of transgenic organisms are continually being updated so that this list only gives an idea of some of the information that is considered important at one point in time

Table 11.3 Some examples of field trials using transgenic Rhizobia

Bacterium	Method of application	Test country
Sinorhizobium meliloti	Alfalfa spray inoculation	USA
S. meliloti	Alfalfa seed coating	USA
Bradyrhizobium japonicum	Soybean seed coating	USA
S. meliloti	Alfalfa seed coating	Ireland
Rhizobium etli	Bean seed coating	USA
S. meliloti	Alfalfa seed coating	Spain
Rhizobium leguminosarum	Pea seed coating	UK, Germany, France
S. meliloti	Mixed into soil	Germany
R. leguminosarum	Pea seed coating	Italy
Rhizobium galegae	Goat's rue inoculation	Finland

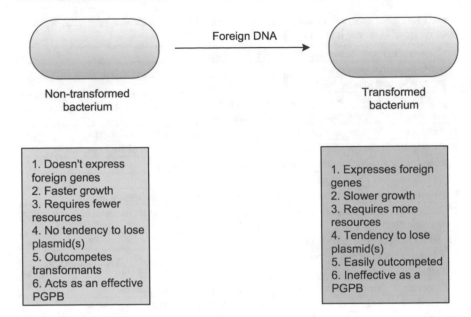

Fig. 11.3 Schematic representation of some of the changes to a bacterium that may occur following the introduction and expression of foreign genes

when the indigenous soil population of *S. meliloti* was high, the transgenic inoculant strain was found in only 6% of nodules. In fact, in several other studies where rhizobia strains had been genetically modified in an effort to improve their symbiotic traits, in the field, those strains appeared to be less competitive than indigenous strains. While this result may be reassuring to those individuals who fear that introduced transgenic strains will take over from naturally occurring strains in the environment, it may not bode well for using genetically manipulated strains to improve crop yields in the field. It has been suggested that the decreased competitiveness of transgenic rhizobial strains may be, at least in part, a consequence of the metabolic load created by the expression of the introduced transgene(s).

11.2   Metabolic Load

The introduction and expression of foreign DNA in a host bacterium typically changes the metabolism of the bacterium with the result that its environmental functioning may be altered to a significant extent (Fig. 11.3). This phenomenon, which is a multifaceted biological response, is due to a metabolic load (also called a metabolic burden or metabolic drain) that is imposed upon the host by the foreign DNA and the proteins/functions that it encodes. A metabolic load can occur for a variety of reasons, including the following:

- An increasing copy number of the foreign gene(s) requires increasing amounts of cellular energy for its replication and maintenance. If the foreign gene(s) is on a plasmid, the transformed bacterium will also need to expend energy and resources to maintain and replicate the plasmid, with larger plasmids and multi-copy number plasmids requiring more of the cell's resources.
- Overproduction of both the introduced encoded target protein(s) as well as any marker proteins (often a part of the vector) may deplete the cellular pools of certain aminoacyl-tRNAs (or even certain amino acids) and/or drain the host cell of its energy (in the form of ATP or GTP). Protein synthesis is an extremely energy and resource intensive process. Thus, the synthesis of foreign proteins may impair the ability of the host bacterium to perform its normal metabolic functions, including processes related to its ability to stimulate plant growth.
- When a foreign protein is overexpressed and then exported from the cytoplasm to the cell membrane, the periplasm, or the external medium, it may "jam" cellular export sites and thereby prevent the proper localization of other, essential host cell proteins, again impairing the bacterium's PGPB activity.
- Host cells with unusual metabolic features, such as an exceptionally high natural rate of respiration, e.g., *Azotobacter* spp., are more likely to be affected by these perturbations than are other host cells. Thus, the introduction of foreign genes into *Azotobacter* spp. and the subsequent expression of those genes can readily debilitate this bacterium, e.g., making it difficult for the transgenic bacterium to fix nitrogen.

One of the most commonly observed consequences of a metabolic load is a decrease in the rate of cell growth (especially on minimal media) after the introduction of foreign DNA. Sometimes, a metabolic load may result in plasmid-containing cells losing all or a portion of the plasmid DNA—if the introduced gene is on the plasmid that is lost, this completely negates the effect of the genetic modification. Even in the laboratory, in the presence of selective pressure, all or part of a recombinant gene may be deleted from the plasmid. Since cells growing in the presence of a metabolic load generally have a decreased level of energy for cellular functions, energy-intensive metabolic processes, such as nitrogen fixation and protein synthesis, are invariably adversely affected by a metabolic load. A metabolic load may also lead to changes in the host cell size and shape and to increases in the amount of extracellular polysaccharide produced by the bacterial host cell. This additional extracellular carbohydrate can cause the cells to stick together thereby decreasing their ability to bind to plant roots and seeds.

When a particular aminoacyl-tRNA becomes limiting, as can become the case when a foreign protein or selectable maker (of a different genus and species) is overexpressed, there is an increased probability that an incorrect amino acid will be inserted in place of the limiting amino acid. In addition, translational accuracy, which depends upon the availability of GTP as part of a cellular proofreading mechanism, may be further decreased as a consequence of a metabolic load from foreign-protein expression. This frequency of errors may diminish the usefulness of the proteins encoded by the host bacterium potentially abolishing its usefulness as a PGPB.

The extent of the metabolic load may be reduced by avoiding the use of (generally multi-copy) plasmid vectors, instead integrating the introduced foreign DNA directly into the chromosomal DNA of the host organism (Fig. 11.4). Plasmids are often unstable, especially the relatively small plasmids that have been developed to be practical for cloning, and there is a risk of losing the plasmid, and thus the introduced trait, in the absence of selective pressure (which is generally not possible to apply in the field). With an integrated cloned gene, without a plasmid vector, foreign genes are stably maintained; problems caused by metabolic load are usually avoided because a single copy of the foreign gene per chromosome typically results in only a low to moderate level of foreign gene expression and the transformed host cell does not waste its resources synthesizing unwanted and unneeded selectable marker encoded proteins. In addition, when the codon usage of the foreign gene is different from the codon usage of the host organism, depletion of specific aminoacyl-tRNA pools may be lessened by chemically synthesizing the target gene to better reflect the codon usage of the host organism.

Incorporation of a foreign gene into a bacterial chromosome is generally achieved by use of either transposition of mobile genetic elements or by homologous recombination. Naturally occurring mobile genetic elements, or transposons, can be modified to carry a target gene along with the essential transposase gene (responsible for inserting the transposon into the chromosomal DNA) between two short terminal sequences required for transposition into a bacterial chromosome. However, transposons by their nature are generally unstable, being able to move within the genome and insert into new positions irrespective of DNA sequence. In addition, they are readily transferred to other bacteria in the rhizosphere.

Stable incorporation of foreign genes into the chromosome by homologous recombination relies on hybridization and strand exchange between two homologous sequences of DNA (Fig. 11.4). First, a reporter (marker) gene, contained on a nonreplicating plasmid, is inserted into a nonessential region of the chromosomal DNA by homologous recombination. Transformants with the marker gene are selected based on the activity encoded by the marker gene. The plasmid is lost as it cannot replicate in the host cells. Then, the same nonessential chromosomal DNA sequence, also contained on a nonreplicating plasmid, is disrupted with the foreign (target) gene under the control of an appropriate promoter. Following transformation of a PGPB containing the marker gene with this construct, one can select for the replacement of the marker gene by the target gene on the chromosome by assaying for the loss of expression of the marker gene.

11.3   Patenting

For a company endeavoring to produce PGPB products for economic gain, it is important that the results of its research efforts (which are often quite expensive) can be legally protected from competitors. One strategy that meets this objective is for

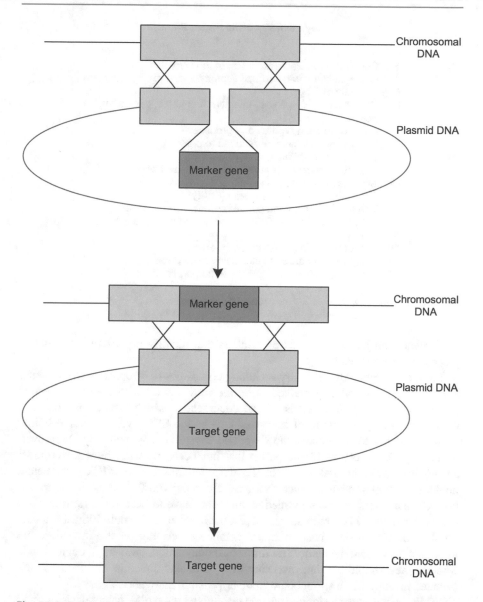

Fig. 11.4 Schematic representation of a two-step process for the integration of a foreign (target) gene into the chromosomal DNA of a host PGPB

the government to grant inventors exclusive rights to the novel products or processes that they develop. These intellectual property rights include trade secrets, copyrights, trademarks, and patents. Trade secrets comprise private information about specific technical procedures and formulations that a company wishes to keep secret from all others. Copyrights protect the authorship of published works from unauthorized use.

Key Elements of the Patent Process

1.   Contains a detailed description of the invention
2.   Patent holder given exclusive commercial rights for 20 years
from the date of filing
3.   Patent awarded to first-to-invent or first-to-file, depending on
jurisdiction
4.   An invention is a product or a process
5.   The invention should be reduced to practice
6.   The invention should be **novel**
7.   The invention must contain a step that is **not obvious** to
workers who are skilled in the field of the invention
8.   The invention should be useful (**utility**)
9.   The application should contain a description sufficiently
detailed to allow persons skilled in the field to implement it
(**feasibility**)
10. Products of nature are not patentable
11. Patents are awarded to indivuals not corporations
12. Patent rights may be assigned to corporations

Fig. 11.5 An overview of some key elements of the patent process

Trademarks can be either words or symbols that identify a particular product or process of one company.

For the biotechnology industry including companies developing new and better PGPB, patents are often perceived to be the most important form of intellectual property (Fig. 11.5). A patent is a legal document in which the government of a particular country (or group of countries such as the EU) gives the patent holder exclusive rights to implement the described invention commercially. Moreover, depending on the claims of the patent that has been approved by a government agency (in the United States the Patent and Trademarks Office or PTO), the patent holder can develop other products that are directly derived from the original invention. Once a patent has been granted, competitors have to license the right to use a patented invention in order to develop a product based on it. By definition, a patent is a public document that must contain a detailed description of the invention that informs others about the nature and limits of the invention, allowing them to decide whether they should continue working in a particular direction or try to use the patented invention as a springboard to other possible innovations.

While patents serve a similar function throughout the world, patent decisions and laws can vary significantly from one country to another. The duration of the exclusive rights of a patent is 20 years from the date that the application is filed, in all countries. In the discussion of patents and patenting that follows key terms that are specific to the patenting process are italicized.

In the United States the applicant who was the *first-to-invent* is awarded the patent. For an individual to prove that he/she have the first date of invention, it is essential keep detailed signed (including witnessed) and dated notebooks detailing the work in the invention or discovery process. When more than one person or group tries to patent an invention, it is necessary for the applicants to prove that they developed the claimed invention first. In almost all other countries, a patent is awarded to the applicant who filed first (*first-to-file* principle). It often takes several years following the filing of the initial patent application before a patent is granted. During that time, the patent application is scrutinized by examiners from a government agency (patent office) to ensure that it meets all of the established criteria for the patent to be granted. Currently, approximately 200,000–300,000 US patents are issued every year (although prior to 2010 only about half this number were issued per year). Approximately 15–25,000 of the patents that are issued each year are in the general area of biotechnology, with between 5 and 25 patents per year relating to some aspect of plant–microbe interaction. Since the beginning of the US patent system, which began at the end of the eighteenth century, more than 10 million US patents have been issued. Because the holding of a patent can be of considerable economic value, it is not a trivial matter to receive one. For this reason, both the patent application and the invention must meet a very strict set of criteria.

Generally, for either an invention (a *product* or a *process*) to be patentable, it must satisfy a number of requirements:

(A) The invention, after having been shown to work (i.e., *reduced to practice*) must be *novel* meaning that the invention does not exist as another patent that is held by someone else in another country; is not an existing product or process; and, outside the United States, has not appeared in some published form before the submission of the patent application. In the United States, an inventor has 1 year following a scientific publication describing the invention in which to apply for a patent.
(B) A patent cannot be granted for something that was merely previously unknown, i.e., a discovery; rather, the invention must contain, as judged by the patent office, an *inventive step* that was *not obvious* to other workers who are skilled in the field of the invention.
(C) The invention must be *useful* in some way, whether it is a process, a compound, a microorganism, or a multicellular organism.
(D) Every patent application must contain a *description of the invention* that is sufficiently thorough and detailed that a person knowledgeable in the same field can implement it.

A patent cannot be granted for anything that is considered to be a *product of nature*. This prevents individuals from gaining a monopoly for something that occurs naturally and has merely been discovered (and not invented) and therefore belongs to the public. This constraint is often quite contentious. In the United States, according to the Supreme Court, virtually "anything under the sun that is made by man" is patentable. However, in many other jurisdictions, this is not the case. Often companies and individuals try to skirt this constraint. For example, instead of applying for a patent on human insulin, a company might apply for a patent on the production of human insulin in a bacterial cell.

In the United States, a patent application is dealt with as follows. The application is typically prepared by an expert, normally a patent lawyer, in consultation with the inventor/scientist. The application is required to have a title; an abstract briefly describing the nature of the application; a section on the background of the invention that includes an extensive description of the current *state of the art* in the field of the invention; a detailed summary of the invention with, if considered helpful, figures and tables (similar to what might be included in a scientific publication); sections that explain the nature of the invention and describe how the invention works; and, finally, a list of *claims* about the invention and how the invention may be used. The application is sent to the PTO, where it is reviewed by an examiner for *novelty*, *nonobviousness*, *utility*, *feasibility*, and general acceptability as a patentable invention.

If a patent examiner agrees that the invention meets all the criteria for patentability, then a patent is awarded. Protection of patent rights is the responsibility of the patent holder, and generally that means bringing a lawsuit(s) against those who are presumed to be infringing on the patent (i.e., using the protected invention without explicit permission). For an individual or company to use a patented invention they must often pay a licensing fee (an agreement between a patent holder and another individual or company who is authorized to use the invention in exchange for an agreed payment). When any disputes arise regarding patent infringement or if a person or company feels that an awarded patent is inappropriate the courts and not the patent office adjudicate these disputes.

If a patent examiner rejects a patent application, then the applicant can appeal the decision to a Patent Appeals Board. If this appeal is turned down, then the decision can be challenged legally. There are a number of cases in which the stakes are considered to be so high that costly court battles go on for many years.

Product patents and process patents are the two major categories of patents. As the names suggest, product patents deal with homogeneous substances, complex mixtures, and various devices while process patents include preparative procedures, methodologies, or actual uses (Table 11.4). The patenting of microbiological and biotechnological inventions is based on the historical experience of the agricultural, fermentation, pharmaceutical, and medical industries. For example, in 1873, Louis Pasteur, arguably the father of modern microbiology, received two patents (US patents #135,245 and #141,072) for a process for fermenting beer. These patents included the living organism (i.e. the particular strains of yeast) used in the process. Thus, based on this historical precedent, there should be no particular impediment to the patenting in the United States of living organisms. Nevertheless, more recently when a scientist attempted, for the first time, to patent in the United States a genetically modified bacterium the case became extremely controversial. In that instance, Dr. Ananda Chakrabarty, who at the time worked for the General Electric Corporation, engineered various strains of *Pseudomonas* spp. by transferring (by bacterial conjugation) different naturally occurring plasmids, each of which carried the genes for a separate hydrocarbon degradative pathway, from one bacterium to another. The net result of this work was that Dr. Chakrabarty and his colleagues eventually developed a genetically modified bacterium which was

Table 11.4 Some examples of PGPB product patents and process patents

Category	Examples
Product patents	
Substance	Cloned genes, recombinant proteins, plasmids, promoter, ribosome binding sites, peptides, oligonucleotides, plant- or bacterial-modifying chemicals
Composition of matter	Selected bacterial strains, transgenic bacteria, encapsulated bacteria (including mixtures of bacteria), bacterial mutants
Devices	DNA sequencing apparatus, PCR apparatus, DNA microarray apparatus, specialized fermenters
Process patents	
Process of preparation	Means of encapsulating bacteria, unique growth media fermentation of selected or engineered bacteria, genetically engineering bacterial strains with unique properties
Method of working	Diagnostic procedures, methods for measuring metabolite production from bacteria and/or plants
Use	Applying PGPB in the field, using bacteria for environmental cleanup

capable of breaking down a number of different components of crude oil. Notwithstanding its potential usefulness in cleaning up oil spills, the US PTO rejected the patent application for the final engineered bacterium on the grounds that microorganisms are products of nature and, as living things, are not patentable. Following an unsuccessful appeal to the US Board of Patent Appeals and Interferences, the US Court of Customs and Patent Appeals overturned the original decision. Following protracted legal maneuvering, in 1980, in the case of Diamond vs. Chakrabarty, in a 5 to 4 landmark decision, the US Supreme Court decided that this engineered microorganism was patentable according to the US Patent Statute, arguing that "a live, human-made microorganism is patentable subject matter...as a manufacture or composition of matter."

The argument against patenting this genetically engineered microorganism tended to center on how the organism was developed. In the past, induced mutation followed by selection for novel properties was an acceptable way to create a patentable living organism. However, genetic engineering was considered by some to be "tampering with nature." Consequently, it was argued that no inventor should benefit from manipulating "products of nature." This position was not upheld. Thus, in the United States from 1980 onward, and later in other countries, organisms, regardless of the means that were used to develop them, must be judged by the standard criteria of *novelty*, *nonobviousness*, and *utility* to determine if they are patentable. In his decision, the Chief Justice of the US Supreme Court, Warren Burger, argued that Congress had intended patentable subject matter to "include anything under the sun that is made by man," he concluded that, "Judged in this light, respondent's micro-organism plainly qualifies as patentable subject matter. His claim is ... to a non-naturally occurring manufacture or composition of matter—a product of human ingenuity."

11.3.1 Patenting in Different Countries

Generally, the rights given by a patent extend only throughout the country in which the patent application was filed. In late 2012, the European Parliament agreed on legislation to enable the establishment of a "unitary patent." Since 2016, this type of patent is valid in participating member states of the EU and replaces the need for validation of a European patent in each of the individual countries concerned. The unitary patent will mean that applicants will pay a single renewal fee, and a single court will deal with all patent issues within the EU. In addition, unitary patents will be accepted in English, French, or German with no further translation into any other languages required. An EU unitary patent confers rights on its owner in each of the countries that are part of this process. Importantly, "European patents shall be granted for any invention, in all fields of technology, providing that they are new, involve an inventive step, and are susceptible of industrial application." Moreover, Morocco, Moldavia, Tunisia, and Cambodia have since opted to join the unitary patent system.

Some countries, notably the United States, permit patent claims to be relatively broad so that it may be possible to claim chemicals or microorganisms or devices that are similar but not identical to the ones described in detail in a particular patent application. Other countries, such as Japan, permit only relatively narrow claims and require each claim to be more extensively documented and proven. As a result, between countries there are somewhat divergent views about what is or is not a patentable invention. As well, patent protection might require numerous smaller and more specific patents in Japan to protect the same invention(s) that might be described in a single a broad omnibus US patent.

11.3.2 Patenting DNA Sequences

At present, isolated nucleic acid sequences, whether DNA, RNA, or cDNA, and proteins are generally patentable. Although the vast majority of these sequences are found naturally in organisms, purification from their natural state is considered sufficient to render them patentable. Since the US Supreme Court decision in the case of Diamond vs. Chakrabarty in 1980, thousands of patent applications for whole genes have been approved by patent offices throughout the world. In the United States, almost 20% of human genes have been patented. In this regard, many of the patented gene sequences are used as diagnostic probes.

During the course of the sequencing of the human genome, researchers isolated and determined the sequence of hundreds of human cDNAs. In some instances these cDNAs were of sufficient length to encode a complete protein while in other instances the isolated cDNAs encoded only a portion of a protein. Beginning in 1991, several attempts were made to patent these cDNAs notwithstanding the fact that researchers did not know what protein(s) they encoded. In these patent applications, it was argued that any human cDNA would (by definition) encode a human protein and that such a protein would likely be an important component of human development or metabolism. However, these patent applications were all rejected based on the fact that the utility of these DNA sequences had not been demonstrated.

One problematic case of patenting a human gene revolves around the company Myriad Genetics (based in the state of Utah in the United States) who claimed methods to detect mutations in the human genes *BRCA1* and *BRCA2* to diagnose a predisposition to breast cancer. The objections to the patent were based on the fact that genes occur naturally in every human. In addition, it was suggested that (in this case) patenting would constitute an obstacle to biomedical research worldwide. Another objection to this patent related to the fact that the discovery of the relevance of these genes to breast cancer was funded by the public.

Following a complex series of court cases and considerable legal maneuvering, the case of Myriad Genetics came before the US Supreme Court. On June 13, 2013, the US Supreme Court unanimously ruled that, "A naturally occurring DNA segment is a product of nature and not patent eligible merely because it has been isolated," thereby invalidating Myriad's patents on the *BRCA1* and *BRCA2* genes. However, the Court also held that manipulation of a gene to create something not found in nature, such as a strand of synthetically produced cDNA ("complementary DNA is patent eligible because it is not natural occurring"), could still be eligible for patent protection. There is now, admittedly, some confusion within both the scientific and legal communities regarding precisely what is and what is not patentable. Since this court decision, the US Patent and Trademarks Office uses the guideline for nature-based products as "whether it possesses markedly different characteristics from its closest naturally occurring counterpart....or whether the product claim covers significantly more." Therefore, even a synthetically created product may be patent ineligible if it does not differ from it natural counterpart.

11.3.3 Patenting Multicellular Organisms

The patenting of multicellular organisms raises a number of ethical and social concerns. However, there is nothing intrinsically new about the patenting of living material. As mentioned above, Louis Pasteur patented living microorganisms more than 100 years ago. Moreover, specific laws that give plant breeders the right to own various plant varieties have been enacted in the United States and elsewhere. The transgenic mouse ("OncoMouse") that carries a gene that makes it susceptible to tumor formation has been a precedent-setting case in many jurisdictions to determine whether genetically modified animals (in addition to microorganisms and plants) are patentable. Currently, patenting of genetically modified animals is sanctioned in many developed countries, including the United States, the EU, Japan, Australia, and New Zealand. However, in other countries such as Canada, it is still not possible to patent a transgenic animal.

11.3.4 Patenting the Plant Microbiome

In nature, seeds are commonly associated with various microorganisms. Thus, treatment of seeds with one or more microorganisms might be viewed as identical to what occurs in nature and therefore not eligible for patenting. However, a plant–microbe combination would be patentable if either the seed or the microbe has been altered in some way or if the particular seed–microbe combination does not occur in nature. Moreover, patent claims for methods used to produce microbes may be patentable. And, of course, customized combinations of microbes (selected to yield optimal performance) are also patentable (as this is seen to be similar to personalized medicine). Although isolated naturally occurring bacteria are not patentable, under the current laws it is nevertheless still possible to patent some of the uses of PGPB.

11.3.5 Patenting and Fundamental Research

Before the development of modern biotechnology, patenting and patent enforcement were rarely of interest to academic researchers working in the biological sciences. However, since the decision handed down by the US Supreme Court in 1980 in the case of Diamond vs. Chakrabarty, there is a view held by some within the academic scientific community that patents and the consequences of patenting may be detrimental to what are perceived to be established scientific values. Prior to 1980, university-based scientific research was largely perceived to be an open system with a free exchange of ideas and materials through publications and personal communications. However, with the development of the science of biotechnology and the subsequent monetization of biological science, some scientists feel that the integrity of scientific inquiry has become secondary to self-interest. At the present time, public recognition and potential financial gain from innovations have become prime motivations for many individual scientists. In the past, there was a tendency to avoid secrecy in basic research. On the other hand, today, biological scientists are often advised by patent lawyers (many of whom are in the employ of universities) to keep their work secret (at least) until a patent application is filed. In the past, it was considered to be standard practice for scientists to readily share both naturally occurring and mutant strains of organisms with one another. Today, it is not uncommon for researchers to insist that "Material Transfer Agreements" that clearly spell out the ownership of the organisms being shared, and the exclusive rights to develop these organisms for commercial purposes, are signed before any microbial strains or isolated genes are shared with colleagues.

As a consequence of the chronic financial constraints that have been imposed on basic research in many developed countries, nonprofit institutions, and especially universities, have sought additional forms of revenue. In most universities, as a condition of employment, faculty members are required to assign the rights of their inventions to their university employer, with some portion of the financial reward (royalty income) that might accrue from the successful development and

A method for replicating a biologically functional DNA, which comprises: transforming under transforming conditionscompatible unicellular organisms with biologically functional DNA to form transformants; said biologically functional DNA prepared in vitro by the method of: (a) cleaving a viral or circular plasmid DNA compatible with said unicellular organism to provide a first linear segment having an intact replicon and termini of a predetermined character; (b) combining said first linear segment with a second linear DNA segment, having at least one intact gene and foreign to said unicellular organism and having termini ligatable to said termini of said first linear segment, wherein at one of said first and second linear DNA segements has a gene for a phenotypical trait, under joining conditions where the termini of said first and second segments join to provide a functional DNA capable of replication and transcription in said unicellular organism; growing said unicellular organisms under appropriate nutrient conditions; and isolating said transformants from parent unicellular organisms by means of said phenotypical trait imparted by said biologically functional DNA.

Fig. 11.6 The first claim of US patent #4,237,224, granted to Stanly Cohen and Herbert Boyer in 1980 and entitled "Process for producing biologically functional molecular chimeras."

commercialization of any scientific invention often being shared by the university with the scientist inventor. Licensing fees and royalties from patents are currently seen by university administrators as a potential source of income. For example, US patent #4,237,224 (Fig. 11.6) was granted to Stanley Cohen (of Stanford University) and Herbert Boyer (then working at the University of California at San Francisco) in 1980 for the process of recombinant DNA technology for both the use of viral and plasmid vectors and the cloning of foreign genes. During the lifetime of this patent, from 1980 to 1997, about $45 million in royalty payments were earned for Stanford University and the University of California. These funds accrued from the yearly licensing fees charged by these two universities to the companies using this technology as part of the companies' efforts to develop commercial and profitable products. In addition, scientists and engineers employed by the Massachusetts Institute of Technology typically file more than 100 patents annually in all research fields (including many in biological science) and generate about $5.5 million per year from licensing the patent rights to these inventions.

One notable exception to the requirement to assign patent rights to the university employer of many researcher has been promulgated by the University of Waterloo in Canada. There, official University of Waterloo policy states that "it is University policy that ownership of rights in IP (intellectual property) created in the course of teaching and research activities belong to the creator(s)." In this model of dealing with intellectual property, it is up to the inventor to decide how to develop their invention, either on their own or in partnership with the university. As a direct consequence of this (some would say enlightened) policy, more than 600 spin-off companies have been developed as a result of research at the University of Waterloo with a tremendous benefit to the local economy and often a benefit to the university as well. The vast majority of research-active universities have established patent policies and have technology offices that facilitate both patenting and the transfer of technology, at a price, to industry. Clearly, entrepreneurial activity has become a fact of life at most research-intensive universities throughout the world. The challenge is to prevent this legitimate technology transfer function from dominating all aspects of academia.

Questions
1. What factors need to be considered before deliberately releasing PGPB to the environment?
2. Why are transgenic PGPB, including rhizobia, often found to be not especially competitive with indigenous soil bacteria?
3. What is metabolic load and how might it affect the functioning of a PGPB?
4. How can one produce transgenic PGPB while keeping the effect of metabolic load to a minimum?
5. How can foreign genes be integrated into the chromosomal DNA of a host PGPB?
6. What is a patent and why is it useful?
7. What are the criteria for deciding whether something is patentable?
8. What sorts of changes might one need to make when patenting the same invention in two different jurisdictions such as the United States and Japan?
9. Why is it difficult to patent complete or partial DNA sequences when the function that these sequences encode is unknown?
10. Why is patenting of scientific discoveries good (or bad) for the further development of science?
11. Is it better for a university or for a researcher/inventor to own the rights to inventions made during the course of research activities at the University?

Bibliography

Aboy M, Crespo C, Liddell K, Liddicoat J, Jordan M (2018) Was the *Myriad* decision a 'surgical strike' on isolated DNA patents, or does it have wider impacts? Nat Biotechnol 36:1146–1149

Alexander M (1985) Ecological consequences: reducing the uncertainties. Issues Sci Technol 1:57–68

Bashan Y (1998) Inoculants of plant growth-promoting bacteria for use in agriculture. Biotechnol Adv 16:729–770

Bruto M, Prigent-Combaret C, Luis P, Moenne-Loccoz Y, Muller D (2014) Frequent, independent transfers of a catabolic gene from bacteria to contrasted filamentous eukaryotes. Proc R Soc Lond B 281:20140848

Caulfield T, Cook-Deegan RM, Kieff FS, Walsh JP (2006) Evidence and anecdotes: an analysis of human gene patenting controversies. Nat Biotechnol 24:1091–1095

De Leij FAAM, Sutton EJ, Whipps JM, Fenlon JS, Lynch JM (1995) Field release of a genetically modified *Pseudomonas fluorescens* on wheat: establishment, survival and dissemination. Bio/-Technology 13:1488–1492

Dolgin E (2015) Safety boost for GM organisms. Nature 517:423

Eriksson D (2018) Recovering the original intentions of risk assessment and management of genetically modified organisms in the European Union. Front Bioeng Biotechnol 6:52

Glick BR (1995) Metabolic load and heterologous gene expression. Biotechnol Adv 13:247–261

Glick BR, Skof YC (1986) Environmental implications of recombinant DNA technology. Biotechnol Adv 4:261–277

Hirsch PR (2004) Release of transgenic bacterial inoculants – rhizobia as a case study. Plant Soil 266:1–100

Hofmann A, Fischer D, Hartmann A, Schmid M (2014) Colonization of plants by human pathogenic bacteria in the course of organic vegetable production. Front Microbiol 5:191

Holman CM (2007) Patent border wars: defining the boundary between scientific discoveries and patentable inventions. Trends Biotechnol 25:539–543

Jackman P, Gaszner M (2018) Patenting the plant microbiome. Indust Biotechnol 14:123–125

Johnson E (1996) A benchside guide to patents and patenting. Nat Biotechnol 14:288–291

Lynch JM (1990) The potential for gene exchange between rhizosphere bacteria. In: Fry JC, Day MJ (eds) Bacterial genetics in natural environments. Chapman & Hall, London, pp 172–181

Ma W, Zalec K, Glick BR (2001) Biological activity and colonization pattern of the bioluminescence-labeled plant growth-promoting bacterium *Kluyvera ascorbata* SUD165/26. FEMS Microbiol Ecol 35:137–144

Mandell DJ, Lajoie MJ, Mee MT, Takeuchi R, Kuznetsov G, Norville JE, Gregg CJ, Stoddard BL, Church GM (2015) Biocontainment of genetically modified organisms by synthetic protein design. Nature 518:55–60

Nascimento FX, Rossi MJ, Soares CRFS, McConkey BJ, Glick BR (2014) New insights into 1-aminocyclopropane-1-carboxylate (ACC) deaminase phylogeny, evolution and ecological significance. PLoS One 9:e99168

Natsch A, Troxler J, Défago G (1997) Assessment of risks associated with the release of wild-type and genetically modified plant growth promoting rhizobacteria. In: Ogoshi A, Kobayashi K, Homma Y, Kodama F, Kondo N, Akino S (eds) Plant growth-promoting rhizobacteria: present status and future prospects. OECD, Paris, pp 87–92

Prosser JI (1994) Molecular marker systems for detection of genetically engineered microorganisms in the environment. Microbiology 140:5–17

Stritzler M, Soto G, Ayub N (2018) Plant growth-promoting genes can switch to be virulence factors via horizontal gene transfer. Microb Ecol 76:579–583

Tiedje JM, Colwell RK, Grossman YL, Hodson RE, Lenski RE, Mack RN, Regal PJ (1989) The planned introduction of genetically engineered organisms: ecological considerations and recommendations. Ecology 70:298–315

Wilkins BM (1990) Factors influencing the dissemination of DNA by bacterial conjugation. In: Fry JC, Day MJ (eds) Bacterial genetics in natural environments. Chapman & Hall, London, pp 22–30

Wilson M, Lindow SE (1993) Release of recombinant organisms. Annu Rev Microbiol 47:913–944

Wöstemeyer J, Wöstemeyer A, Voight K (1997) Horizontal gene transfer in the rhizosphere: a curiosity or a driving force in evolution? Adv Bot Res 24:399–429

Printed in the United States
by Baker & Taylor Publisher Services